50 YEARS OF OCEAN DISCOVERY

NATIONAL SCIENCE FOUNDATION

1950 — 2000

Ocean Studies Board
National Research Council

NATIONAL ACADEMY PRESS
Washington, D.C.

NATIONAL ACADEMY PRESS • 2101 Constitution Ave., N.W. • Washington, DC 20418

NOTICE: The project that is the subject of this report was approved by the Governing Board of the National Research Council, whose members are drawn from the councils of the National Academy of Sciences, the National Academy of Engineering, and the Institute of Medicine. The members of the committee responsible for the report were chosen for their special competencies and with regard for appropriate balance.

This report was supported by a grant from the National Science Foundation. The views expressed herein are those of the authors and do not necessarily reflect the views of the sponsor.

Library of Congress Catalog Number 99-67879

International Standard Book Number 0-309-06398-1

Additional copies of this report are available from:

National Academy Press
2101 Constitution Avenue, NW
Box 285
Washington, D.C. 20055
800-624-6242
202-334-3313 (in the Washington Metropolitan area)
http://www.nap.edu

Preface

Oceanography is a relatively young branch of science. The Challenger Expedition in 1872-1876 began the exploration of the deep sea but it was during World War II that the significance of this knowledge became of national importance. The U.S. Office of Naval Research, formed soon after the war, sponsored basic research as well as work relevant to the Navy. The National Science Foundation (NSF), created in 1950, gradually augmented and diversified funding in oceanography until, now, NSF is the major federal supporter of marine science.

During the first few decades after WWII there was a tremendous efflorescence in our understanding of fundamental processes in the sea; in ocean circulation, plate tectonics, and the biochemical basis for the productivity of marine life. Oceanography expanded rapidly and then, of necessity, grew more slowly until it is now a mature discipline. So this is a very suitable point to look back at our achievements and the major role of NSF, as well as looking to the future of this field.

The United Nations declared 1998 the International Year of the Ocean and NSF, along with other federal agencies, celebrated the U.S. commitment to the oceans. As part of these activities and to commemorate, ahead of time, the NSF Fiftieth Anniversary, NSF's Division of Ocean Sciences approached the National Research Council's Ocean Studies Board (OSB) to plan and execute a symposium on Fifty Years of Ocean Discovery. The OSB formed a planning committee composed of scientists and managers who have observed NSF's role in U.S. ocean sciences from outside and inside NSF. The following members of the planning committee did a great service to the oceanography community in planning and participating in this symposium: Feenan Jennings (National Science Foundation and Texas A&M University, retired), John Knauss (University of Rhode Island), Marcia McNutt (Monterey Bay Aquarium Research Institute), Walter Munk (Scripps Institution of Oceanography), Andrew Solow (Woods Hole Oceanographic Institution), Sandra Toye (National Science Foundation, retired), Karl Turekian (Yale University), and Robert Wall (National Science Foundation and University of Maine,

retired). The planning committee put together a two-and-one-half-day symposium to highlight the achievements of ocean sciences and the individuals who made the advances over the past 50 years. We emphasized the development of the institutional structures within NSF required to work successfully with the expanding ocean science community. We ensured discussion of the major issues that emerged in recent decades; especially the balance of large and small science. We wanted a glimpse into the future of ocean sciences through the ideas of the next generation.

These topics, the speakers, and a highly participatory audience of almost 400 made the symposium an exciting and enjoyable event. A poster session highlighted the history of individual institutions and some of the major ocean science programs. An exhibit of photos of marine organisms provided a backdrop for a reception at the historic National Academy of Sciences building. NSF sponsored participation in the symposium by ocean science students nominated by their institutions. There was a genuine feeling of celebration among several generations of ocean scientists coming together to discuss the past and future of our field.

This volume contains the presentations from the symposium speakers. Nearly all of them were directly involved in the events they discuss and this gives an immediacy to their stories. The appendices provide other information that we hope will serve those who delve into this period of the history of our science and NSF's role in it. We are grateful to Dr. Gary Weir and Dr. David van Keuren for checking the historical accuracy of several of the historical papers from the first day of the symposium. Videotapes of the symposium are also available from the National Science Foundation and the Ocean Studies Board. On behalf of the planning committee, I wish to express our appreciation for NSF's sponsorship of this activity, and thank the OSB staff, especially Ed Urban and Ann Carlisle, for all their work. Above all we are grateful to the many individuals who took such a full part in this event and made it a very memorable occasion.

John Steele
Chair, Planning Committee

Contents

Keynote Lecture

The Emergence of the National Science Foundation as a Supporter of Ocean Sciences in the United States

John A. Knauss

Graduate School of Oceanography, University of Rhode Island, and
Scripps Institution of Oceanography, University of California, San Diego

ABSTRACT

U.S. oceanography grew rapidly after World War II, and in the years immediately after the war, the Office of Naval Research, which began in 1946, provided most of the support and much of the leadership. The National Science Foundation (NSF) began in 1950, but for a number of years its support of oceanography was marginal except for biological oceanography. This began to change with the International Geophysical Year of 1957-1958, and by the time the International Decade of Ocean Exploration (IDOE) began in 1970, NSF had in place the organizational structure necessary to become the dominant player. The timing was excellent, because 1970 was the year of the Mansfield Amendment, which limited military support of science in universities to those programs of military relevance. NSF also housed the National Sea Grant College Program for a brief period, from its formation in 1967 to its transfer in 1970 to the newly established National Oceanic and Atmospheric Administration. NSF policies have significantly influenced the course of science in the United States. Two policies specific to oceanography have contributed much to its strength and vitality. They are the NSF policy that assigns and supports ships through individual academic oceanographic institutions rather than through a single organization, and NSF's development of a support structure that allows for and encourages large, multi-investigator, multi-institutional programs, a type of program that came to flower with the IDOE and continues today.

The long period of growth of American oceanography began with World War II. The war provided a jump start to a field that until then had few practitioners in the United States and little in the way of support. Harry Hess, who skippered a destroyer, used its echo sounder to explore the bottom of the Pacific Ocean and discovered flat-topped seamounts, *guyots*. Harold Sverdrup and Walter Munk developed the techniques for calculating the strength and time of arrival of ocean swell on landing beaches. Athelstan Spilhaus, then at Woods Hole, held the patent for the development of the mechanical bathythermograph, a device used to determine the range limitations of sonar, but also a device that taught us about the seasonal thermocline, and was later used by Fritz Fuglister and his Woods Hole colleagues for tracking the cold wall of a meandering Gulf Stream.

But as important as World War II was in providing opportunities for those few who were already engaged in marine science, I believe its most important oceanography legacy was that it introduced the study of the oceans to scientists from a variety of backgrounds who found themselves working at either Scripps or Woods Hole. Carl Eckart, Russel Raitt, Brackett Hersey, Allyn Vine, and a number of others never returned to their original disciplines. For anyone interested in this transition (at least for non-biological oceanography), I highly recommend the *Transactions of the American Geophysical Union* (AGU). The difference between the *AGU Transactions* of just before and immediately after World War II is remarkable.

Because of the Office of Naval Research (ONR), the transition from wartime to peacetime science was smooth. ONR began in 1946, the year after the war ended. Many, of course, left Scripps and Woods Hole, the two big centers of World War II ocean research, to return to their earlier careers, but for those who remained, ONR was there to provide a wide range of support. The Bureau of Ships and other naval operations groups would continue to supply significant funds for a variety of research activities related to their military mission, but ONR allowed Scripps and Woods Hole to broaden their agendas.

And ONR ensured that oceanography would be supported elsewhere. By the time the National Science Foundation (NSF) arrived on the scene four years later in 1950, ONR was supporting oceanography at the University of Washington, whose program was essentially put on hold during World War II, and at a number of new centers: Texas A&M, the Chesapeake Bay Institute of the Johns Hopkins University, the University of Miami, and, when Maurice Ewing ended his association with Woods Hole, the Lamont Geological Laboratory of Columbia University. Later came Oregon State University and the Universities of Rhode Island and Hawaii.

Many have noted how ONR was the template from which many of NSF's policies and practices were formed. What is sometimes forgotten is the wide range of basic science that ONR supported in those early days. Navy support for oceanography is obvious, but in the beginning ONR provided funding for research ranging from cosmic rays and white dwarf stars, to the structure of protein and the biochemistry of muscle,[1] and to developments in nuclear physics and high-speed computing.[2] In the days before there was an NSF or a National Institutes of Health (NIH), ONR was the primary source of federal funding for all basic research supported by the federal government, more than a thousand projects at more than 200 institutions in 1948. The total budget? Less than 30 million dollars per year.[3]

Roger Revelle, still in uniform, was the first head of ONR's Geophysics Branch whose mandate included meteorology, oceanography, geography, geology, and geophysics. Revelle had returned to Scripps as director by the time I came to work in ONR in 1949. I was the sole program officer for oceanography, and my academic preparation was an undergraduate wartime degree in meteorology, which included a single course in oceanography. But little in the way of scientific expertise was required. In 1949 the principal investigator of each of our contracts was the laboratory director. Each proposal might list a number of individual projects, but the director had the freedom to move funds and scientists from one project to another and to undertake new initiatives. All that was required was a brief quarterly progress report. Our two largest contracts were with Scripps and Woods Hole. When I first arrived, each was for $125,000 a year.

ONR's role as the sole federal support for oceanographic basic research ended in 1950 with the formation of the National Science Foundation. Alan Waterman, ONR's chief scientist, became NSF's first director, and in time others moved from ONR to NSF. But unlike fields of science less key to the Navy's primary mission, ONR maintained its dominant role in oceanography for a number of years. It was

ONR, not NSF, that underwrote the development of programs at Oregon State University and the University of Rhode Island some years after the formation of NSF.[4] And it was ONR that underwrote the development of manned submersibles, first the support of Jacques Piccard's bathyscaph *Trieste*, and later the construction of Woods Hole's *Alvin*.

When did NSF become the dominant player in support of oceanography? One can look at budgets, and I have, but all who have had any intimacy with the federal budget know that interpretation is not easy. It is difficult enough to assign categories in a contemporary budget. It is almost impossible to reconstruct the actual division of funds after 20 or more years. Major budget items can be tucked away in categories that can be easily overlooked by those doing historical research. For example, one federal report for fiscal year 1969 shows the Navy's contractual oceanographic program 40 percent larger than that of NSF; another shows them essentially equal.[5]

One must also distinguish between the various types of oceanography. ONR provided relatively little support for biological oceanography, and although the Atomic Energy Commission began supporting a wide range of oceanography, including biological oceanography, in the mid-1950s, as near as I can judge, NSF was the primary source of funds in this field from the beginning. Its total support of biological oceanography was larger than that for all other fields of oceanography in 1962.[6]

My own sense is that the passing of the torch from ONR to NSF began with the International Geophysical Year (IGY) of 1958 and was completed about the time the International Decade for Ocean Exploration (IDOE) began in 1970.[7] ONR

---

[1] Pfeiffer, J. 1949. The Office of Naval Research. *Scientific American* 180(2):11-15.

[2] Moss, M. 1986. Interview with Marvin Moss, Director, Office of Naval Research. *Naval Research Reviews* 38(3):38-41.

[3] See Pfeiffer, reference 1.

[4] Wayne Burt, one of the first of the Scripps postwar Ph.D.s was able to convince the administration of Oregon State University to begin an oceanography program on the basis of a promise by ONR to provide a research vessel, the *Acona* (John Byrne, personal communication).

[5] For example: Table 9, page I-19, in *Science and Environment*, panel report, No. 1 of the 1969 *Report of the Commission on Marine Science, Engineering and Resources* (Stratton Commission) shows the 1968 NSF ocean science budget as $19.2 million for 1968. Table A-2, page 21, of *Marine Science Affairs*, the 1969 report to the President from the National Council on Marine Resources and Engineering Development estimates the 1968 NSF ocean science budget as $35 million.

[6] The summary table of an internal NSF document entitled "10-Year Projection of National Science Foundation Plans to Support Basic Research in Oceanography," dated March 27, 1962, estimates that $10 million of the $19.5 million total oceanography budget for fiscal year 1962 went to support "biological oceanography." The percentage for biology may actually be higher since one might assume that some biological oceanography was supported in two other programs of that table: "Antarctic Program" and "International Activities."

[7] Lambert, R.B., Jr. 1998. Emergence of Ocean Science Research in NSF, 1951-1980. *Marine Technology Society Journal* 32(3):68-73. Figure 1. The NSF IGY budget for oceanography for the three year period, 1956-1958 was significantly larger than the entire NSF ocean budget from its beginning in 1952 through fiscal year 1959. The IDOE program that began in 1970 more than doubled the annual NSF non-biological oceanography budget.

was the dominant player in academic oceanography at the start of the IGY; it was no longer as we moved into the IDOE. However, it is important to recognize that during this period of transition, and continuing today, NSF and ONR supported parallel and joint programs. It is a measure of the skill and common purpose of the program managers in both agencies that those doing the research could mostly ignore the details of the funding as they went about their research, and it is sometimes difficult to remember today which agency supported which parts of which program.

This transition period when primary support of oceanography passed from ONR to the National Science Foundation coincides with the development of the necessary oceanography infrastructure within NSF. The National Science Foundation has from the beginning been organized mostly along disciplinary lines. That NSF did not recognize oceanography as a separate discipline in 1950 is no surprise. The only Ph.D. granting institution at that time was Scripps. With the active assistance of the Office of Naval Research the number of degree-granting institutions began to grow. By 1960 there were a half dozen.[8] But most who called themselves oceanographers during this period had earned their degrees in other disciplines, and many continued to question whether a degree in such an ill-defined field as oceanography was the best training. The organizational structure of NSF reflected this uncertainty. As Sandra Toye writes in her administrative history of ocean science in the National Science Foundation later in this volume, "For oceanography, an inherently interdisciplinary field, NSF's early organizational choices created problems that would not be fully rectified for 25 years." A few marine biologists found a home in one or another section of the Biological and Medical Science Division, and it was relatively easy for marine geology and geophysics to find a home in the Earth Sciences Program, but there was no obvious home for physical and chemical oceanography.[9]

The International Geophysical Year of 1957-1958 brought new money and new prominence to oceanography and the rest of the Earth sciences, and the National Science Foundation structure slowly changed to meet these challenges. The reconstituted Earth Science Section established in 1962 had four programs, one of which was Oceanography under the direction of John Lyman. The Oceanography Facilities program led by Mary Johrde (primarily ship support) was added in 1967, and a Biological Oceanography program headed by Ed Chin was formed in 1968 in the Biological and Medical Science Division.[10] As part of a significant reorganization of NSF in 1970, biological oceanography was transferred to the Ocean Science Research Section to join the rest of the oceanographic disciplines. Ship support, polar programs, the Deep Sea Drilling Program, and the International Decade of Ocean Exploration were made a part of a new Directorate of National and International Programs.[11]

By 1970 the National Science Foundation had an administrative structure adequate to the challenge of the rapidly expanding field of oceanography. The timing was excellent; 1970 was also the year of the Mansfield Amendment which forbade the Department of Defense to fund projects in basic science unless they were closely related to a military function or operation. The 20-year-long passing of the torch from ONR to NSF for primary responsibility for the support of oceanography was now complete.

In 1966, while NSF was still grappling with how to integrate oceanography into its organization, Congress created the National Sea Grant College Program and placed it in the National Science Foundation. Senator Pell, who introduced the first Sea Grant legislation, was not certain NSF was the best home. For a time he even considered the Smithsonian Institution. Noting that the Smithsonian had served as the nineteenth century launch pad for both fisheries and the weather service, he thought it might serve a similar role for Sea Grant until such time as a better fit could be found within the administration. However, the Smithsonian did not rise to the challenge during legislative hearings, and Sea Grant went to NSF almost by default.[12] Sea Grant was about applied research, and it included research in economics and the other social sciences. It had an educational component, and perhaps most critically, it had a significant public outreach program patterned after the very successful agricultural extension service. Sea Grant was not an easy fit in the National Science Foundation of the 1960s. The 1969 report of the Commission on Marine Science, Engineering and Resources (the Stratton Commission) recommended the bringing together of various ocean-oriented agencies within the federal government into a new National Oceanic and Atmospheric Administration (NOAA). NOAA was established in 1970 and the Sea Grant program was a part of it.

Patterns of support established by National Science Foundation during this period have done much to shape the development of oceanography in this country and the way it is practiced today. The most obvious, and probably the most important, is peer-reviewed science proposals, which is prac-

---

[8]As reported in the 1962 NSF 10-year projection, reference 6 above, the number of Ph.D. oceanography degrees awarded by these institutions was estimated at no more than about nine a year during the late 1950s.

[9]Some program managers were more sympathetic than others to a field far removed from their own area of interest. I remember claims during this period, claims that I like to believe were apocryphal, that one's chance of gaining NSF support was dependent upon the guy in the mail room since it was his responsibility to decide on which desk to drop an oceanography proposal.

[10]Lambert, reference 7.

[11]Ibid.

[12]I worked closely with Senator Pell on the development of the Sea Grant program and remember his concern about placing Sea Grant in NSF and his search for alternatives.

ticed throughout NSF. Elsewhere in this volume one can find examples and discussion of oceanographic achievements resulting from such support. In the remainder of this paper I wish to concentrate on two NSF policies that I believe have done much to shape in a very positive way the structure of oceanography within our universities. The two are ship operations and the large, multi-investigator, multi-institutional program. The first was set in motion by ONR and later bought into by the National Science Foundation. It is now the UNOLS (University-National Oceanographic Laboratory System) (see Byrne and Dinsmore paper later in this volume). The second began in 1970 with NSF's International Decade of Ocean Exploration. There were multiship, multi-investigator, multi-institutional programs before the IDOE (e.g., Operation Cabot, the multiship study of the Gulf Stream in 1950 and the International Indian Ocean Expedition that began a dozen years later), but I expect most would agree that this type of program gained full expression with the IDOE. These multi-investigator programs have long outlasted the original source of funding. The IDOE folded its administrative tent in 1981, but IDOE-like programs continue to constitute a significant share of the NSF oceanography budget.[13]

SHIP SUPPORT

Nowhere is cooperation between the National Science Foundation and the Office of Naval Research better seen than in the construction and support of research ships. And nowhere, I suspect, have the two agencies had to work harder to bring in line their differing modes of operation. Prior to World War II, only Woods Hole had a research vessel, the *Atlantis*, that could venture far from shore for any length of time. After the war the mark of an academic oceanographic research program was an oceangoing ship. Many were surplus World War II vessels, modified with varying degrees of success to perform oceanographic research. Their names are familiar to all of this period: first *Crawford, Horizon, Vema, Spencer F. Baird*; later *Chain, Argo, Trident, Pillsbury, Yaquina, Alaminos,* and others.[14] It was not until the Navy made available to Lamont one its first AGOR vessels, the *Robert D. Conrad*, in 1962, and NSF built the *Atlantis II* for

Woods Hole in 1963, that the academic research fleet began to acquire ships designed for the task.[15]

I firmly believe the ship support practices that evolved in the United States after World War II were critically important to the development of oceanography in this country. The system that has developed, whereby the oceangoing research fleet is operated by the university research establishment is unique. I have sometimes wondered how this decision was made. How did it come about that each of the major oceanographic institutions operated one or more research vessels capable of operating far from home port? Was it a conscious decision, the pros and cons carefully weighed and thoroughly thought through, or did it just happen? I believe it was the latter. Woods Hole had a ship, the *Atlantis*, built and supported before World War II with funds from the Rockefeller Foundation. If Scripps was going to carry out the Marine Life Research Program designed by Sverdrup before he returned to Norway at the end of World War II, it would need some vessels to conduct the monthly surveys. Among its first acquisitions was the 143-foot, 900-ton seagoing tug *Horizon*, capable of working in the open ocean for weeks at a time. One reason Maurice Ewing left Woods Hole to found the Lamont Geological Observatory was that the Woods Hole director, Columbus Islen, could not guarantee the ship time Ewing wanted for his worldwide geological and geophysical surveys. He got it with *Vema*, a former 700-ton, 200-foot yacht built in 1923.[16] These early decisions at Scripps and Lamont were made before there was an NSF. Perhaps this is all it took; once Lamont, Scripps, and Woods Hole had their vessels, the pattern was established. If you were going to be an oceanographic research institution, you needed a research ship.

As ONR began the practice of supporting a growing number of oceanographic research institutions, it also did its part in providing these institutions with supporting research vessels. In time, the Universities of Hawaii, Miami, Rhode Island, and Washington, along with Texas A&M University and Oregon State University, all had major oceangoing facilities. NSF bought into the practice and slowly became the dominant player in terms of determining how these vessels were to be used and supported.

It did not have to be this way. One could imagine NSF and ONR following the route NSF established in the support of the meteorology departments of this country. It could have established the ocean equivalent of the National Center for Atmospheric Research in Boulder, Colorado which, in the beginning at least, supported the airplanes and other large field equipment required by academic meteorologists. For many years the U.S. practice of having its major academic oceanographic institutions operate an oceangoing research fleet was unique. Even today, the practice is rare.

---

[13]Lambert reference 7. Purdy, G.M., M.R. Reeve, D.F. Heinrichs, and M.A. Booth. 1998. A question of balance: Funding basic research in the ocean sciences. *Marine Technological Society Journal 32(3):91-93*. During the IDOE decade of 1971-1981 the total IDOE budget was comparable to the funds provided individual investigators (Figure 1 in Lambert). Although the ratio of total support for IDOE-like programs to individual investigator projects has fluctuated significantly between the IDOE and the present, the ratio is once again approaching unity.

[14]Nelson, S.B. 1971. *Oceanographic Ships, Fore and Aft*. Office of the Oceanographer of the Navy. U.S. Government Printing Office, Washington, D.C.

[15]Ibid.
[16]Ibid.

I am convinced that the spawning of a number of major university oceanographic institutions, each operating one or more large ships and other facilities, has been a major reason for U.S. world leadership in oceanography. Managing a large research vessel can expand one's intellectual horizons. Someone on your staff has an interest in the flux of ions across the air-sea interface, which requires taking your ship to a variety of locations; then comes the question of what happens when these chemical constituents leave the boundary layer; and five years later, as happened to me at the University of Rhode Island, you suddenly find yourself with an atmospheric chemistry program larger than that in any meteorology department in the United States at the time.

It can also add another dimension of experience. Finding ways to repair an engine in a foreign port some 4,000 miles away, or getting a crew member out of jail in the same port, broadens your range of experience and increases your confidence. I expect there was never any doubt in the minds of the directors of Scripps, Woods Hole, Lamont, and the University of Miami (at that time the big four of oceanographic institutions) that they were quite capable of overseeing the work of Global Marine and the conversion and operation of its oil drilling vessel the *Glomar Challenger* for the Deep Sea Drilling Program. Later, the informal consortium was formalized as the Joint Oceanographic Institutions (JOI), Inc. and participation in the ocean drilling program became international, but major operational responsibility still rests in the different JOI institutions. I know I am not alone when I claim that the Ocean Drilling Program (as it is now called) is one of the most successful, as well as one of the longest-running, singularly focused international oceanographic research programs. The leadership, including the formal direction of the program, has come from the major university oceanographic institutions of the United States.

If somehow the decision to operate major oceanographic facilities had gone the other way—if the responsibility for ship operations had been vested in a central organization, for example, the marine equivalent of NCAR—then I expect buoy arrays, submersibles, and similar facilities would have been housed there also. If this had happened I believe ocean science would be weaker in the United States than it is today.

However, having said this, I believe it is the National Science Foundation, and not the major oceanographic institutions, that deserves much of the credit for ensuring that the academic ship operation program works as well as it does and continues to be acceptable to the ever-growing group of scientists who need to find a way to work at sea. In the beginning, each institution ran its ship program differently. What might be available as ship support or a ship's scientific equipment at one institution would not necessarily be available at another. One needed to know the details. As a consequence the scientist in charge of a seagoing program generally came from the same institution as the ship. As the family of seagoing oceanographers grew, not all made their careers at ship-operating institutions. The UNOLS program does much to guarantee uniformity of support among the ships of the fleet. The ships are different and some programs can only be accommodated on certain vessels, but the number of unpleasant surprises concerning vessel furnished facilities and support has been significantly reduced, if not entirely eliminated. You no longer need be a part of a ship-operating institution to conduct a major research program requiring a research vessel.[17]

BIG PROGRAMS

If the early ship support practices of ONR and NSF that evolved into UNOLS did much to define the early development of academic oceanographic institutions in the United States, the NSF support of multi-institutional, multi-investigator programs through the International Decade of Ocean Exploration did much to define the practice of oceanography in the 1970s. These large programs, each with its own catchy acronym, established a pattern of doing oceanography that continues today. How the IDOE came to be and was organized is described in the paper by Jennings later in this volume.[18] I wish to note the contribution of these programs to what might be called the sociology of oceanography. Just as I believe the operation of oceangoing ships did much to define the oceanographic institutions for the first half of NSF's 50 years, so I believe NSF's sponsorship of large multi-investigator programs has done much during the last 25 years to develop a level of cultural sophistication among the oceanographic community found in relatively few others fields of science.

Oceanographic field work is expensive and often frustrating. Many of the most interesting problems are best attacked by a multiprobe program. Those who succeed at sea with their observational and experimental programs soon learn that it can be dangerous to put all of one's effort into a single approach or a single instrument. We also learn that serendipity often plays a larger role than we wish to admit in whatever successes we achieve. By 1970, many of those with extensive seagoing experience were ready to embrace the concept of a multifaceted, multi-investigator approach to oceanographic field work. We believed it was the most cost-effective way to attack problems that were often not as well defined as we liked to suggest in our proposals. Multi-investigator programs followed naturally, and several were in the planning stage at the start of the IDOE: the Geochemical Ocean Sections Study (GEOSECS), the Mid-Ocean Dynamic Study (MODE), and the Climate: Long-Range Inves-

---

[17]For a discussion of UNOLS and its establishment, see Byrne and Dinsmore, this volume.

[18]Jennings, F.D. and L.R King. 1980. Bureaucracy and science: The IDOE in the National Science Foundation. *Oceanus*. 23(1):12-19; see also Jennings, this volume.

tigation Mapping and Predication (CLIMAP) program.[19] These large multi-investigator programs that reached full flower in the IDOE have continued to constitute a significant fraction of NSF's ocean budget.[20]

One reason these large programs continue is because they generate interesting, often exciting, science, and they appear to be cost-effective. I also believe that an unanticipated positive contribution of these programs is that they teach a degree of cooperation, mutual respect for different approaches and personalities, and an understanding of institutional complexities. Scientists who participate in—some might suggest those who survive—these large multi-investigator, multi-institutional programs generally come away with a deeper understanding of their colleagues and the various ways of achieving their own goals.

I do not wish to imply that scientific cooperation is unique to oceanography. Joint investigations and joint papers are the rule, not the exception, in science; and much organization and cooperation is required to gain the maximum effectiveness from such large pieces of equipment as telescopes, satellites, high-energy machines, or deep-sea drilling vessels. Committees of the National Research Council and others are well equipped to outline important problems that require attention. But I do believe it is relatively rare in science to not only have a problem first defined by a committee, but then to have a committee outline the approach, determine what kinds of scientific specialists are required to successfully implement the approach, provide a steering committee to ensure that there are no significant holes in the combined proposals, make certain that deadlines are met, and do all of the other chores necessary to ensure that the whole of the multi-investigator program is greater than the sum of its parts.

Joint programs of this kind are not to everyone's taste, nor do I expect many to make a career of participating in such programs one after another, but I do believe that large multi-investigator programs contribute to the education of those who participate and that oceanography is a stronger field today because so many have been associated with at least one such program. I have no proof, and others may disagree, but it is my sense in talking to colleagues who have been so involved, that they have a deeper understanding of what it takes to mount a successful science program (whether it be single or multi-investigator, small science or large science). They know how to go about it, and they have a confidence that they can succeed. I believe NSF's multi-investigator, multi-institutional programs, which started with the IDOE and continue today, have contributed significantly to all of oceanography, both big science and small.

In summary, oceanography in this country was jump-started during World War II, and the Office of Naval Research was there immediately after the war to support and to expand the field. The formation of the National Science Foundation in 1950 began a slow, 20-year passing of the torch of primary support from ONR to NSF that was completed in 1970 with the start of the IDOE and the passage of the Mansfield Amendment. Both agencies deserve major credit for ensuring that the transition went as smoothly as it did. For those of us who were supported during this period it was sometimes difficult to remember which agency was responsible for which support. There were few hiccups as we transitioned from one grant or contract to another.

Fifty years of partnership between those who practice science and those who support science result in cultural patterns that we sometimes take for granted. As I reflect on how oceanography has been supported and is supported in other parts of the world, I see several examples of NSF's way of doing things that I believe have been important to the development of oceanography in this country. The first, of course, is the peer-reviewed grant proposal system, which pervades all of NSF and has been widely discussed for many years. Two others are peculiar to oceanography and may be less obvious, but I believe each has played an important role in the development of both oceanographic institutions and oceanographers in this country. They are institutional support of ships through the UNOLS program and the sponsorship of multi-investigator, multi-institutional programs that began with the IDOE. As NSF looks back on 50 years of support for oceanography, it has much of which to be proud.

---

[19]Ibid.
[20]See Lambert and Purdy et al., references 7 and 13.

Landmark Achievements of Ocean Sciences

Achievements in Biological Oceanography

RICHARD T. BARBER AND ANNA K. HILTING

Duke University, NSOE Marine Laboratory

INTRODUCTION

For the Ocean Studies Board's examination of achievements of the National Science Foundation (NSF) in ocean research, we have been asked "to focus on the landmark achievements in biological oceanography in the past 50 years, the individuals involved, the new technology and ideas that made these achievements possible, how one discovery built on the foundations of earlier ones, discoveries made at the intersections of disciplines, and the role that NSF programs and institutional arrangements had in making these achievements possible."

The period addressed, the first 50 years of the National Science Foundation, has been a heady time for biological oceanography, and identification of landmark achievements of this period is a fitting tribute for the 1998 International Year of the Ocean. The pace of biological investigation of the ocean quickened as this period began in the 1950s. To appreciate the magnitude of the acceleration, a brief look at biological oceanography in the first half of this century is useful.

Before World War II, biological oceanography had two main themes. The first and more important by far was as a handmaiden to fisheries science. Where were exploitable resources, why did they vary so in abundance, and how could more resources be found? These were the demanding questions asked of biological oceanography. These questions, particularly the old question of why recruitment to exploitable fish stocks varies so much from year to year and from place to place (Hjort, 1926), have proved to be profoundly complex questions that biological oceanography still struggles with today (cf. the current Global Ocean Ecosystems Dynamics [GLOBEC] program).

The second major theme was discovery per se. Exotic and strange animals, and to a lesser degree plants, commanded interest among scientists as well as the general public; they illustrated the marvels of adaptation and evolution. Exploration for its own sake has always motivated biological oceanographers, but as the discipline matured, this motivation became less fashionable. NSF has never supported biological oceanographers for the sole purpose of looking for strange new organisms, yet discovery is what makes biological oceanography so much fun.

After World War II the climate of oceanographic research was different. One of the authors (RTB) spent an undergraduate summer in Woods Hole in the mid-1950s, a time when the overriding impression was that there were many exciting questions and unlimited opportunities. Ideas newly aired then were the ecosystem constructs of the Odum brothers (Odum and Odum, 1955), the quantitative and predictive plankton studies of Riley (1946), and the elegant ecology and evolution theories of Hutchinson (1961); the brilliance of these ideas inspired students and researchers. The relative merits of applied versus basic research became a topic of frequent discussion. At the same time, the distance between biological oceanography and fisheries science widened here in the United States until, by the 1960s, there was little significant intellectual exchange between the two disciplines. A few iconoclastic individuals argued for studying ocean biology whether or not an application for the new knowledge could be envisioned, implying that knowledge per se was good.

In addition to knowledge for the sake of knowledge, there arose in the postwar period the specter of pollution and the notion that it was necessary to know how the ocean functioned to avoid inadvertently destroying it. All of these needs or objectives—food from the ocean, discovery, knowledge for its own sake, and the custodial sense arising from pollution concerns—provided motivations to expand and reshape the field. But by far the most important impetus driving the expansion of biological oceanography was the cornucopia of new resources available for science.

Financial resources that flowed to American science as a result of *Sputnik*, the scientific race with the Soviets, and the Cold War trickled down (or up) to biological oceanography. It should be noted that NSF made a conscious decision to support biological oceanography in the 1950s because it foresaw that biological oceanography was unlikely to receive support elsewhere, including the Office of Naval Research. These new resources drove the increase in basic, NSF-supported, research during this period. Responding to the increased supply of resources by recruiting young scientists in large numbers, biological oceanography became a discipline of its own and settled into a steady rate of progress and expansion. In this paper we identify nine landmark achievements of biological oceanography of the past 50 years and discuss who made them, their sequelae, and NSF's role in making them possible.

PROCEDURE

This review got its start in March 1998 at an NSF-sponsored retreat where a group of about 50 people pondered the future of biological oceanography; the exercise was called OEUVRE (Ocean Ecology: Understanding and Vision for Research). In considering what's exciting for the future of biological oceanography, there was a thorough and wide-ranging discussion of achievements of the past two or three decades. We have used the OEUVRE report liberally; its results are presented in the paper by Peter Jumars later in this volume.

Next we queried about 150 practicing biological oceanographers on their opinions of the landmark achievements of the past 50 years. Almost everyone responded to our query, and it was fascinating to see this thoughtful self-evaluation of our discipline. Organizing and collating the many replies was educational, but this informal survey did not lend itself to quantitative analysis. We also looked at citation indices (McIntosh, 1989; Parsons and Seki, 1995) but did not use this information because biological oceanography was not a specific category. In the end we made a subjective selection which was, for the most part, consistent with the suggestions provided by the community. We thank the respondents and acknowledge how much we learned from their replies, but we absolve them from responsibility for the following.

Because neither of the authors has formal training as a historian, we are in every sense amateurs at writing history. Our strongest, or perhaps weakest, characteristic is a passionate interest in biological oceanography and its history. Another important weakness is that we are practicing biological oceanographers. It is unrealistic to expect an objective history of baseball from players who are in the middle of a playoff game. Our paper is very subjective—interesting and informative, we hope, but not necessarily objective.

Selecting achievements to include was not difficult; the agonizing aspect was what to leave out. In biology there are many kinds of achievements. In this short paper we do not do justice to the diversity of biological oceanography. Also, as any NSF program manager in biological oceanography will tell you, there is no tidy framework for organizing the different parts of biological oceanography. Our list is therefore eclectic as well as subjective.

TWO WONDERFUL ACCIDENTS: VENTS AND OCEAN COLOR

We begin with two landmark achievements that more or less fell into the laps of biological oceanographers.

Chemosynthetic Hydrothermal Vent Communities (Plate 1)

This is an easy landmark to start with because it has all the dramatic elements of discovery. We may no longer set out on voyages of discovery, but in the past 50 years the pace of biological discovery has been awesome. In 1976, when geologists discovered the hydrothermal vents, biological oceanography received a much-appreciated jolt of intellectual stimulation (Corliss et al., 1979). The existence of a new kind of ecosystem with dramatic new biochemical adaptation fueled the imagination of everyone. The names associated with this pioneering work on chemosynthesis are a cross section of the gentry of biological oceanography. Cavenaugh, Childress, Grassle, Jannasch, Karl, Lutz, and Somero were early leaders in this work, but the list soon expanded to include several dozen individuals (see references below). From this work we learned how organisms adapt biochemically to temperature extremes and lack of oxygen, a line of investigation that has led to the discovery of active microbes deep in the Earth. This work also provides a rational organizing paradigm for the search for life on other celestial bodies.

What is amazing about the discovery of chemosynthetic ecosystems is that, once discovered, they have turned up everywhere in the ocean: on the continental shelves and slopes, in the deep sea, and at plate margins and ridge crests (Van Dover, 1990, 1998, 1999). They are hot vents or cold seeps; their reducing power comes from hydrogen sulfide or methane. Chemosynthetic ecosystems even exist on whale carcasses (Smith et al., 1989).

The mystery is how we overlooked these ubiquitous ocean ecosystems for so long, and we wonder what other surprises the ocean holds.

NSF's Biological Oceanography Program has been the lead agency in support of this work, and *Alvin* support by NSF made rapid progress possible. The discovery, response by scientists, and response by NSF provide a model of science at its best.

Chemosynthetic Hydrothermal Vent Communities References[1]

1979   Corliss, J.B., J. Dymond, L.I. Gordon, J.M. Edmond, R.P. von Herzen, R.D. Ballard, K. Green, D. Williams, A. Bainbridge, K. Crane, and T.H. van Andel. 1979. Submarine thermal springs on the Galapagos Rift. *Science* 203:1073-1083.

1979   Jannasch, H.W., and C.O. Wirsen. 1979. Chemosynthetic primary production at East Pacific sea floor spreading centers. *BioScience* 29:592-598.

1980   Karl, D.M., C.O. Wirsen, and H.W. Jannasch. 1980. Deep-sea primary production at the Galapagos hydrothermal vents. *Science* 207:1345-1347.

1981   Cavanaugh, C.M., S.L. Gardiner, M.L. Jones, H.W. Jannasch, and J.B. Waterbury. 1981. Procaryotic cells in the hydrothermal vent tube worm *Riftia pachyptila* Jones: Possible chemoautotrophic symbionts. *Science* 213:340-341.

1981   Felbeck, J., J.J. Childress, and G.N. Somero. 1981. Calvin-Benson cycle and sulphide oxidation enzymes in animals from sulphide-rich habitats. *Nature* 293:291-293.

1983   Arp, A.J., and J.J. Childress. 1983. Sulfide binding by the blood of the hydrothermal vent tube worm *Riftia pachyptila. Science* 219:295-297.

1984   Lutz, R.A., R.D. Turner, and D. Jablonski. 1984. Larval development and dispersal at deep-sea hydrothermal vents. *Science* 226:1451-1454.

1985   Paull, C.K., B. Hecker, R. Commeau, R.P. Freeman-Lynde, C. Neumann, W.P. Corso, S. Golubic, J.E. Hook, E. Sikes, and J. Curray. 1985. Biological communities at the Florida escarpment resemble hydrothermal vent taxa. *Science* 226:965-967.

1985   Grassle, J.F. 1985. Hydrothermal vent animals: Distribution and biology. *Science* 229:713-717.

1985   Okutani, T., and K. Egawa. 1985. The first underwater observation on living habitat and thanatocoenoses of *Calyptogena soyoae* in bathyal depth of Sagami Bay. *Venus: Japanese Journal of Malacology* 44:285-289.

1989   Smith, C.R., H. Kukert, R.A. Wheatcroft, P.A. Jumars, and J.W. Deming. 1989. Vent fauna on whale remains. *Nature* 341:27-28.

1990   Van Dover, C.L. 1990. Biogeography of hydrothermal vent communities along seafloor spreading centers. *Trends in Ecology and Evolution* 5:242-246.

1998   Van Dover, C.L. 1998. Vents at higher frequency. *Nature* 395:437-439.

1999   Van Dover, C.L. 1999. Deep-sea clams feel the heat. *Nature* 397:205-220.

Ocean Color—Seeing the Ocean for the First Time

The Coastal Zone Color Scanner (CZCS) launched in 1978 showed biological oceanographers the patterns, variability, complexity, and coherence of ocean biology for the first time. Biological oceanography became a global discipline in a single step. It is, of course, somewhat facetious to call this new satellite-based remote sensing capability an "accident." Far-sighted individuals such as Giff Ewing and Charlie Yentsch kept prodding the National Aeronautics and Space Administration (NASA) in the right direction; they provided an accurate vision of what could be. However, the

real drivers in the early days were the spirit of NASA, its engineers, and their unquenchable drive to build whatever could be built and flown on satellites. Our reading of the event is that NASA was looking for challenges, and the quantitative assessment of ocean surface chlorophyll and related pigments by reflected light was a challenge they took on with enthusiasm. Ironically, biological oceanographers don't even know the names of these creative NASA engineers who built the CZCS, but that doesn't reduce our debt to them.

The first CZCS data of reflected light that became available in the late 1970s started a scramble to put together systems to process and interpret this new kind of data. The key algorithms produced at the University of Miami (Gordon and Clark, 1980; Gordon et al., 1983) were the "open sesame" that permitted biological oceanographers to see the ocean for the first time (see references below). As CZCS images flooded into our consciousness it became obvious that we needed to train a cohort of biological oceanographers who would know how to use the new technology. This hard work has paid off. When a new, much improved U.S. ocean color satellite Sea-Viewing Wide Field of View Sensor (SeaWiFS) (Plate 2) flew in August 1997, the community was ready. As a result, the pace of biological oceanography has quickened all around the globe.

The space-based analysis of chlorophyll concentration based on ocean color revealed (1) oceanography's chronic problem of undersampling; (2) dominance of mesoscale physical processes in determining the spatial distribution of phytoplankton; (3) effect of topography on biomass; (4) complexity of the seasonal progression of phytoplankton blooms; and (5) magnitude of interannual variability. Space-based analysis changed not only our perception of the ocean, but also our ideas of what constitutes good biological oceanography. Of the various landmark achievements mentioned here, this is one that profoundly affects all biological oceanographers and indeed each citizen of the planet. Having seen the totality of the oceans, mankind can no longer maintain the concept of discrete or isolated components of the ocean.

NASA, of course, was the major patron of this work, but NSF has been and remains an important supporter of the synthesis and interpretation of this exciting new way to view the ocean. This NASA-NSF cooperation is an example of science support at its best.

Ocean Color References

1980   Gordon, H., and D.K. Clark. 1980. Atmospheric effects in the remote sensing of phytoplankton pigments. *Boundary-Layer Meteorol.* 18:299-313.

1983   Gordon, H.R., D.K. Clark, J.W. Brown, O.B. Brown, R.H. Evans, and W.W. Broenkow. 1983. Phytoplankton pigment concentrations in the Middle Atlantic Bight: Comparison of ship determinations and CZCS estimates. *Applied Optics* 22:3929-3931.

1986   Esaias, W.E., G.D. Feldman, C.R. McClain, and J.A. Elrod. 1986. Monthly satellite-derived phytoplankton pigment distribution for the North Atlantic Ocean basin. *Eos, Trans., AGU* 67:835-837.

---

[1]References are given in chronological, rather than alphabetical, order to emphasize the progression of the discoveries in each landmark area.

1987   Yoder, J.A, C. McClain, J. Blanton, and L. Oey. 1987. Spatial scales in CZCS-chlorophyll imagery of the southeastern U. S. continental shelf. *Limnol. Oceanogr.* 32:929-941.

1989   Feldman, G., N. Kuring, C. Ng, W. Esaias, C.R. McClain, J. Elrod, N. Maynard, D. Endres, R. Evans, J. Brown, S. Walsh, M. Carle, and G. Podesta. 1989. Ocean color, availability of the global data set. *Eos, Trans., AGU* 70:634-641.

1993   Sullivan, C.W., K.R. Arrigo, C.R. McClain, J.C. Comiso, and J. Firestone. 1993. Distributions of phytoplankton blooms in the Southern Ocean. *Science* 262:1832-1837.

1993   Yoder, J.A., C.R. McClain, G.C. Feldman, and W.E. Esaias. 1993. Annual cycles of phytoplankton chlorophyll concentrations in the global ocean: A satellite view. *Global Biogeochem. Cycles* 7:181-193.

Global Productivity and Productivity Regimes— The Stepchildren of Ocean Color

Soon after Steeman-Nielsen (1952) introduced the radioactive carbon tracer method to measure primary productivity, biological oceanographers began to use the new productivity observations to speculate about the existence of differing oceanic productivity regimes and to estimate global productivity (Ryther, 1959). Two signal achievements in the estimation of global productivity were Ryther's synthesis (1969) dealing with productivity in different oceanic regimes and the synthesis by Koblentz-Mishke et al. (1970) of all the available radiocarbon productivity data. Both contributions advanced biological oceanography, but undersampling compromised both efforts.

Global CZCS chlorophyll coverage provided a way to break out of this sampling limitation using the productivity-chlorophyll-light relationship described first by Ryther and Yentsch (1957). High-resolution spatial and temporal patterns of phytoplankton biomass permitted objective estimates of global primary productivity (see references below) as well as the size and seasonal variability of the various productivity regimes or biogeochemical provinces of the world ocean (Longhurst, 1998). Arguably the most important scientific contributions of the satellite ocean color breakthrough to date have been improved estimates of global productivity and the birth of an objective ecological geography of the sea.

Global Productivity and Productivity Regimes References

1952   Steemann-Nielsen, E. 1952. The use of radioactive carbon (^{14}C) for measuring organic production in the sea. *J. Cons. Int. Explor. Mer* 144:38-46.

1957   Ryther, J.H., and C.S. Yentsch. 1957. The estimation of phytoplankton production in the ocean from chlorophyll and light data. *Limnol. Oceanogr.* 2:281-286.

1959   Ryther, J.H. 1959. Potential productivity of the sea. *Science* 130:602-608.

1969   Ryther, J.H. 1969. Photosynthesis and fish production in the sea. *Science* 166:72-76.

1970   Koblentz-Mishke, O.J., V.V. Volkovinsky, and J.G. Kabanova. 1970. Plankton primary production of the world ocean. Pp. 183-193 in W.S. Wooster (ed.), *Scientific Exploration of the South Pacific.* National Academy of Sciences, Washington, D.C.

1982   Smith, R.C., R.W. Eppley, and K.S. Baker. 1982. Correlation of primary production as measured aboard ship in southern California coastal waters and as estimated from satellite chlorophyll images. *Mar. Biol.* 66:281-288.

1985   Eppley, R.W., E. Stewart, M.R. Abbott, and V. Heyman. 1985. Estimating ocean primary production from satellite chlorophyll, introduction to regional differences and statistics for the Southern California Bight. *J. Plankton Res.* 7:57-70.

1988   Platt, T., and S. Sathyendranath. 1988. Oceanic primary production: Estimation by remote sensing at local and regional scales. *Science* 241:1613-1620.

1992   Balch, W.M., R. Evans, J. Brown, G. Feldman, C. McClain, and W. Esaias. 1992. The remote sensing of ocean primary productivity: Use of new data compilation to test satellite algorithms. *J. Geophys. Res.* 97:2279-2293.

1995   Longhurst, A., S. Sathyendranath, T. Platt, and C. Caverhill. 1995. An estimate of global primary production in the ocean from satellite radiometer data. *J. Plankton Res.* 17:1245-1271.

1996   Antoine, D., J.M. Morel, and A. Morel. 1996. Oceanic primary production. 2. Estimation at global scale from satellite (Coastal Zone Color Scanner chlorophyll). *Global Biogeochem. Cycles* 10:57-69.

1997   Behrenfeld, M.J., and P.G. Falkowski. 1997. Photosynthetic rates derived from satellite-based chlorophyll concentration. *Limnol. Oceanogr.* 42:1-20.

1998   Longhurst, A. 1998. *Ecological Geography of the Sea.* Academic Press, San Diego, California. 398 p.

1999   Esaias, W.E., R.L. Iverson, and K. Turpie. 1999. Ocean province classification using ocean color data: Observing biological signatures of variations in physical dynamics. *Global Change Biology,* in press.

1999   Iverson, R.L., W.E. Esaias, and K. Turpie. 1999. Ocean annual phytoplankton carbon and new production, and annual export production estimated with empirical equations and CZCS data. *Global Change Biology,* in press.

FOUR SPECIFIC BREAKTHROUGHS

Deep-Sea Diversity

The discovery of high biological diversity in the deep sea in the late 1960s changed the way deep-sea biology was viewed, and sparked theoretical debates on how diversity is maintained in a large, monotonous environment such as the deep sea (see references below). The diversity analyses, set in motion in the 1960s by Howard Sanders and Bob Hessler, were followed up by Paul Dayton, Fred Grassle, Gil Rowe, and Pete Jumars. This work was enhanced by the availability of the submersible *Alvin*, which gave researchers direct observation and the ability to do *in situ* benthic experiments. The skill these early workers gained in using *Alvin* for diversity and metabolic studies made it possible for them to shift rapidly to work on the hydrothermal vents soon after their discovery in 1976.

Alvin changed our perception of the deep sea just as the CZCS satellite changed our perception of the surface ocean. Images of the seafloor—particularly the monotonous, soft-sediment abyssal regimes—documented how different the deep-sea environment is from any other that ecologists have visited.

The discovery of high diversity in the deep sea was critically important to the evolution and maturation of biological oceanography because it provided scientific respectability to this expensive research. Deep-sea animals were found to be interesting and sometimes weird, as *National Geographic* articles frequently reminded us, but of what relevance was deep-sea ecology? The discussion of diversity thrust deep-sea research into a mainstream ecology debate that was important and exciting. This development was pivotal because NSF is most comfortable supporting hypothesis-driven research on questions that are significant to mainstream science. After Howie Sanders and Bob Hessler published on deep-sea diversity (Sanders, 1967; Hessler and Sanders, 1969; Sanders and Hessler, 1967; Dayton and Hessler, 1972; Grassle and Sanders, 1973), there were abundant hypotheses to be tested, and tested they were. The Biological Oceanography Program at NSF was, and still is, the major supporter of this work.

Deep-Sea Diversity References

1967   Hessler, R.R., and H.L. Sanders. 1967. Faunal diversity in the deep-sea. *Deep-Sea Res.* 14:65-78.

1968   Sanders, H.L. 1968. Marine benthic diversity: A comparative study. *Am. Natur.* 102:243-282.

1969   Rowe, G.T., and R.J. Menzies. 1969. Zonation of large benthic invertebrates in the deep-sea off the Carolinas. *Deep-Sea Res.* 16:531-537.

1969   Sanders, H.L., and R.R. Hessler. 1969. Ecology of the deep-sea benthos. *Science* 163:1419-1424.

1972   Dayton, P.K., and R.R. Hessler. 1972. Role of biological disturbance in maintaining diversity in the deep sea. *Deep-Sea Res.* 19:199-208.

1973   Grassle, J.F., and H.L. Sanders. 1973. Life histories and the role of disturbance. *Deep-Sea Res.* 20:643-659.

1976   Jumars, P.A. 1976. Deep-sea species diversity: Does it have a characteristic scale? *J. Mar. Res.* 34:217-246.

New and Regenerated Productivity

This landmark achievement had its origin in a *Limnology and Oceanography* publication by Dugdale and Goering (1967) that introduced a deceptively simple notion: primary productivity in the ocean can be divided into the portion that uses locally recycled nutrients (regenerated production) and the portion that uses nutrients newly transported into the euphotic zone (new production), usually by the physical processes of mixing and upwelling. Dugdale and Goering's exciting and powerful concept was presented in very basic terms and specifically included "new" nutrients entering from the atmosphere, a process that was not considered important in 1967 but is now known to be significant.

The new production concept, together with the Dugdale (1967) paper on nutrient uptake dynamics in the same issue of *Limnology and Oceanography*, provided biological ocean-ography with the mathematical formalism needed for rigorous, quantitative modeling of ocean productivity and biogeochemical fluxes. (See also Eppley et al., 1969; MacIsaac and Dugdale, 1969.) This formalism fueled the explosive growth of modeling described in the modeling section later in this paper.

Eppley and Peterson (1979) further developed the concept by arguing that at steady state the magnitude of new production is equal to the export flux of particulate organic matter out of the euphotic zone to the ocean interior. Together, the Dugdale and Goering (1967) and Eppley and Peterson (1979) papers have impressive citation index scores. At the ages of 31 and 19 years, respectively, they are cited more now than they were in their first decades. They are like fine wines. Significantly, Eppley and Peterson (1979) estimated global new production to be about 4 petagrams per year and suggested for the first time that this number approximates the sinking flux of organic carbon and, hence, the rate at which the deep sea sequesters atmospheric carbon dioxide. This number has proved very durable; it is still used in global biogeochemical budgets.

As a consequence of the work of Dugdale and Goering (1967) and Eppley and Peterson (1979), a link was forged between physical and biological oceanography. The new concept required that physical processes of mixing and upwelling be an integral part of ecosystem models dealing with new production, fish production, or export of organic material from the surface layer. Ocean physics and biology were formally wed by this landmark achievement.

The technological advance that made progress on new production possible was the use of a stable isotope tracer ^{15}N and a mass spectrometer to measure it precisely. The ^{15}N tracer method was a logical development of Steemann-Nielsen's (1952) breakthrough use of ^{14}C as a tracer of carbon fixation.

Dugdale and Goering's work was supported by the NSF International Indian Ocean Expedition and its successor program, the Southeastern Pacific Expedition, using the NSF ship *Anton Bruun* for focused biological oceanography. The NSF decision to fund this vessel specifically for biological oceanography was a decision that had positive long-range consequences for the field. In addition to expeditionary support from NSF, laboratory support came from the Atomic Energy Commission (AEC), now the Department of Energy (DOE). In the 1960s and 1970s, NSF and AEC had a productive partnership, with NSF providing focused investigator and expedition awards and AEC providing block grants to support research groups. Dugdale and Goering went to sea with John Ryther's AEC-supported research group at the Woods Hole Oceanographic Institution. Eppley was a member of J.D.H. Strickland's AEC group at Scripps Institution of Oceanography and headed this group later during its most productive years. However, NSF was the lead agency responsible for this breakthrough and the agency should take great pride in this landmark achievement.

New and Regenerated Productivity References

1967   Dugdale, R.C., and J.J. Goering. 1967. Uptake of new and regenerated forms of nitrogen in primary production. *Limnol. Oceanogr.* 12:196-206.

1967   Dugdale, R.C. 1967. Nutrient limitation in the sea: Dynamics, identification and significance. *Limnol. Oceanogr.* 12:655-695.

1969   MacIsaac, J.J., and R.C. Dugdale. 1969. The kinetics of nitrate and ammonia uptake by natural populations of marine phytoplankton. *Deep-Sea Res.* 16:47-58.

1969   Eppley, R.W., J.N. Rogers, and J.J. McCarthy. 1969. Half-saturation constants for uptake of nitrate and ammonium by marine phytoplankton. *Limnol. Oceanogr.* 14:912-920.

1979   Eppley R.W., and B.J. Peterson. 1979. Particulate organic matter flux and planktonic new production in the deep ocean. *Nature* 282:677-680.

How Zooplankton Swim, Feed, and Breed

Zooplankton live in a medium that, to them, is viscous and structured (Koehl, 1993). The process of capturing food does not involve passive sieving as much as it involves purposeful ingestion of particular food targets (see references below). Phytoplankton and other tasty prey items leave a chemical trail in the viscous water, and zooplankton follow such trails to find and eat a particular victim. Data suggest that the same process is at work in finding mates (Howlett, 1998; Yen et al., 1998). This view of the zooplankton world strains our credulity: because of our size, we cannot easily comprehend the low Reynolds number world in which zooplankton—especially copepods and smaller—swim, feed, and breed. This work showed us a world that is very common on our planet, but beyond our ken.

This new understanding has come in large part from the intellectual prodding of a single individual, Rudy Strickler, although he has had some very capable collaborators such as Mimi Koehl, Gus Paffenhöfer, Holly Price, Jeanette Yen, and others. A fascinating thing about this breakthrough is that it was immediately adopted by the field and entrained into the mainstream ideas. Zooplankton "gurus" such as Bruce Frost, Charlie Miller, Mike Roman, Sharon Smith, and Peter Wiebe had prepared the way for rapid assimilation of these new ideas by arguing that zooplankton feeding is selective and purposeful. Miller, in particular, had long emphasized that copepods fed in a viscous medium. Strickler's innovative high-speed movies of live copepod feeding showed how selectivity is realized. The technical breakthrough that made this advance possible was microcinematography. In this case, live copepods superglued to a dog hair on a microscope slide were filmed with a high-speed, strobe movie camera focused on the tethered animal. Innovation has many faces. NSF was the major source of support for this innovative work, which is continuing at an accelerated pace, but the Office of Naval Research (ONR) has also been a significant patron.

How Zooplankton Swim, Feed, and Breed References

1977   Hamner, P., and W.M. Hamner. 1977. Chemosensory tracking of scent trails by the planktonic shrimp *Acetes sibogae australis*. *Science* 195:886-888.

1977   Rubenstein, D.I., and M.A.R. Koehl. 1977. The mechanisms of filter feeding: Some theoretical considerations. *Amer. Nat.* 111:981-994.

1980   Alcaraz, M., G.A. Paffenhöfer, and J.R. Strickler. 1980. Catching the algae: A first account of visual observations on filter feeding calanoids. Pp. 241-248 in W.C. Kerfoot (ed.), *Evolution and Ecology of Zooplankton Communities*. University Press of New England, Biddefort, Maine.

1981   Koehl, M.A.R., and J.R. Strickler. 1981. Copepod feeding currents: Food capture at low Reynolds number. *Limnol. Oceanogr.* 26:1062-1073.

1982   Paffenhöfer, G.A., J.R. Strickler, and M. Alcaraz. 1982. Suspension-feeding by herbivorous calanoid copepods: A cinematographic study. *Mar. Biol.* 67:193-199.

1982   Strickler, J.R. 1982. Calanoid copepods, feeding currents, and the role of gravity. *Science* 218:158-160.

1983   Price, J.J., G.A. Paffenhöfer, and J.R. Strickler. 1983. Modes of cell capture in calanoid copepods. *Limnol. Oceanogr.* 28:116-123.

1993   Koehl, M.A.R. 1993. Hairy little legs: Feeding, smelling, and swimming at low Reynolds number. *Contemp. Math.* 141:33-64.

1998   Howlett, R. 1998. Sex and the single copepod. *Nature* 394:423-425.

1998   Yen, J., M.J. Weissburg, and M.H. Doall. 1998. The fluid physics of signal perception by mate-tracking copepods. *Phil. Trans. R. Soc. Lond.* 353:787-804.

Iron Hypothesis

The iron issue is an example of classic science progress:

a. There was a nagging question.
b. A tentative explanation was advanced.
c. Available data did not support the explanation.
d. The data, however, were suspect.
e. A technical (analytical) breakthrough was made.
f. The new data suggested an hypothesis.
g. The hypothesis was tested and confirmed.
h. Textbooks had to be revised.

For the iron issue the nagging question was: Why do excess plant nutrients persist in the surface ocean in certain regions such as the Antarctic, equatorial Pacific, and Northeast Pacific? For 50 years there had been speculation that iron limitation might be a factor, but measurements showed there was abundant iron in seawater.

John Martin set out to improve the analytical chemistry of iron, and when he had done so, he found that iron was much less abundant in the ocean than previously thought. Martin's innovations in iron chemistry alone would have earned him a place in history, but John Martin continued his quest with great zest. In a 1990 paper in *Paleoceanography*, he published the "Iron Hypothesis," which proposed that glacial-interglacial changes in atmospheric carbon dioxide were driven by variations in dryness, dust, and iron that forced variations in new production and, hence, atmospheric carbon dioxide drawdown in Antarctic waters.

The rest of this story is well known. John Martin gained considerable media attention with the radical notion that iron addition in the Southern Ocean could be used to "engineer down" atmospheric carbon dioxide. This marked the first time that biological oceanography per se commanded prime-time media attention, and it was no surprise that Martin's proposed iron enrichment method to draw down atmospheric CO_2 met with considerable negative publicity and was unpopular with biological oceanographers, environmentalists, and federal agencies. Martin himself kept his radical notion, which he always mentioned with a playful grin, separate from his serious determination to test the Iron Hypothesis. His critics did not or would not recognize this distinction.

John H. Martin died in June 1993, but his iron hypothesis was tested successfully in an *in situ* transient iron enrichment experiment in September 1993 (Martin et al., 1994) and again in May 1995 (Coale et al., 1996). It has now become evident that iron is a limiting or regulating nutrient in many marine and freshwater habitats for many organisms, not just primary producers. At a recent American Society of Limnology and Oceanography (ASLO) meeting on aquatic sciences more than 50 papers referred to iron effects. As with the ubiquity of chemosynthetic ecosystems, the question is, How could we have missed the importance of iron for so long?

Martin's proposed research to test the iron hypothesis with an *in situ* transient iron addition in the equatorial Pacific Ocean was controversial from the start (Chisholm, 1995). There was significant opposition because of worries that confirmation of the hypothesis would lead immediately to reckless climate engineering. Furthermore, no one had ever modified and marked a patch of open-ocean water, and many oceanographers were dead certain that it couldn't be done. Two courageous program managers, Ed Green of ONR and Neil Anderson of NSF, devised a Byzantine funding arrangement to get Martin's experiment done despite their agencies' aversion to controversy. Without heroic efforts by these two individuals, the rapid progress in testing the Iron Hypothesis would not have taken place. It is regrettable that at present there are no *in situ* iron experiment projects under way by U.S. investigators; fortunately, other countries are forging ahead boldly with work in the Antarctic and North Pacific oceans.

Iron Hypothesis References

1990   Martin, J.H. 1990. Glacial-interglacial CO_2 change: The iron hypothesis. *Paleoceanography* 5:1-13.

1994   Martin, J.H., et al. 1994. Testing the iron hypothesis in ecosystems of the equatorial Pacific Ocean. *Nature* 371:123-129.

1995   Chisholm, S.W. 1995. The iron hypothesis: Basic research meets environmental policy. *Reviews of Geophysics* 33:1277-1288.

1996   Coale, K.H. et al. 1996. A massive phytoplankton bloom induced by an ecosystem-scale iron fertilization experiment in the equatorial eastern Pacific Ocean. *Nature* 383:495-501.

INDIVIDUAL INVESTIGATORS VERSUS TEAMS

Work on the preceding achievements was set in motion and doggedly pursued by individual investigators: deep-sea diversity by Howard Sanders and Bob Hessler; new and regenerated productivity by Dick Dugdale and Dick Eppley; zooplankton milieu by Rudy Strickler; and the iron hypothesis by John Martin. Of course, science in general (and oceanography, in particular) is a team activity, and these individuals had important and essential collaborators, but for the breakthroughs described here, these individual investigators were key to the achievement. In this context, these achievements are quite unlike the first two—the discovery of vents and the gaining of a global perspective through satellite imagery—and the following three, all of which were set in motion by teams.

THE MOST FAR REACHING ACHIEVEMENT

Recognizing the Microbial Character of the Pelagic Foodweb

Over the past 25 years our vision of the pelagic foodweb structure has changed dramatically. We now view the traditional "diatom-copepod-fish" foodweb as a relatively minor component. The foodweb consistently present in all oceanic habitats is based on pico- and nanoplankton-sized autotrophs and heterotrophs, which are efficiently grazed by flagellates and ciliates. The pelagic foodweb is microbe-centric. ("Microbe" in this context means small autotrophs, heterotrophs, and mixotrophs, and refers to both prokaryotes and eukaryotes.) Pioneering work by Malone (1971) introduced these ideas regarding picoplankton productivity and micrograzer regulation, but it was not until the late 1970s that this revolution gathered momentum.

The microbial revolution was the easiest achievement to select. In our informal survey it was by far the first choice for inclusion as a landmark achievement, and it was the accomplishment that one of the authors (RTB) suggested at the OEUVRE meeting as the major advance of the past 20 years. There is wide consensus that the microbial revolution is of paramount importance for biological oceanography. It is a revolution still in progress and it appears to be different things to different people (Azam, 1998; Steele, 1998).

In 1974, Larry Pomeroy's paper titled "The Ocean's Food Web: A Changing Paradigm" foretold the microbial revolution by asking a logical sequence of questions:

• Do small autotrophs carry out a major portion of oceanic primary production?

• Is nonliving organic matter, both dissolved and particulate, an important link in oceanic foodwebs?

• Do protist grazers such as ciliates and flagellates play a major role in grazing the autotrophic and heterotrophic microbes?

• Is leakage during feeding an important source of new dissolved organic material for heterotrophic microbes?

• Do microbes carry out the bulk of the respiration in the oceanic foodweb?

• Is recycling by the microbial foodweb a significant fate for newly produced organic matter?

At the time he asked them, Pomeroy's questions were unanswerable because of technical constraints. The saga of the microbial loop tells how one after another methodological advance allowed Pomeroy's questions to be answered. Hobbie et al. (1977) developed the fluorescent staining technique that permitted rapid counting and discrimination of bacteria, protozoa, and phytoplankton. The bacteria numbers found were high, but relatively constant. Bacterial production measured by Azam et al. (1983) was surprisingly high. Landry and Hassett (1982) and Fenchel (1982) found that protistan micrograzers provided the grazing mortality that held bacteria and picoautotrophs to relatively constant values. Rapidly growing micrograzers keep up with increases in growth rate of their bacterial and phytoplankton prey but never "overgraze" the prey because of threshold effects that make it unprofitable for micrograzers to feed when prey density drops below a given value.

The next step was to identify the source and magnitude of organic substrates for the heterotrophs. Measurement of dissolved organic carbon (DOC) was in disarray in 1974 when these questions were posed, but with a strong community effort supported by NSF, the DOC problem was painstakingly solved (Williams and Druffel, 1988; Peltzer and Brewer, 1993; Sharp, 1993). The presence of rapid DOC recycling was confirmed and other questions relative to DOC and bacterial production were rapidly solved (Ducklow and Carlson, 1992; Hansell et al., 1993).

In the mid-1980s, the new technology of flow cytometry enabled Chisholm et al. (1988, 1992) to discover a novel picoplankter that is now considered the most abundant autotroph in the world. How could we have overlooked these abundant organisms for so long?

Further work on micrograzer rates (Landry and Hassett, 1982; Landry et al., 1997) showed that grazer control of the pico- and nanophytoplankton was the norm and recycling by the microbial foodweb is a significant fate for primary production in the open ocean. Hard work and technical breakthroughs have confirmed most of the suggestions of Pomeroy (1974). Plate 3a shows how Steele (1998) entrained these ideas into a model of the pelagic foodweb; Plate 3b shows another representation of the concept. The Biological Oceanography Program at NSF was the major patron of the work that led this revolution. The response of NSF to the microbial revolution showed that this agency could adapt rapidly to a changing paradigm.

Recognizing the Microbial Character of the Pelagic Foodweb References

1971 Malone, T. 1971. The relative importance of nanoplankton and netplankton as primary producers in tropical oceanic and neritic phytoplankton communities. *Limnol. Oceanogr.* 16:633-639.

1974 Pomeroy, L.R. 1974. The ocean's food web, a changing paradigm. *BioScience* 24:499-504.

1977 Hobbie, J.E., R. J. Daley, and J. Jasper. 1977. Use of nucleopore filters for counting bacteria by fluorescence microscopy. *Appl. Env. Microbiol.* 33:1225-1228.

1980 Fuhrman, J.A., J.W. Ammerman, and F. Azam. 1980. Bacterioplankton in the coastal euphotic zone: Distribution, activity and possible relationships with phytoplankton. *Mar. Biol.* 60:201-207.

1981 Williams, P.J. Le B. 1981. Incorporation of microheterotrophic processes into the classical paradigm of the planktonic food web. *Kieler Meeresforschung* 5:1-28.

1982 Fenchel, T. 1982. Ecology of heterotrophic microflagellates. IV. Quantitative occurrence and importance as bacterial consumers. *Mar. Ecol. Prog. Ser.* 9:35-42.

1982 Fuhrman, J.A., and F. Azam. 1982. Thymidine incorporation as a measure of heterotrophic bacterioplankton production in marine surface waters: Evaluation and field results. *Mar. Biol.* 66:109-120.

1982 Landry, M.R., and R.P. Hassett. 1982. Estimating the grazing impact of marine micro-zooplankton. *Mar. Biol.* 67:283-288.

1983 Azam, F., T. Fenchel, J.G. Field, J.S. Gray, L.A. Meyer-Reil, and T.F. Thingstad. 1983. The ecological role of water-column microbes in the sea. *Mar. Ecol. Prog. Ser.* 10:257-263.

1988 Chisholm, S.W., R.J. Olson, E.R. Zettler, R. Goericke, J.B. Waterbury, and N.A. Welschmeyer. 1988. A novel free-living prochlorophyte abundant in the oceanic euphotic zone. *Nature* 334:340-343.

1992 Chisholm, S.W. et al. 1992. *Prochlorococcus marinus* nov. gen. nov. sp.: An oxyphototrophic marine prokaryote containing divinyl chlorophyll *a* and *b*. *Arch. Microbiol.* 157:297-300.

1992 Ducklow, H.W., and C.A. Carlson. 1992. Oceanic bacterial production. *Advances in Microbial Ecology* 12:113-181.

1995 Landry, M.R., J. Kirshtein, and J. Constantinou. 1995. A refined dilution technique for measuring the community grazing impact of microzooplankton, with experimental tests in the central equatorial Pacific. *Mar. Ecol. Prog. Ser.* 120:53-63.

1997 Landry, M.R., R.T. Barber, R.R. Bidigare, F. Chai, K.H. Coale, H.G. Dam, M.R. Lewis, S.T. Lindley, J.J. McCarthy, M.R. Roman, D.K. Stoecker, P.G. Verity, and R.T. White. 1997. Iron and grazing constraints on primary production in the central equatorial Pacific: An EqPac synthesis. *Limnol. Oceanogr.* 42:405-418.

1998 Azam, F. 1998. Microbial control of oceanic carbon flux: The plot thickens. *Science* 280:694-696.

1998 Steele, J.H. 1998. Incorporating the microbial loop in a simple plankton model. *Proc. Roy. Soc. Lond.* B 265:1771-1777.

Dissolved Organic Carbon (DOC) References:

1988 Williams, P.M., and E.R.M. Druffel. 1988. Dissolved organic matter in the ocean: Comments on a controversy. *Oceanography* 1:14-17.

1993 Hansell, D.A., P.M. Williams, and B.B. Ward. 1993. Comparative analyses of DOC and DON in the Southern California Bight using oxidation by high temperature combustion. *Deep-Sea Res.* 40:219-234.

1993   Peltzer, E.T., and P.G. Brewer. 1993. Some practical aspects of measuring DOC-sampling artifacts and analytical problems with marine samples. *Marine Chemistry* 41:243-252.

1993   Sharp, J. 1993. The dissolved organic carbon controversy: An update. *Oceanography* 6:45-50.

TWO NEW AVENUES TO UNDERSTANDING

The six achievements described above were revolutionary in that they each overturned an old consensus and forced a new reality suddenly onto center stage. Revolutions are fun, particularly for the young at heart, but they are not the only route to scientific progress. The achievements discussed next are evolutionary, rather than revolutionary, in that they consist of steady, stepwise increases in knowledge and understanding. In addition, they involve many individuals; the advance cannot be credited to any one person.

Modeling

A subtle, but pervasive, achievement of biological oceanography is that modeling has become a mainstream activity; it permeates so much of our work that graduate students in the discipline assume it is integral to biological oceanography. Modeling was at one time an esoteric craft practiced by a gifted few; now it is the norm. Today's biological oceanography graduate student is more likely to have a model than a microscope.

The evolution from Gordon Riley's original models, which were "run" by hand calculation, according to one enduring myth of biological oceanography, to the numerous coupled global ocean-atmosphere-biota models now running is marked by steady advances. A select number of contributors after Riley made improvements, added complexity, and incorporated more sophisticated forcing. The line from Riley (1946) led through John Steele (1959 and 1974), whose slim volume *The Structure of Marine Ecosystems* (1974) enticed mathematicians, physicists, and physical oceanographers to try their hand at the new craft. Even today one usually finds Steele's volume on the shelves of individuals recruited to biological modeling from the physical sciences.

With new talent entering the field, modeling gathered momentum in the 1970s and 1980s (Walsh, 1975; Jamart et al., 1977; Steele and Frost, 1977; Wroblewski, 1977; Evans and Parslow, 1985; Hofmann, 1988). Genealogies of modeling accomplishments in biological oceanography, impossibly difficult to construct, would be marked by lots of branching and fusion. One important milestone, the Fasham Model (Fasham et al., 1990), was an upper-ocean ecosystem model that was widely distributed by its generous originators. Dozens, if not hundreds, of researchers adapted the Fasham Model to their own ends; this was the code that caused a bloom of biological oceanography models in small computers around the world. One particularly influential application of the Fasham Model that demonstrated the power of physical-biological models was a seasonal North Atlantic ecosystem study by Sarmiento et al. (1993).

Biological oceanography modeling is at the forefront of modeling in a number of areas: the use of data assimilation, coupled physical-biological models, single-species population models, ecosystem models, and the use of massively parallel supercomputers to simulate biogeochemical processes in general circulation models (Hofmann and Lascara, 1998).

The growth of modeling is aptly demonstrated in Brink and Robinson (1998), *The Sea*, Volume 10, which has three chapters dealing with various aspects of interdisciplinary modeling of the coastal ocean. Together, these three chapters have 371 references. The growth of this area of biological oceanography exceeds the assimilative capacity of a single individual.

NSF programs such as GLOBEC and the Joint Global Ocean Flux Study (JGOFS) are making a significant investment in modeling, but there persists some uncertainty about the best way to manage this powerful new research activity to ensure that the sum of its parts will be realized.

Modeling References

1946   Riley, G.A. 1946. Factors controlling phytoplankton populations on Georges Bank. *J. Mar. Res.* 6:54-73.

1949   Riley, G.A., H. Stommel, and D.F. Bumpus. 1949. Quantitative ecology of the plankton of the western North Atlantic. *Bull. Bingham Oceanog.* 12(3):1-169.

1959   Steele, J.H. 1959. The quantitative ecology of marine phytoplankton. *Biol. Rev.* 34:129-158.

1974   Steele, J.H. 1974. *The Structure of Marine Ecosystems*. Harvard University Press, Cambridge, Mass. 128 pp.

1975   Walsh, J.J. 1975. A spatial simulation model of the Peru upwelling ecosystem. *Deep-Sea Res.* 22:201-236.

1977   Jamart, B.B., D.F. Winter, K. Banse, G.C. Anderson, and R.K. Lam. 1977. A theoretical study of phytoplankton growth and nutrient distribution in the Pacific Ocean off the northwest U.S. coast. *Deep-Sea Res.* 24:753-773.

1977   Steele, J.H., and B.W. Frost. 1977. The structure of plankton communities. *Phil. Trans. Roy. Soc. Lond.* 280:485-534.

1977   Wroblewski, J.J. 1977. A model of phytoplankton plume formation during variable Oregon upwelling. *J. Mar. Res.* 35:357-394.

1985   Evans, G.T., and J.S. Parslow. 1985. A model of annual plankton cycles. *Biological Oceanography* 3:327-347.

1987   Frost, B.W. 1987. Grazing control of phytoplankton stock in the open subarctic Pacific Ocean: A model assessing the role of mesozooplankton, particularly the large calanoid copepods, *Neocalanus* spp. *Mar. Ecol. Prog. Ser.* 39:49-68.

1988   Hofmann, E.E. 1988. Plankton dynamics on the outer southeastern U.S. continental shelf. III. A coupled physical-biological model. *J. Mar. Res.* 46:919-946.

1990   Fasham, M.J.R., H.W. Ducklow, and S.M. McKelvie. 1990. A nitrogen-based model of plankton dynamics in the oceanic mixed layer. *J. Mar. Res.* 48:591-639.

1993   Sarmiento, J.L., R.D. Slater, M.J.R. Fasham, H.W. Ducklow, J.R. Toggweiler, and G.T. Evans. 1993. A seasonal three-dimensional ecosystem model of nitrogen cycling in the North Atlantic euphotic zone. *Global Biogeochem. Cycles* 7:417-450.

1998   Brink, K.H., and A.R. Robinson (eds.). 1998. *The Sea, Vol. 10,*

The Global Coastal Ocean Processes and Methods. John Wiley & Sons, Inc., New York, 604 pp.

1998   Hofmann, E.E., and C.M. Lascara. 1998. Overview of interdisciplinary modeling for marine ecosystems. Chapter 19, pp. 507-540 in K.H. Brink and A.R. Robinson (eds.), *The Sea, Vol. 10, The Global Coastal Ocean Processes and Methods.* John Wiley & Sons, Inc., New York.

1998   Steele, J.H. 1998. Incorporating the microbial loop in a simple plankton model. *Proc. Roy. Soc. Lond.* B 265:1771-1777.

In Situ Observations and Experiments

When Jacques Cousteau first lured people under the sea, marine biologists joined the activity with enthusiasm. Scientific advances from these new *in situ* observations remained modest until biologists ventured into the pelagic realm. Once there, they found a world that had no counterpart in the mangled samples harvested by nets or water collection bottles (Alldredge, 1972; Madin, 1974; Hamner, 1975). Transparent and iridescent organisms, large and small, were abundant (Hamner et al., 1978). Organic aggregates were ubiquitous, and these large, gossamer structures were found to have very high rates of microbial activity (Silver and Alldredge, 1981; Caron et al., 1982). The aggregates appear to be self-contained biospheres with populations of producers and consumers living together. *In situ* observations by divers, submersibles, and remotely operated vehicles (ROVs) revealed a great diversity of large planktonic organisms, particularly cnidaria, ctenophores, and salps (Robison, 1995) (Plate 4). Some of these are so delicate that they disintegrate in the wake of a swim fin; others are as tough as shoe leather. *In situ* observations showed that the pelagic realm is anything but barren or boring.

A characteristic that is very much a part of being a biologist is the inclination to give nature a gentle prod and watch the response. Connell (1961) and Paine (1966) used manipulation of intertidal communities to establish the hugely successful field of experimental marine ecology. From this work we have learned many rules about how communities are structured. Thirty years after Connell and Paine, Martin's successful *in situ* open-ocean experiment was carried out by adding iron to a 64-km^2 patch of the equatorial Pacific and following the enriched patch for about 10 days (Martin et al., 1994). Interest in the confirmation of the Iron Hypothesis overshadowed the demonstration by this work that open-ocean experiments can be done. Just as *in situ* observations have revealed a biology that bottles and nets cannot capture, *in situ* experiments in the open ocean will reveal how intact, pelagic communities respond to environmental variations. When *in situ* ocean experimentation is coupled with *in situ* sensors and data assimilation (von Alt and Grassle, 1992), our discipline will have reached the end of its adolescence. Experimental intertidal marine ecology is very much a mainstream research activity; experimental biological oceanography is still only a glimmer in the eye of a few visionaries.

Both *in situ* ocean observations and *in situ* ocean experiments are unorthodox by the standards of traditional biological oceanography. However, the exciting new insights that resulted from work in both areas showed that NSF was wise to support this unconventional and risky research.

In Situ Observations References

1972   Alldredge, A.L. 1972. Abandoned larvacean houses: A unique food source in the pelagic environment. *Science* 177:885-887.

1974   Madin, L.P. 1974. Field observations on the feeding behavior of salps (Tunicata: Thaliacea). *Mar. Biol.* 25:143-148.

1975   Hamner, W.M. 1975. Underwater observations of blue-water plankton. Logistics, techniques, and safety procedures for divers at sea. *Limnol. Oceanogr.* 20:1045-1051.

1978   Hamner, W.M., L.P. Madin, A.L. Alldredge, R.W. Gilmer, and P.P. Hamner. 1978. Underwater observations of gelatinous zooplankton: Sampling problems, feeding biology and behavior. *Limnol. Oceanogr.* 20:907-917.

1981   Silver, M.W., and A.L. Alldredge. 1981. Bathypelagic marine snow: Vertical transport system and deep-sea algal and detrital community. *J. Mar. Res.* 39:501-530.

1982   Caron, D.A., P.G. Davis, L.P. Madin, and J. McN. Sieburth. 1982. Heterotrophic bacteria and bacterivorous protozoa in oceanic aggregates. *Science* 218:795-797.

1995   Robison, B.H. 1995. Light in the ocean's midwaters. *Scientific American* (July):60-65.

In Situ Experiments References

1961   Connell, J.H. 1961. The influence of interspecific competition and other factors on the distribution of the barnacle *Chthamalus stellatus*. *Ecology* 42:710-723.

1966   Paine, R.T. 1966. Food web complexity and species diversity. *Am. Nat.* 100:65-75.

1992   von Alt, C.J., and J.F. Grassle. 1992. LEO-15—An unmanned long term observatory. *Proc. Oceans '92* 2:829-854.

1994   Martin, J.H., et al. 1994. Testing the iron hypothesis in ecosystems of the equatorial Pacific Ocean. *Nature* 371:123-129.

1996   Coale, K.H., et al. 1996. A massive phytoplankton bloom induced by an ecosystem-scale iron fertilization experiment in the equatorial eastern Pacific Ocean. *Nature* 383:495-501.

EPILOGUE

Reading over our selection of landmark achievements, we note with chagrin that we have failed to cite the achievements of the one individual, Alfred Redfield, who was most responsible for the dramatic advance of biological oceanography in the past 50 years. His groundbreaking work gave biological oceanographers both the Redfield Ratio and Redfield's Rule (Redfield, 1958). We acknowledge that all of the biological oceanographers cited in this paper had the advantage of standing on Redfield's broad shoulders.

We have also failed to cite the work of a series of exceptionally productive biological oceanographers who were multi-faceted leaders. Mikhail Vinogradov, David Cushing, Gotthilf Hempel, Ramon Margalef, Akihiko Hattori, Achim Minas, André Morel, and Takahisa Nemoto are individuals whose overarching leadership left an indelible mark on biological oceanography. These individuals all led expeditions, directed laboratories, made important scholarly contributions, and at the same time were mentors to a generation of talented biological oceanographers. The significant contributions of these individuals to biological oceanography will have to be recognized at a future opportunity.

References Not Mentioned Under Specific Landmarks

1926   Hjort, J. 1926. Fluctuations in the year classes of important food fishes. *J. Cons. Int. Explor. Mer.* 1:1-38.

1955   Odum, H.T., and E.P. Odum. 1955. Trophic structure and productivity at a windward coral reef community on Eniwetok Atoll. *Ecol. Monogr.* 25:291-320.

1958   Redfield, A.C. 1958. The biological control of chemical factors in the environment. *Amer. Scientist* 46:205-221.

1961   Hutchinson, G.E. 1961. The paradox of the plankton. *Amer. Nat.* 95:137-145.

1989   Cushing, D.H. 1989. A difference in structure between ecosystems in strongly stratified waters and in those that are only weakly stratified. *J. Plankton Res.* 11:1-13.

1989   McIntosh, R.P. 1989. Citation classics of ecology. *The Quarterly Review of Biology* 64:31-49.

1995   Parsons, T.R., and H. Seki. 1995. A historical perspective of biological studies in the ocean. *Aquat. Living Resour.* 8:113-122.

Achievements in Chemical Oceanography

John W. Farrington

Woods Hole Oceanographic Institution

PREFACE

The charge given to me by the steering committee is as follows: focus on *landmark achievements* in chemical oceanography over the past 50 years, the *individuals involved*, the *new technology* and *ideas* that made these achievements possible, *how one discovery built on the foundations of earlier ones, discoveries made at the intersections of disciplines*, and the *role that NSF programs and institutional arrangements* had in making these achievements possible.

I am honored to have the opportunity to share my views on this topic of achievements in chemical oceanography since the 1950s. Given the credentials and landmark (should we call them "seamark" or "channel buoy?") contributions of the others, it is clear that I am a substitute for those much more qualified to satisfy the charge of the committee. For various reasons, those more qualified were not available to write this paper. I suspect that I am the substitute because someone on the organizing committee obtained a copy of my undergraduate transcript and learned that my grades in history and political science were reasonable and certainly much better on the average than my grades in the sciences and math. The committee must also have learned how excited and enthusiastic I am about the study of the oceans and about scientific research and education in general.

I have an apology. The space allocated for this paper is limited, and there is an abundance of significant contributions by individuals and groups deserving of explicit recognition—more than can be incorporated into this paper. Admittedly important areas of research—marine biochemistry, natural product chemistry, and contributions of marine isotopic chemistry to paleoclimate and paleoceanographic studies—that could be thought of by many as marine geochemistry or marine chemistry are not included because of space and time limitations and because they seemed to be beyond the charge given to me. I have continued to revise the paper after initial oral presentation. Those who view the videotape of the presentation and compare it with this written version will note a few significant additions. I sought advice on this paper from several colleagues, but I did not conduct a systematic survey by questionnaire. In hindsight, the lack of a more systematic survey may have been a mistake, but I have had the good fortune in my career to have met and listened to many of the chemical oceanographers and marine chemists in the United States and elsewhere. Their papers, lectures, seminars, and informal conversations inform this paper. I remain less than fully satisfied with the completeness of this paper and have yielded reluctantly to personal limits of scholarship and the requirement to submit the paper in written form by a deadline.

For the readers who are expecting a mention of their favorite element, I regret that limited space precludes a full exploration of the oceans using the periodic table of elements as a guide, although in my opinion that would be fascinating. While delivering the oral version of this paper, I wore a tie that incorporated the periodic table of elements to celebrate the event. Thus, I can assure you that all the elements were close to my heart. (There was about an equal outburst of groans and laughter after this statement at the talk.) When I quote references from the years before the present, I use the language of those times and I do not correct statements to the gender-neutral language of our times.

INTRODUCTION

The organizing committee scheduled this paper between "biological oceanography" and "physical oceanography" in the symposium program. Many of the significant achievements in chemical oceanography through the 1950s might best be described as applications of chemistry to understanding biological and physical processes in the oceans. The same can be stated about chemistry applied to understanding geological processes. There is merit in organizing the study of the oceans in some manner, and doing so using the fundamental, underpinning science disciplines as a template has advantages. However, I submit that we must keep foremost as the guiding principle of our endeavors that advancing knowledge of the oceans was the central objective of the research discussed at this symposium.

I found it difficult to define the boundaries of chemical oceanography when preparing this review. Early in the process of preparing this paper, I realized that this was not an important aspect of the undertaking. The record of accomplishments using chemistry to understand the oceans and oceanic processes involves research efforts by individuals and groups who may be primarily self-identified or generally recognized as physical oceanographers, biological oceanographers, or marine geologists. My colleague, Dr. James R. Luyten, Senior Associate Director and Director of Research at Woods Hole Oceanographic Institution (WHOI), brought to my attention a recent editorial in *Science* "How to Change the University" (Hazzaniga, 1998). A quote from this editorial is thought-provoking and has implications in the world of research and scholarship in general: "The modern university is partitioned along academic lines that no longer truly reflect today's intellectual life." (p. 237). Perhaps this was what the organizing committee for this symposium had in mind when it set forth the charge of "discoveries at the intersections of disciplines."

I believe that those of us studying the oceans should continue to be vigilant and take heed that we do not allow organizational boundaries among or within disciplines to frustrate significant advances in our knowledge of the oceans. Sverdrup, Johnson, and Fleming (1942) with their powerful, wide-ranging (and now venerable) text set the example for us to follow:

> Oceanography embraces all studies pertaining to the sea and integrates the knowledge gained in the marine sciences that deal with such subjects as the ocean boundaries and bottom topography, the physics and chemistry of sea water, the types of currents, and the many phases of marine biology. The close interrelation and mutual dependence of the single marine sciences have long been recognized. (Sverdrup, Johnson, and Fleming, 1942, p.1)

This is the appropriate place to acknowledge the lasting contributions of Richard H. Fleming, Professor of Oceanography, University of Washington, and the co-author of *The Oceans*, who was a leader in pioneering studies setting the scene for the post-1950 studies of the chemistry of the oceans.

A detailed assessment of progress in chemical oceanography for the past three decades—essentially for the 1970s, 1980s, and 1990s—was assembled recently in an effort funded by the National Science Foundation called *Futures of Ocean Chemistry in the United States*—an effort with the clever acronym FOCUS. Many excellent chemical oceanographers, marine chemists, and geochemists contributed to the FOCUS report and it is available on the World Wide Web (FOCUS, 1998). The 1970s through 1990s received an extensive treatment by these experts. I concen-

trate my effort here on the 1950s, 1960s, and early 1970s because this is the time period during which several of the most important contributions and activities occurred over the past 50 years. In fact, I will go back a bit before 1950 to set the scene, and then provide a thread of continuity from this paper to the FOCUS report by a limited discussion of important research efforts from the 1970s into the 1990s.

The Ocean Studies Board of the National Research Council organized this event, and it is appropriate to note that staff of this board and its predecessor boards and committees have provided invaluable service at the interface between the scientific community in general and the federal agencies since the early 1950s. Richard C. Vetter was a key staff person for these boards and committees, serving as Executive Secretary of the Committee on Oceanography during a significant portion of this time. When Dick retired, he advertised on an OMNET (electronic mail) bulletin board that he had a small collection of reports, books, and news clippings to be made available to anyone who would pay for the shipping and promise to keep the collection together and make it available to students in particular, as I recall. I was fortunate to be the selected recipient. This collection contained a copy of the June 1, 1964, weekly professional magazine of the American Chemical Society, *Chemical and Engineering News*. A part of a featured special report was "*Chemistry and the Oceans.*"

There is an interesting statement in that report: "Chemical oceanography is an old science recently revitalized." (p. 6A) Some may question this since several folks think of oceanography as a relatively young science. Many chemical oceanography texts—for example, a compilation of papers *Chemical Oceanography* edited by J.P. Riley and G. Skirrow (1965); *The Sea*, Volume 5: *Marine Chemistry*, edited by Professor Edward D. Goldberg (1974) and used as keystone learning and reference guides by my generation of chemical oceanographers; and the recent text of Professor Michael E.Q. Pilson (1998)—provide guidance about the history of chemical oceanography and marine chemistry. Wallace (1974) provides a very thorough and highly recommended review of the history of chemical analysis of seawater up to the mid-1900s and then continues with a thorough review as these analyses pertained to chlorinity and salinity determinations well into the 1960s. Comprehensive reviews of various topics in chemical oceanography have been assembled by leading researchers in the volumes of *Chemical Oceanography* edited by J.P. Riley and R. Chester beginning in 1975 (Riley and Chester, 1975).

CHEMICAL OCEANOGRAPHY PRIOR TO 1950

In the next three paragraphs, I paraphrase or quote from the texts cited above (Riley and Skirrow, 1965; Goldberg, 1974; Wallace, 1974; Pilson, 1998).

Aristotle expounded on the possible origins of the salt in the sea (Wallace, 1974). Since Aristotle's contribution, eminent scientists and chemists of their times have made significant contributions to understanding the chemistry of the oceans. Among them, during the 1600s and 1700s, were Robert Boyle and his "Observations and Experiment About the Saltness of the Sea" (1674), "which, in the opinion of several (modern) writers, established him as the founder of the science that is now referred to as chemical oceanography" (Wallace, 1974, p. 1). Others from those years included Edmund Halley, Count Luigi Ferdinando Marsigili, Antoine Lavoisier, and Joseph Louis Gay-Lussac. From the late 1700s through the 1800s, Alexander Marcet, Johann Forchhammer, and William Dittmar undertook painstaking analyses of seawater, which provided the heralded "Marcet's Principle," or the constancy of ratios of several major ions in seawater. Wallace (1974, p. 121) states, "Dittmar's report on the chemistry of the 77 water samples of the 'Challenger' expedition represents the most extensive seawater analysis performed before or since." The importance of knowing the density of seawater drove a significant part of chemical oceanography during the period of 1900 to 1950 to focus on salinity measurements or surrogates, mainly chlorinity, and to affirm the constancy of the ratios of the major ions of seawater. In addition, measurements of nutrients, dissolved oxygen, and the components of the carbonate system and alkalinity were pursued. It was this combination of understanding biological systems, refining and confirming the chlorinity-salinity-density relationships, and the beginnings of the understanding of distinctive chemical compositions for distinguishing water masses that characterized chemical studies of the oceans at that time.

During the 1920s, analytical methods for nutrient substances—mainly compounds of nitrogen and phosphorus—began to appear and to be improved. Individuals such as Atkins, Harvey, and Cooper and the organizing activities of the International Council for the Exploration of the Sea were important in this effort. Harvey's (1928) book *The Biological Chemistry and Physics of Seawater* captures chemical oceanography of the time as it was involved with biological productivity.

Overlapping in this time frame, in the 1930s, V.M. Goldschmidt, the renowned geochemist, and his school conducted their research on crustal abundances and ionic potential classifications. Goldschmidt and his group also initiated their studies of the mass balances and geochemical cycles of elements, including the oceans in their research (e.g., Goldschmidt, 1933, 1937). Also during this time Buch of Finland and others initiated studies of the physical chemistry of carbon dioxide in seawater. Wattenberg on the *Meteor* expedition drew attention to the fact that some areas of the ocean were supersaturated while others were undersaturated with respect to calcium carbonate.

Elizabeth Noble Shor, in her historical account of the Scripps Institution of Oceanography, quotes Norris Rakestraw: "One of the most striking observations of marine biology is the fact that some parts of the ocean are fertile while other parts are quite barren. There must be chemical factors which determine fertility, and an explanation of this was perhaps the first serious question which oceanographers asked the chemist. In the year 1930 there were probably no more than a dozen professional chemists in the world who were actively interested in the ocean, and practically every one of them was trying to answer this question" (Shor, 1978, p. 321).

The first chemical laboratory at the Scripps Institution of Oceanography was founded by Erik G. Moberg in 1930 (Shor, 1978). This was the beginning of a tradition of excellence in chemical oceanography and marine chemistry that continues to the present. Further north on the U.S. West Coast, Thomas G. Thompson at the University of Washington labored to improve the analyses of seawater during the 1920s to 1940s. People from those times who should know (NAS, 1971a) described Thompson's laboratory as follows: "For some years this laboratory was the most productive center for chemical oceanography in the United States." (p. 10) Beginning in the 1930s, chemical work began at the Woods Hole Oceanographic Institution (WHOI) with the efforts of Redfield, Seiwell, and Rakestraw, who also conducted research on the questions of the interaction of the biology and chemistry of the sea. Rakestraw later moved to the Scripps Institution of Oceanography.

J.P. Riley (1965) notes that as early as 1935 to 1937, fluorimeteric determinations of uranium in seawater coupled with other observations of the low concentrations of radium in seawater, led to the observation that uranium and radium-226 were in disequilibrium in seawater. The explanatory hypothesis was removal of thorium-230 from the water and its incorporation into sediments.

By 1940, the complexion of chemical oceanography had changed notably. Marine geology, or the geological aspects of oceanography, had been developing through the previous decades, and it had become quite evident that the chemistry of seawater was fundamentally involved in sedimentation phenomena.

Another major division appeared—marine geochemistry—concerned not merely with the use of chemistry to solve geological problems, but also with the part the ocean plays in the broad, general weathering cycles. Since most of the chemical elements have been found in seawater, the chemist is provided with an endless number of problems concerning the source, speciation, function, and significance of these elements and their interactions." (NAS, 1971a, Chapter 1, pp. 10-11).

The World War II years provided a focus for further understanding salinity and the major chemical components contributing to salinity because of its relationship to sound transmission in the sea. At the Woods Hole Oceanographic Institution, Alfred C. Redfield and his former graduate student at Harvard University, Bostwick H. Ketchum, conducted extensive research on antifouling paints with great success. During the past several years in the United States, there has often been a vigorous debate about definitions and values pertaining to "basic" and "applied" research. It is interesting that Redfield noted in a taped interview conducted in 1973 by his daughter, Elizabeth R. Marsh, "I learned one thing from [work during the war on] the paint thing, and that was that it was pretty good fun on an applied problem. Because if you had an applied problem which couldn't be solved by existing engineering principles, it meant that you didn't know what the fundamental problems were" (Marsh, 1973).

After World War II, the Office of Naval Research (ONR) continued a strong interest in ocean research, including the chemistry of the oceans (Anderson, 1973). Although this symposium was focused on the National Science Foundation and oceanography, it is important to acknowledge that ONR funding of chemical oceanography and marine chemistry was critical in the years following World War II, especially the 1950s and 1960s, as NSF funding in this arena was initiated and then increased.

Research that has had a major influence in chemical oceanography and marine geochemistry was W.F. Libby's discovery of radioactive carbon produced in the atmosphere from cosmic rays. Continuing the strong connection between biological or ecological considerations and ocean chemistry, the renowned limnologist and ecologist G. Evelyn Hutchinson wrote a provocative paper "The Problems of Oceanic Geochemistry" (Hutchinson, 1947).

The 1950s was a period of intensification of the more traditional (at the time) and mainly descriptive studies of nutrients, dissolved oxygen, and major and minor elements in general. A summary of this particular research focus is found in the report of a meeting convened by the National Academy of Sciences (NAS, 1959) on the physical and chemical properties of seawater. One glimpse of the thinking at that time is provided by this exchange, which can be found in the discussion section of the report. Professor W.T. Holsar of the Institute of Geophysics at the University of California at Los Angeles asked the question, "Can different oceans be characterized by differences in chemical composition?" Professor Richard H. Fleming of the University of Washington answered, "Yes. If a sample labeled only by depth is presented to a chemist, he can, by analyzing chlorinity, calcium, alkalinity and nutrients, distinguish whether it is from the Atlantic, Pacific or Indian Oceans." (p. 95)

An important event of the 1950s in chemical oceanography, as it was for oceanography in general, was the International Geophysical Year (IGY) of 1957-1958. From the perspective of the present, I believe that in addition to obtaining valuable data, the IGY expeditions provided experience with intensive water sampling and chemical analyses of large numbers of samples and experience with multinational collaboration. Chemical measurements were made as "routine" aspects of the hydrographic sections. Experience gained from "catching water" during the hydrographic casts of these expeditions was translated directly into practical improvements for water sampling—hydrographic casts—of the 1960s, which in turn, made possible the significant progress to come in the 1970s and later.

Early National Science Foundation Grants

The 1950s were the formative years of the National Science Foundation. I thank Dr. Michael Reeve of the Ocean Sciences Division of the National Science Foundation and his staff for making available a compilation of grants during the early years of NSF (Reeve, 1998). I have selected all the grants whose titles indicate that they pertain to marine chemistry, geochemistry, and chemical oceanography in some manner (Table 1). Many of the recipients of these grants are widely recognized today as leaders of the 1950s through the present in geochemistry, geology, and chemical oceanography or marine chemistry. Many other prominent chemical oceanographers and marine geochemists were fully funded by the Office of Naval Research and by the Atomic Energy Commission and thus may not have had the time or the inclination to submit a proposal to NSF in the early days of the 1950s.

As far as can be determined, the first NSF grant (listed under Earth Sciences) that could be described as focused in some area of chemical oceanography was awarded to T.J. Chow and T.G. Thompson of the University of Washington, "Distribution of Some Minor Elements in Seawater" (Table 1). Things picked up in 1954 and through the IGY, NSF's Earth Sciences funded research that would have profound effects on our knowledge of the chemistry of the sea and still influence our research today. As the titles of the grants in Table 1 indicate or hint, this research involved one of the major intellectual forces and practical applications of chemistry to the oceans in chemical oceanography and marine geochemistry of the past 50 years—radioactive isotope and stable isotope chemistry analyses of seawater and sediment samples to elucidate physical, biological, geochemical, and biogeochemical processes in the oceans.

Descriptive Chemical Oceanography Shifts Toward Quantifying Rates

The decade also heralded a significant move from the use of chemical measurements for descriptive oceanography to the initiation of the use of chemical measurements to quantify rates of oceanic processes. These were the early career years of several scientists who would make significant contributions to marine chemistry and chemical oceanography and the use of chemistry to understand and quantify oceanic processes: Harmon Craig and Edward D. Goldberg of the Scripps Institution of Oceanography, Wallace Broecker of Columbia University and Lamont-Doherty Geological Observatory, and Karl Turekian of Yale University, among others.

TABLE 1   NSF Grants Related to Marine Geochemistry and Chemical Oceanography in the 1950s

Year	Grant
1953	T.J. Chow and T.G. Thompson (University of Washington). *Distribution of some minor elements in seawater.*
1954	C. Urey (University of Chicago). *Isotopic abundances relating to geochemical research.* D.B. Erickson (Columbia University). *Lithological and micropaleontological investigation of the Atlantic Ocean.* J.L. Kulp (Columbia University). *Time relations of ocean floor sediments.* M.L. Keith (Pennsylvania State Univ.). *Fractionation of stable isotopes in geological processes.*
1955	J.L. Kulp (Columbia University). *Carbon-14 dating of archeological and anthropological specimens.* W.H. Dennen and E. Mencher (Massachusetts Institute of Technology) *Geochemical investigations of sedimentary rocks.* D.W. Hood (Texas A&M Univ.). *Calcium carbonate solubility equilibrium in sea water.*
1956	E.S. Barghoorn (Harvard University). *Organic residues in fossil sediments.* H.B. Moore (University of Miami). *Oxygen-density relationships and phosphate control of Caribbean waters.* H.D. Holland (Princeton University). *Radiation damage measurements as a guide to geologic age.* V.T. Bowen (Woods Hole Oceanographic Institution). *Research instrumentation for sampling water at all depths.* E.S. Devey (Yale University). *Radiocarbon dating and other forms of geochronometry.*
1957	J.L. Kulp (Columbia University). *Isotope geology of strontium and rubidium.* F.F. Koczy (University of Miami). *Distribution of radioactive elements in the oceans.* E.K. Ralph (University of Pennsylvania). *Half-life of carbon-14.* K.O. Emery and A. Hancock (University of Southern California). *Rate of deposition of sediments off Southern California.*
1958	B.B. Benson (Amherst College). *Oxygen isotope variations in ocean water.* C.C. Patterson and T.J. Chow (California Institute of Technology). *Lead isotopes in the oceans.* W.S. Broecker (Columbia University). *Radiocarbon age determinations.* T.G. Thompson (University of Washington). *Organic compounds in sea water.* K.K. Turekian (Yale University). *Crustal abundance of nickel, cobalt and chromium.* **Geophysical Year Related Grants**: *Radiocarbon or Radiochemistry.* (University of California-Scripps Institution of Oceanography, Columbia University, Texas A&M University, Woods Hole Oceanographic Institution) *Procurement of Equipment for Carbon Dioxide Measurement.* (University of Washington)

SOURCE: Reeve (1998).

Some publications of the 1950s would have major, lasting impact on chemical oceanography and marine geochemistry. W.W. Rubey published his very influential contribution *Geologic History of Sea Water: An Attempt to State the Problem* (Rubey, 1951). In the same year, Urey et al. (1951) reported on their result of measuring paleotemperatures using stable isotopes of oxygen. Harmon Craig wrote *The Geochemistry of Stable Carbon Isotopes* (Craig, 1953), setting the scene for many studies using stable isotopes of carbon. Edward D. Goldberg wrote *Marine Geochemistry 1: Chemical Scavengers of the Sea* (Goldberg, 1954), setting the scene for many studies to follow related to particle scavenging of chemicals in seawater. V.M. Goldschmidt, continuing his pioneering efforts in geochemistry over two decades, published his highly acclaimed book, *Geochemistry* (Goldschmidt, 1954). Goldberg and Arrhenius (1958) published their paper on residence times of elements in the oceans. According to Goldberg (1965), the important concept of residence times for elements in the oceans, as estimated from inputs from rivers (and the atmosphere) and removals to sediments (and assuming steady state conditions), was introduced by Barth (1952). As a harbinger of things to come in the 1960s and later with respect to the utilization of the uranium decay series to quantify several processes in the ocean, Goldberg and Koide (1958) published a paper about ionium-thorium chronology in sediments.

Initiation of Modern Studies of the Oceans' Role in the Carbon Dioxide-Climate Concerns

As noted previously, much of chemical oceanography in the decades prior to 1940 had been focused on biologically related problems. One of the other areas of interest was the exchange of carbon dioxide between the sea and the atmosphere, including the physical chemistry of carbon dioxide and its solution in seawater (NAS, 1971a). The role of the oceans in the cycle of carbon and particularly the carbon dioxide exchange between the ocean and the atmosphere was identified as a major research focus at the Scripps Institution of Oceanography and championed by the Scripps' director Roger Revelle, beginning in the 1950s. Not only did Revelle recognize the significance of the atmosphere-ocean exchange of carbon dioxide and its relationship to climate issues, he participated personally in the research, and he recruited a diverse group of talented chemists and geochemists to conduct research on the problem, as has been chronicled by Shor (1978). As one example, Revelle brought Charles David Keeling (Keeling, 1958) to Scripps in 1956 and encouraged him to study carbon dioxide in the atmosphere-ocean system (Keeling, 1968). Clearly, one of the most influential papers pertaining to chemical oceanography and oceanography in general of the 1950s, and in all of the literature up to the present in oceanography, is the paper by Revelle and Suess (1957), "Carbon Dioxide Exchange Between the Atmosphere and Ocean and the Question of an Increase of Atmospheric CO_2 During Past Decades."

Roger Revelle had a significant influence on the future of chemical oceanography and marine geochemistry research, and ocean and environmental sciences in general, so much so that it is difficult to capture in words. The closest to an accurate description is that of MacLeisch (1982-1983) who identifies Roger Revelle by the apt designation "Senior Senator of Science." Roger Revelle received the National Medal of Science of the United States in 1990.

Tracers, Ocean Circulation and Mixing, and Global Biogeochemical Cycles

A very influential paper in chemical oceanography and marine geochemistry presented at the International Oceanographic Congress, and subsequently published in 1961, was by Broecker, Gerard, Ewing, and Heezen (1961) "Geochemistry and Physics of Ocean Circulation." This paper was similar to the paper published in 1960 by the same authors in the peer-reviewed journal literature (Broecker et al., 1960) and was largely the outcome of the Ph.D. thesis research of Wallace S. "Wally" Broecker, completed in 1957 at Columbia University, "Application of Radiocarbon to Oceanography and Climate Chronology" (Broecker, 1957)—the launch of a truly illustrious career by arguably one of the most influential, scholarly geoscientists of his times. Wally Broecker received the National Medal of Science of the United States in 1996 and the Blue Planet Award for his many and diverse scientific contributions.

Three additional points are worthy of mention about this specific contribution by Broecker et al. (1961). First, there is evidence of the success of an earlier NSF investment in the establishment of radiocarbon measurement capability through grants to Kulp in 1954 and 1955 (Table 1). Second, NSF continued to invest in the early career development of Wally Broecker as evidenced by its grant to him in 1958 (Table 1). Third, this was the ocean science and geoscience communities' introduction to the powerful reasoning and explanatory teaching style of Wally Broecker. Readers are invited to compare the reasoning and analogies in Broecker et al. (1961) to that found in the later influential texts *Chemical Oceanography* (Broecker, 1974), *Tracers in the Sea* (Broecker and Peng, 1982), and *How to Build a Habitable Planet* (Broecker, 1985).

Karl K. Turekian was also a graduate student of Professor Kulp at Columbia University, at the same time as Wally Broecker, and they collaborated on some projects (e.g., Broecker et al., 1958). Karl has been influential in many ways in his career. McElway (1983) wrote a profile of Karl Turekian, "Academic Gladiator," in which he captured the Karl Turekian I know: wide-ranging intellect, superb teacher, scrappy debater, eclectic in his significant contributions to Earth sciences—including chemical oceanography and marine geochemistry—through the use of geochemical mea-

surements of various types. Karl's earlier publications indicated the breadth and depth of contributions to come (e.g., Turekian, 1955, 1957, 1958; Turekian and Kulp, 1956). Karl's book *Oceans* (Turekian, 1968) provided many of the undergraduates and beginning graduate students of my generation with a concise, readable, important introduction to marine sciences. Karl's influence can be found in some of the most important areas of chemical oceanography from the 1950s to the present as well as in much research on global biogeochemical cycles.

Physical Chemistry of Seawater and Lars Gunnar Sillen

Other aspects of the chemistry of the oceans were receiving increased attention. In 1959, Professors Gustaf Arrenhius and Edward D. Goldberg invited Professor Lars Gunnar Sillen, one of the world's foremost inorganic chemists of the time, to give a lecture at the International Oceanographic Congress in New York, between August 31 and September 12. His paper, (another very influential paper from this decade) "The Physical Chemistry of Seawater" (Sillen, 1961) was published in the proceedings of the Congress edited by Dr. Mary Sears. Goldberg (1974) quotes from Sillen and I repeat Goldberg's quote here:

> . . . it may be worthwhile to try to find out what the true equilibrium would be like, and that one might learn from a comparison with the real system. We shall often find that sufficient data are lacking to make the discussion very precise. Neither the laboratory data on chemical equilibria (needed for the model) nor the geochemical data (for the real system) are always as accurate as one might wish. Still, it may be worth while to try this approach. (p. ix)

The process described by Sillen of attempting to define equilibrium or steady-state conditions from fundamental chemical principles and laboratory experiments and then comparing the resulting chemical distributions, including detailed chemical speciation, with actual measured distributions in the oceans, is at the heart of much chemical oceanography and marine chemistry research of the past three decades and at present.

Nuclear Weapons Test Fallout: Environmental Quality and Tracers in the Sea

The initiation of nuclear weapons testing in the Pacific Ocean in the 1950s by the United States, and elsewhere by other members of the "nuclear weapons club," was accompanied by concern for the fate and effects of several radioactive elements and led to an intensification of research concerned with "biogeochemical" cycling in the oceans (NAS, 1957). Much funding was provided from the Atomic Energy Commission to understand many aspects of oceanic processes, including chemical processes. There were major concerns about the ultimate exposure of marine life and the critical pathways back to people through the sea. Of course, the introduction of radioactive elements from the weapons testing that continued through the mid-1960s also provided tracers that would become important in verifying and contributing to advances in our understanding of oceanic processes. In one sense, this was an experiment, albeit an experiment that rational scientists would not design and execute deliberately. However, oceanographers would have been remiss in not taking advantage of the tracers introduced by nuclear weapons testing.

Among those responding to this important challenge was Vaughan T. Bowen, a zoologist who received his Ph.D. at Yale University, studying with Professor G. Evelyn Hutchinson. Bowen had been recruited to the Woods Hole Oceanographic Institution by Alfred Redfield and had been studying the distribution of major and minor elements in marine organisms. Under his leadership, Bowen's research group began conducting research on the biogeochemical cycles of radioactive elements entering the oceans, an effort that Bowen pursued until his retirement in the mid-1980s. Bowen and Sugihara (1957) were among the first in the world to publish data on strontium-90 activity in seawater according to compilations prepared in 1971 (NAS, 1971b)—the first of many papers from this group to contribute to our knowledge of biogeochemical cycles of artificial radionuclides in the oceans. Koczy (1956) was another of the pioneers studying the geochemistry of radioactive elements in the ocean.

THE 1960S AND INTO THE EARLY 1970S

This was the decade of explosive growth and maturation in chemical oceanography, marine geochemistry, and marine chemistry. The decade began with the publication of the papers by Sillen (1961) and Broecker et al. (1960, 1961), mentioned earlier, followed by the important papers of Garrels and co-workers (Garrels et al., 1961; Garrels and Thompson, 1962). These efforts of Bob Garrels eventually led to the very productive and influential collaboration with Fred MacKensie and to the influential book *Evolution of Sedimentary Rocks* (Garrels and MacKensie, 1971).

In 1963, the paper that summarized the thinking and work of Alfred C. Redfield and coworkers on the influence of the chemical composition of organisms, mainly plankton, on the chemical composition of seawater—the famous Redfield or RKR (Redfield, Ketchum, and Richards) ratio (Redfield et al., 1963) was published. In an interview with his daughter (Marsh, 1973), Redfield attributes the origin of that idea to an earlier paper (Redfield, 1958).

Many more scientists were becoming engaged in analyses of seawater for nutrients and other chemicals. In an influential attempt to codify some of the important lessons learned to date, Strickland and Parsons (1965) published their first manual about seawater analysis, which would be followed by a second edition several years later (Strickland and Parsons, 1972). Francis Richards summarized the state of knowledge and importance of studying anoxic basins (Richards, 1965), stimulating several expeditions in future years to the Cariaco Trench and Black Sea (and several fjords) to study the details of biogeochemistry at the interface of oxic and anoxic waters and in anoxic waters.

Scholarship contributions are at the heart of the intellectual enterprise. In addition, organizational leaders with vision, who are also excellent scientists in their own areas of expertise, are important to move fields of research forward. John M. Hunt is this type of person. In 1964, John Hunt was hired away from Carter Oil Company (a subsidiary of Standard Oil of New Jersey) to head the newly formed Chemistry and Geology Department at the Woods Hole Oceanographic Institution. John would chair this department, and later the separate Chemistry Department, for a decade. John made his most important scholarly research contributions in the field of petroleum geochemistry (Dow, 1992). Of equal importance, John Hunt had a lasting impact on marine chemistry, geochemistry, and chemical oceanography through his efforts to build the Chemistry and Geology Department, and later the Chemistry Department, at the Woods Hole Oceanographic Institution with appointments of a diverse group of researchers to yield one of the better marine chemistry and geochemistry departments in the world (Dow, 1992).

Carbon Dioxide, the Carbon Cycle, and Climate

During the 1960s, and continuing to the present, C. David Keeling launched into a time-series measurement of carbon dioxide in the atmosphere (e.g., Keeling, 1973; Keeling et al., 1976a,b). This intense focus by Keeling and collaborators on a time-series certainly numbers among the more important individual research group efforts in marine geochemistry and atmospheric chemistry of the entire period from 1950 to the present. His data, plotted as concentration of carbon dioxide in the atmosphere versus time at the Mauna Loa, Hawaii, sampling station, have become known worldwide among scientists and environmental policy and management practitioners, including heads of state.

From my perspective, Revelle and Suess sounded the alarm about the potentially serious climatic consequences of modern civilization's use of fossil fuels, the resultant increase of carbon dioxide in the atmosphere, and the role of the ocean in the global carbon cycle. Keeling and coworkers provided the data documenting the increase of carbon dioxide in the atmosphere attributable to fossil fuel combustion and limestone use.

Keeling's data also begged the question of understanding the magnitude of the exchange of carbon dioxide between the atmosphere and the ocean. This required not only an understanding of air-sea exchange processes, but also an understanding of the general circulation and mixing time of the ocean. Broecker et al. (1961), as already noted, provided the pioneering effort to use carbon-14 as a tracer to advance knowledge of oceanic mixing times and confirm general circulation patterns.

Elucidation of the details of the carbon dioxide-carbonate system was, and continues to be, a critical area of research throughout the 1950s to the present. Many marine chemists and chemical oceanographers tackled this central problem, as has been documented very nicely by Gieskes (1974), Broecker and Peng (1982), and most recently by Pilson (1998). Biological productivity and remineralization of the biologically produced organic matter as part of the carbon cycle internal to the ocean were the subjects of considerable and important research efforts as reviewed and summarized by one of the main participants (Menzel, 1974).

The details of organic matter composition in seawater and the underlying surface sediments, and by interpretation the processes acting on the organic matter, began to yield to modern analytical organic chemistry methods through the pioneering efforts of Egon Degens at WHOI with his laboratory's studies of amino acids and carbohydrates; Jeffrey Bada and coworker's studies of amino acids at the Scripps Institution of Oceanography; studies of fatty acids by Peter M. Williams of the Scripps Institution of Oceanography; studies of fatty acids and sterols in sediments by Patrick L. Parker and his students at the University of Texas; the research of Max Blumer of the Woods Hole Oceanographic Institution on hydrocarbons and fatty acids in seawater, organisms, and sediments; and efforts of several other scientists (Duursma, 1965; Andersen, 1977; Kvenvolden, 1980, and references therein). I was in the group of marine organic geochemists engaged in our doctoral studies when these pioneering works appeared and they significantly influenced our research.

GEOSECS: The Most Important Chemical Oceanography-Marine Geochemistry Program of the 1950s to 1990s

I am of the opinion that the most important chemical oceanography-marine geochemistry program of the 1950s to the present was initiated in the late 1960s as part of the International Decade of Ocean Exploration: the Geochemical Ocean Sections Study (GEOSECS). One of the main participants, Dr. Peter Brewer, provides an interesting and informative account of GEOSECS in his paper later in this volume, and I will simply add an interesting story about a few of the influences that launched GEOSECS.

Henry Stommel had proposed an elegant theory about the general circulation of the oceans (Stommel, 1957, 1958; Stommel and Arons, 1960a,b; Bolin and Stommel, 1961; Arons and Stommel, 1967). The ability to use tracers such as the carbon-14 activity of the carbon dioxide-carbonate system of seawater to estimate mixing and circulation times had been demonstrated by researchers in the 1960s, follow-

ing the seminal work of Broecker et al. (1960, 1961). In an interview for an *Oceanus* volume in honor of Hank Stommel, Wally Broecker (1992) states that it was Hank Stommel who launched GEOSECS.

> Ed Goldberg (Scripps Institution of Oceanography) and I were attending some sort of meeting at WHOI during the late 1960s. Hank came to us and said that radiocarbon measurements in the sea were of great importance. He went on to gently chastise us (the geochem community) for doing only scattered stations.
>
> What is needed, he said, is a line of stations extending the length of the Atlantic.
>
> Gee, we said, that would cost a million dollars, a sum greater than the entire NSF annual budget for ocean chemistry.
>
> Hank replied, "Well it would be worth more than a million."
>
> He spurred us to propose such a venture. Soon plans were being formulated not only to do carbon-14 but also a host of other chemical and isotopic properties along Hank's Atlantic line. Boosted by the appearance of Department of Energy [initially from ERDA, DoE's predecessor] monies, Hank's dream became a reality that ultimately covered the entire world ocean and cost NSF $25 million. (p. 73)

Harmon Craig (1992), in his letter nominating Henry Stommel for the National Medal of Science, which Hank Stommel received from President George Bush in 1989, states:

> Henry Stommel is the complete scientist, naturalist and sailor, with an eye to every interesting problem and observation that comes along. One of the best examples of this wide-ranging perception of new and interesting developments has been his interest in welding together the tracer geochemistry people and the physical oceanographers for a total look at oceanic circulation and mixing. One result of Henry Stommel's interest in this area was the GEOSECS (Geochemical Ocean Sections Study) Program, which was overwhelmingly considered the best of the NSF-IDOE (International Decade of Ocean Exploration) programs and is the model for the present WOCE (World Ocean Circulation Experiment) initiative, which will expand on and continue the GEOSECS Studies during the next decade.

John Edmond, a major participant in GEOSECS and a marine geochemist who has made several significant contributions to the field, recounts his view of some of the chemical oceanographic achievements that made GEOSECS a possibility (Edmond, 1980). He states, and I paraphrase, that there were several significant efforts and discoveries, such as efforts by Derek Spencer of the Woods Hole Oceanographic Institution and Karl Turekian of Yale University to overcome many obstacles and make oceanic trace-metal profile measurements a practical proposition; the pioneering efforts of Gote Ostlund and Claes Rooth of the University of Miami to measure tritium in the Atlantic Ocean; and measurement of primordial helium in the deep Pacific by Harmon

Craig of the Scripps Institution of Oceanography and Brian Clarke of McMasters University demonstrating that volcanogenic input to the deep ocean was a real thing.

GEOSECS was a monumental undertaking with extensive cruise tracks in the Atlantic, Pacific, and Indian Oceans (see Brewer paper later in this volume). The program had its trials and tribulations in analytical chemistry, among other challenges (harking back to the importance of careful, precise analyses of seawater first highlighted in the 1920s through 1950s) as recounted by Brewer. However, ultimately it was a significant success (Craig, 1972, 1974; Craig and Turekian, 1980; Edmond, 1980) and led to follow-on programs of increasing sophistication and better time and space scale resolution: TTO (Transient Tracers in the Ocean), WOCE (World Ocean Circulation Experiment), and to some extent, the JGOFS (Joint Global Ocean Flux Study) program.

Success in GEOSECS was due to the combined efforts of many people, as is the case for most advances in chemical oceanography and marine geochemistry. The importance of having first-rate sampling systems and onboard analyses systems cannot be overemphasized, and these were provided for GEOSECS by Arnold Bainbridge and the GEOSECS Operations Group. Broecker and Peng (1982) dedicate their book to Arnold Bainbridge in recognition of his important contributions.

Marine Radioactivity and Chemicals of Environmental Concern

Chemical oceanographers, marine chemists and geochemists, physicists, geologists, biologists, and ecologists intensified the investigation of the invasion of radioactive fallout into the oceans and also provided preliminary assessments of the use of artificial radionuclides as tracers of oceanic processes. Radioactivity in the marine environment was assessed in a report of that title published in 1971(NAS, 1971b), although the actual work on the report began in 1967. I highly recommend a careful reading of this report. The breadth and depth of advances in knowledge and the literature cited in this report are impressive. One example of information contained in the report (NAS, 1971b, Chapter 7, Figure 4) should serve to whet the reader's intellectual appetite (Figure 1).

NSF funded much fundamental research in ocean chemistry during the 1960s alone or in partnership with the Office of Naval Research and the Atomic Energy Commission. Understanding the fundamental biogeochemical cycles of chemicals in seawater became the key to assessing some very important societal problems in addition to the role of the ocean in the carbon dioxide-related climate issues and contamination of the oceans by artificial radionuclides. Natural cycles of several elements and compounds were being modified by human activities (SCEP, 1970; NAS, 1971c). Nutrient enrichments in coastal waters were a recognized prob-

lem (NAS, 1971c). People were poisoned with mercury in the Minimata Bay area of Japan. Rachel Carson (1962) documented, in layperson's terms, the promiscuous use of chlorinated pesticides and unintended adverse effects. Shortly thereafter, in the late 1960s, the issue of PCBs (polychlorinated biphenyls) in the environment and potential problems with these chemicals became known. The Santa Barbara oil spill of 1968 captured people's attention for a period of time. Polluted rivers and polluted air were obvious near industrialized areas. The first Earth Day would occur in 1970. All of this is chronicled in a readable book *The Health of the Oceans* by Edward D. Goldberg (1975a).

Earlier in this paper (see Table 1), the NSF grant to Patterson and Chow was noted. They conducted fundamental research about the biogeochemistry of lead in the marine environment. This led to a critically important finding of the evidence of lead input to the oceans from human activities (Chow and Patterson, 1966). Using isotope dilution mass spectrometry, Patterson's laboratory at California Institute of Technology set the standard for analysis for lead in marine (and other) samples. A summary of their work up to 1976 and its influence on the research of others concerned with the biogeochemistry of lead in the environment is found in Patterson et al. (1976). Claire C. Patterson was recognized for his pioneering geochemical research on lead isotopes by the award of the V.M. Goldschmidt Medal in 1980 (Epstein, 1981). I highly recommend reading Patterson's acceptance speech (Patterson, 1981) to those entering an environmental chemistry career.

Max Blumer, organic geochemist at the Woods Hole Oceanographic Institution, had been supported by both NSF and ONR to undertake fundamental investigations of organic compounds in the marine environment. Max focused on hydrocarbons and fatty acids in the contemporary environment and on pigment diagenesis products in ancient sediments. He had developed elegant and sophisticated trace analytical organic chemistry methods and applied them to the analyses of biosynthesized hydrocarbons in marine animals, plants, seawater, and surface sediments in the 1960s (see Farrington, 1978, for a more complete review). In the fall of 1969, the barge *Florida* went aground and spilled No. 2 fuel oil onto the marshes and subtidal area of Buzzards Bay near West Falmouth, Massachusetts. Thus began modern studies of oil pollution in the marine environment. Max Blumer and his laboratory group applied their sophisticated methodology to analyses of surface mud and shellfish— days, weeks, months, and then two years after the visible oil slick had disappeared. They proved beyond reasonable doubt that No. 2 fuel oil persisted long after "conventional wisdom," founded in visual observations, suggested that the oil compounds would be gone from the marine environment (Blumer et al., 1970; Blumer and Sass, 1972a,b). Of equal importance, Max Blumer collaborated with WHOI colleagues in biological oceanography, Howard Sanders and John Teal, and a graduate student advised by Teal, Kathryn

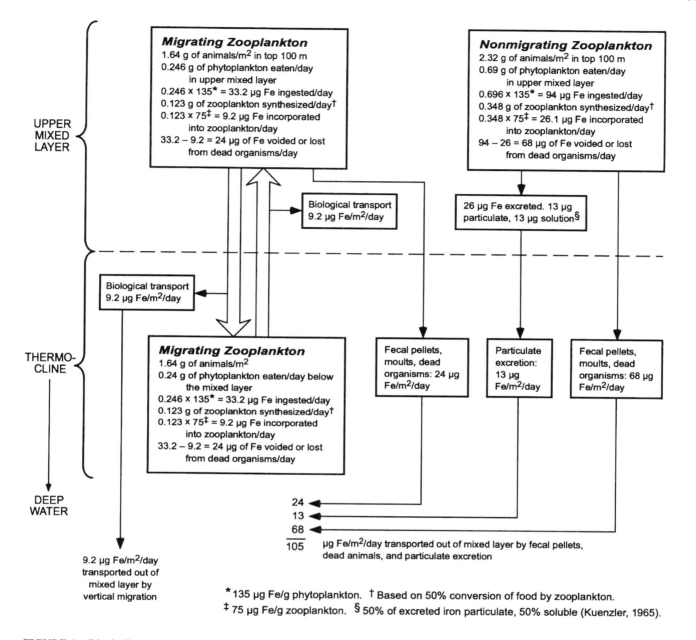

FIGURE 1 Block diagram showing the downward transport of iron by biological mechanisms in phytoplankton and zooplankton in the northeastern Pacific Ocean. SOURCE: Figure 4 in NAS (1971b).

Burns, to demonstrate a linkage between elevated concentrations of No. 2 fuel oil compounds and adverse effects in subtidal, intertidal, and marsh communities of marine organisms (e.g., Burns and Teal, 1971; Burns, 1976; Sanders, 1978; Sanders et al., 1980). Blumer, Sanders, and Teal pioneered modern oil pollution studies, along with colleagues studying the *Arrow* spill in Canada (e.g., Gordon and Michalik, 1971).

The results of these oil spill studies influenced early studies of the magnitude and biogeochemistry of chronic

inputs of petroleum in coastal and estuarine ecosystems (Farrington and Quinn, 1973). Max Blumer shared generously of his knowledge of oil pollution with my Ph.D. thesis advisor James G. Quinn and me during a crucial phase of our studies in 1969 and 1970. I mention this because it is one example of many of how personal communications during scientific meetings and personal visits advance scientific knowledge.

The examples of Patterson's laboratory and of Blumer's research are but two of many examples of NSF-funded basic research providing the background and underpinning for environmental quality research and the application of this research to providing crucial knowledge and solutions to critical environmental quality issues in the marine environment.

National Academy of Sciences Marine Chemistry Panel, 1969-1971

At the turn of the decade, the Marine Chemistry Panel of the Committee of Oceanography of the National Academy of Sciences met and assessed progress and the challenges ahead. The panel membership consisted of a cross section of the intellectual and organizational leaders of marine chemistry in the United States and Canada. Their names and affiliations (at that time) are *Norris Rakestraw*, chairman, retired from Scripps Institution of Oceanography; *Richard Bader* and *John Bunt*, Rosenstiel School of Marine and Atmospheric Sciences, University of Miami; *James Carpenter*, head of the Oceanography Section of the National Science Foundation; *Dayton Carritt*, director of the Institute for Man and His Environment, University of Massachusetts; *Gordon Erdman*, Phillips Petroleum Company Research Center; *Robert Garrells* and *Edward Goldberg*, Scripps Institution of Oceanography; *John Hunt*, Woods Hole Oceanographic Institution; *David Menzel*, Skidaway Institute of Oceanography; *Timothy Parsons*, Institute of Oceanography, University of British Columbia; and *Ricardo M. Pytkowicz*, Oregon State University.

Alfred C. Redfield wrote the dedication to this report at the request of the committee:

> The members of the Committee responsible for this report dedicate it to their chairman, Norris W. Rakestraw, the dean of marine chemists in the United States (NAS, 1971a, p. iii).

The panel's statement in the introduction to the report poignantly outlined the role of chemistry and chemical research in the oceans and, from the perspective of almost three decades later, captured the general framework within which most of the research of the intervening years has been pursued.

> There are several viewpoints from which the chemist can approach the ocean. He can consider the oceans as:
>
> A dynamic mixing system, in which composition changes take place partly from internal processes and partly as a result of the circulation and mixing of water masses
>
> A reservoir that is intermediate between the runoff of components from the continents and exchange reactions with the sediments
>
> A biological system in which virtually all the biochemical changes associated with living organisms take place

> A grand septic tank in which organic materials are decomposed mostly in the near-surface water or at the bottom
>
> A vast chemical system, in which interactions occur among an enormous number of components, both organic and inorganic, ranging in concentration through 12 orders of magnitude
>
> An environment that is being invaded by man through his social, agricultural, and industrial activities
>
> The key to interpreting the past history of the earth and as the custodian of its relics. (p. 5-6)

The panel went on to note:

> It was said by the late Fritz Koczy of the University of Miami: Chemical reactions in the ocean . . . are largely determined by phenomena which occur at interfaces . . . seawater is bounded by two of the most extensive interfaces on earth—the one where it meets the air above, the other where it mingles with the sediment below. (p. 6)

Atmosphere-ocean interactions, in addition to the carbon dioxide exchange, and the chemistry of the surface ocean microlayer have been the subject of much innovative research since that time. Macintyre (1974) and Berg and Winchester (1978) chronicle the important contributions of Robert Duce, John Winchester, Joseph Prospero, William Barger, William Garrett, and others to this effort in the 1960s and early 1970s.

On the subject of sediment geochemistry, Professor Robert A. Berner of Yale University began his research in the 1960s and became one of the more important contributors to our understanding of marine sediment geochemistry (Siever et al., 1961; Berner, 1963, 1964) continuing to the present. Berner's book *Early Diagenesis* (Berner, 1980) captured much of his thinking on the subject and is one of the leading texts on this subject. Many of the researchers from the 1970s through the present, including this author, were strongly and positively influenced by Berner's work.

Many significant sediment geochemistry studies required comparisons of depth profiles of solid phase and pore water geochemistry. For studies of some important geochemical reactions, obtaining samples of pore water from deep ocean samples in a manner that avoided compromising sample integrity due to changes in temperature and pressure when bringing sediment cores up to the surface was a major challenge. The pioneering efforts of Fred Sayles and his coworkers (Sayles et al., 1976) are illustrative of how efforts to develop novel *in situ* sampling instrumentation by excellent scientists and careful analysts has led to significant advances in marine chemistry and geochemistry. Fred Sayles used this instrumentation, and subsequent improved versions, to make significant contributions to understanding fluxes of chemicals at the sediment-water interface (e.g., Sayles, 1979, 1981).

THE 1970S TO THE 1990S[1]

Physical Chemistry and Aquatic Chemistry: Theory and Experimentation

The text, *Aquatic Chemistry* by Werner Stumm and James J. Morgan (1970), which was aimed at the broader arena of the chemistry of all waters, brought a modern and fundamental underpinning of physical chemistry to aquatic chemistry, including chemical oceanography and marine geochemistry. This book became a required text for many of my generation and a standard reference. The text has been published in a second and third edition (Stumm and Morgan, 1981, 1995). Other excellent texts of similar aim have followed in the intervening years, but I believe that the first edition of Stumm and Morgan had a powerful positive influence on the field following in the wake of the Sillen (1961) paper.

In the arena of physical chemistry from the 1970s to the present, perhaps no other chemical oceanographer-marine geochemist has made more significant contributions than Frank Millero of the University of Miami. Early evidence of this was his contribution "Seawater as a Multicomponent Electrolyte" (Millero, 1974).

International Decade of Ocean Exploration and Chemical Oceanography-Marine Geochemistry

The International Decade of Ocean Exploration programs in chemical oceanography and marine geochemistry were more than the flagship GEOSECS effort. Several programs were grouped together under the overarching theme of environmental quality, including GEOSECS (Jennings and King, 1980). For example, the Manganese Nodules Program, which became known as MANOP in Phase II, undertook important research to better understand the geochemistry of manganese nodules—much touted in the late 1960s and early 1970s as a valuable mineral resource (Knecht, 1982).

In 1971 through early 1972, IDOE launched a one-year Baseline Data Acquisition Program for a limited survey of the extent of contamination of the marine environment by chemicals of environmental concern. The broad focus was on chemicals entering the environment as a result of human activities mobilizing both naturally occurring chemicals (e.g., trace metals and petroleum), and chemicals synthesized only by modern industrial processes (e.g., chlorinated pesticides and PCBs). A conference workshop of three days was convened in May 1972, under the leadership of Professor Edward Goldberg to assess what had been learned from the

Baseline effort. I attended this conference, having contributed some of the data as a result of my ongoing postdoctoral research with Max Blumer, initiated in July 1971. Although all of us at the workshop recognized the limitations of the sparse data sets, these data were all we had. Data for trace metals, chlorinated pesticides and PCBs, and the less biodegradable petroleum hydrocarbons could be interpreted in the broadest sense within the framework of lessons learned about biogeochemical cycles from studies of the fallout radionuclides. In the forward to *Radioactivity in the Marine Environment* (NAS, 1971b), Dr. Philip Handler, president of the National Academy of Sciences summarized this point:

> It is particularly appropriate that this contribution to our understanding of the marine environment be available at a time when man is increasingly concerned with the ways in which his own actions may affect his environment. Though this work is specifically addressed to radioactivity in the marine environment, many of the concepts that pertain to our understanding of this problem can be applied effectively to studies of other wastes discharged into the marine environment, including industrial wastes, municipal sewage, pesticides, nutrients, heavy metals and heat. It is perhaps ironic that of the many substances that man has introduced into his environment over the centuries, he understands best and controls most rigorously the radioactive materials that have been produced only during the past quarter century. We are indeed fortunate that our intense concern for public safety and protection from radioactivity since 1950 has stimulated much basic research that can be applied to other serious environmental problems that we are just beginning to recognize. (p. iii)

It was within that type of framework that Ed Goldberg led the "Baseline Conference" to consensus. The fact that we were meeting at the Brookhaven National Laboratory, the Memorial Day weekend was approaching, and Ed controlled the arrival of the buses to the airport provided one impetus for the participants to reach consensus. The consensus was important because Ed Goldberg and a dedicated secretarial staff labored over the weekend to produce the final version of the workshop report and have it printed (Goldberg, 1972). Then Goldberg delivered the report the following week to the First United Nations Conference on the Environment meeting in Stockholm, Sweden, where the report influenced that conference's deliberations on environmental quality concerns and the ocean. In recognition of this and many other research and science-policy interactions, Ed Goldberg shared the Tyler Prize for Environmental Achievement in 1989 for his many contributions to understanding marine environmental quality issues.

The Baseline Surveys (Goldberg, 1972), and also other scientific research and survey data assessed in the very influential Workshop on Critical Problems of the Coastal Zone, under the leadership of Bostwick H. Ketchum, held May 22 to June 3, 1972, in Woods Hole, Massachusetts (funded in part by the National Science Foundation), led to the inescap-

---

[1] See the *FOCUS* (1998) report for an in-depth review and the FOCUS summary in this volume.

able recognition that there were some serious environmental quality problems in the coastal zone. Several of these problems were associated with chemicals of environmental concern, including trace metals, pesticides, petroleum hydrocarbons, and excessive nutrient inputs (Ketchum, 1972). This spawned many environmental quality research efforts in the coastal zone. The origin of the U.S. Environmental Protection Agency's (EPA's) Mussel Watch Program, a prototype monitoring program during 1976-1980 for chemicals of environmental concern in coastal areas (Goldberg, 1975b; Goldberg et al., 1978; Farrington et al., 1983), can be traced directly from the experience in the IDOE Baseline Program and individual investigator research efforts funded by a mix of NSF, ONR, and the Atomic Energy Commission. The current operational National Oceanic and Atmospheric Administration (NOAA) Status and Trends Monitoring Program grew out of the U.S. EPA Mussel Watch Program prototype.

The IDOE-NSF follow-on programs to the Baseline Surveys took two pathways: one mainly biogeochemical and one mainly biological effects. In the first pathway, research on marine pollutant transfer was pursued between 1972 and 1976 and is summarized in the workshop book edited by Windom and Duce (1976). The part of this effort concerned with atmospheric inputs to the oceans eventually evolved under the leadership of Robert Duce, among others, to the SEAREX (Sea-Air Exchange) Program of the 1980s (Duce, 1989) and then to other follow-on programs assessing the atmospheric transport of chemicals to the ocean.

In the second pathway, mesocosms were used in CEPEX (Controlled Pollution Experiment) studies undertaken with large plastic enclosures hung in the sea. Within a few years, CEPEX—and mesocosm experiments at Loch Ewe in Scotland—influenced the development of the MERL (Marine Ecosystems Research Laboratory) mesocosms at the Graduate School of Oceanography, University of Rhode Island, funded by U.S. EPA (Grice and Reeve, 1982). In addition, the effect of pollutants at the organism and tissue levels was pursued in the NSF-funded PRIMA (Pollutant Responses in Marine Animals) program (Jennings and King, 1980).

The Uranium Decay Series and Chemical Oceanography-Marine Geochemistry

A considerable number of scientists in numerous studies since the late 1960s, have utilized the uranium decay series radionuclides (Figure 2) to unravel, quantitatively, processes at the boundaries of the oceans and internal processes in the oceans. So many scientists have used this (e.g., see Broecker and Peng, 1982, and Pilson, 1998 for discussion and references), and the knowledge gained has been so important, that it is important to highlight this as an achievement.

Hydrothermal Vents, Riverine Inputs, Atmospheric Inputs and the Inner Workings—Exchanges with the Atmosphere and Deposition to Sediments

There has been amazing progress in understanding the biogeochemical cycles in the oceans since 1970. I have borrowed a cartoon from Professor Conrad Neumann of the University of North Carolina-Chapel Hill that captures many of the important aspects of biogeochemical cycles of the oceans (Figure 3). A reviewer drew my attention to the fact that my favorite part of biogeochemical cycles—organic matter (dissolved and particulate)—is missing. Nevertheless, the cartoon captures much of what needs to be qualitatively depicted.

The most exciting discovery of the 1970s in oceanography, by almost all accounts, was the discovery of the vents of hot fluids at the ridge crest valleys of the midocean ridge system and the unexpected associated fauna founded in a chemosynthetic food web (Ballard, 1977; Corliss et al., 1979; Edmond, 1982). This has been described in Dick Barber's paper on biological oceanography immediately preceding this paper and in Bob Ballard's paper later in this volume. As pointed out by Corliss et al. (1979) and Edmond (1982) among several others, the vents not only were important from a biological perspective, but provided documentation of what had been suspected from analyses of altered basalt dredged from the ridge crests or obtained by submersible, that the interaction of seawater with hot and warm basalt at the ridge crests had an important influence on overall seawater chemical composition and in balancing global biogeochemical cycles on geologic time scales.

At the other end of the inputs pipeline, the flow of dissolved and particulate material into the oceans via rivers received increased and significant attention from the 1970s through the present (e.g., Martin et al., 1981; Milliman and Meade, 1983). A continuing vexing challenge was to understand the effect of increased salinity on the chemical composition of estuarine water as materials flowed from the fresh river water into the more saline estuaries. Sholkovitz and coworkers carried out a series of elegant experiments titrating river water with seawater and observing the effects on the chemistry and physical chemical forms in the resulting solutions (e.g., Sholkovitz, 1976).

William J. Jenkins began studies in W.B. Clarke's laboratory to measure helium-3:helium-4 ratios and also apply this to measuring tritium (Jenkins et al., 1972; Clarke et al., 1976). Bill Jenkins continues to make major important contributions to oceanography. As the citation for the 1997 Bigelow Medal awarded to Bill Jenkins states:

> The key to Bill Jenkin's success is that he is one of those rare people who can make superb measurements and can also place the data into sound, quantitative models, allowing him to contribute to diverse fields in a unique way. Few scientists have had as much impact and achieved recognition among so many different scientific communities.

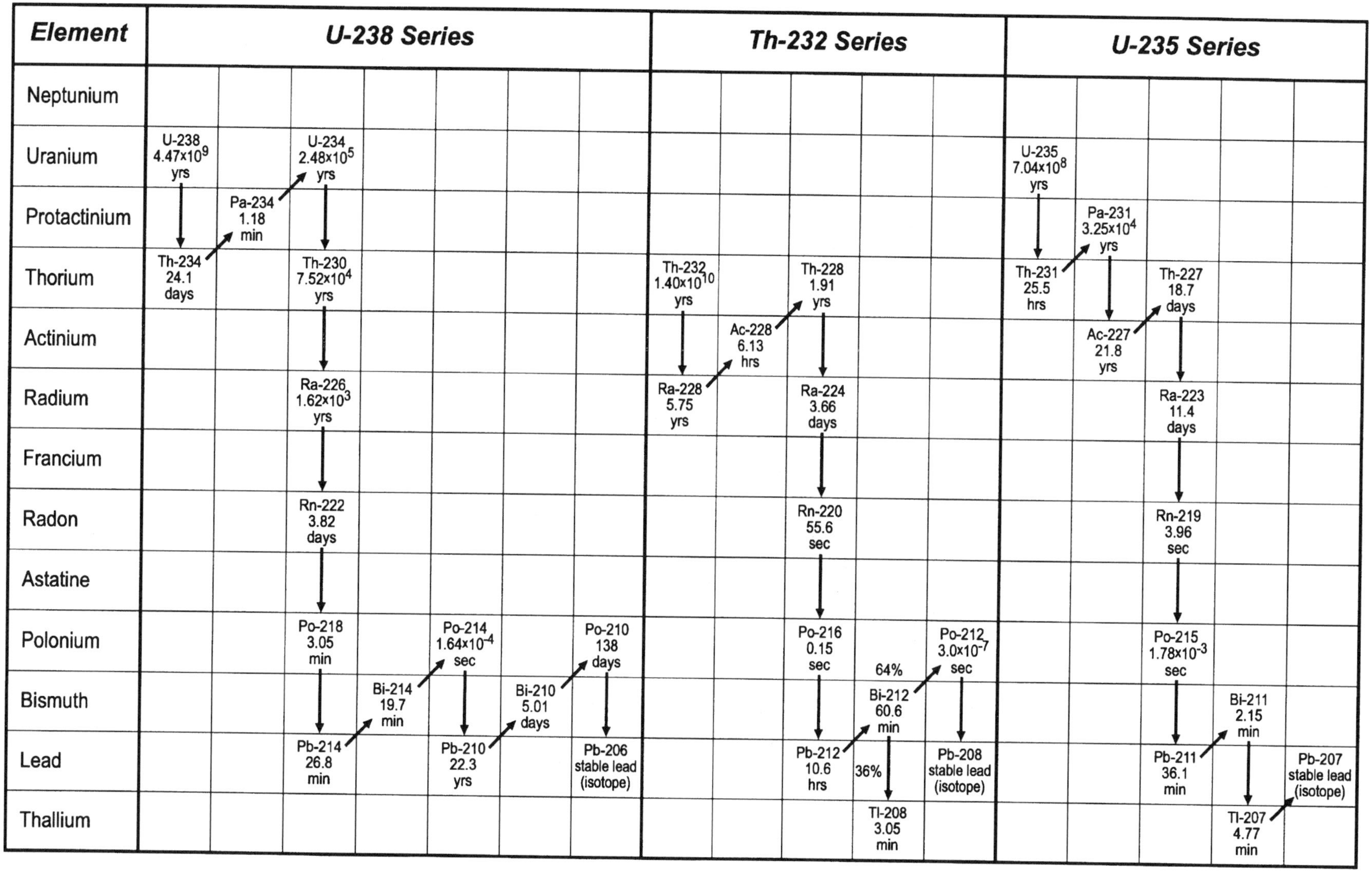

Element	U-238 Series	Th-232 Series	U-235 Series
Neptunium			
Uranium	U-238 4.47×10^9 yrs; U-234 2.48×10^5 yrs		U-235 7.04×10^8 yrs
Protactinium	Pa-234 1.18 min		Pa-231 3.25×10^4 yrs
Thorium	Th-234 24.1 days; Th-230 7.52×10^4 yrs	Th-232 1.40×10^{10} yrs; Th-228 1.91 yrs	Th-231 25.5 hrs; Th-227 18.7 days
Actinium		Ac-228 6.13 hrs	Ac-227 21.8 yrs
Radium	Ra-226 1.62×10^3 yrs	Ra-228 5.75 yrs; Ra-224 3.66 days	Ra-223 11.4 days
Francium			
Radon	Rn-222 3.82 days	Rn-220 55.6 sec	Rn-219 3.96 sec
Astatine			
Polonium	Po-218 3.05 min; Po-214 1.64×10^{-4} sec; Po-210 138 days	Po-216 0.15 sec; Po-212 3.0×10^{-7} sec	Po-215 1.78×10^{-3} sec
Bismuth	Bi-214 19.7 min; Bi-210 5.01 days	Bi-212 60.6 min (64%)	Bi-211 2.15 min
Lead	Pb-214 26.8 min; Pb-210 22.3 yrs; Pb-206 stable lead (isotope)	Pb-212 10.6 hrs; Pb-208 stable lead (isotope)	Pb-211 36.1 min; Pb-207 stable lead (isotope)
Thallium		Tl-208 3.05 min (36%)	Tl-207 4.77 min

FIGURE 2 Uranium and thorium decay series. Reprinted from Figure 4-1 in Broecker and Peng (1982) with permission from Lamont-Doherty Earth Observatory.

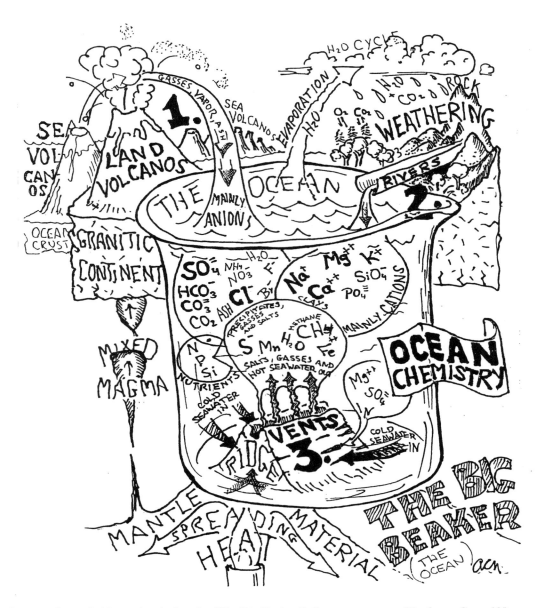

FIGURE 3   Cartoon of oceanic biogeochemical cycles "The Big Beaker."   Cartoon courtesy of Professor Conrad Neumann, University of North Carolina, Chapel Hill.

A pioneering and thoughtful paper by Oliver C. Zafiriou (1977) "previewed" the field of marine photochemistry, stimulating a fresh look at the role of photochemical reactions in the ocean.   Since that time, with the efforts of Zafiriou, Zika, and others, our knowledge of marine photochemistry has expanded rapidly (Zika, 1987).   Marine organic geochemistry moved from descriptive, qualitative studies to become more quantitative and more oceanic process oriented (e.g., see Gagosian, 1983; Farrington, 1987; Lee and Wakeham, 1989; and the review volume edited by Farrington, 1992).

The internal fluxes of materials on particulate matter in the ocean were the subject of significant efforts in chemical oceanography-marine geochemistry.   Honjo, Spencer, and Brewer undertook an effort using large sediment traps to assess the vertical fluxes of large particles in the oceans in their PARFLUX effort (Honjo, 1978; Spencer et al., 1978; Brewer et al., 1980).   Similar efforts were undertaken simultaneously by several other investigators (e.g., Gardner, 1977; Staresinic et al., 1978; Knauer et al., 1979; and reviews by Brewer and Glover, 1987).

A very important small research group effort by Werner

G. Deuser of the Woods Hole Oceanographic Institution, funded by NSF, adopted the Honjo sediment trap design and undertook a pioneering effort to make time-series sediment trap measurements in the Sargasso Sea. Deuser and coworkers documented that there was a seasonal flux of particles to the deep Sargasso Sea (e.g., Deuser and Ross, 1980; Deuser et al., 1981). These oceanic time series measurements built on the Station S measurements off Bermuda, conducted by Hank Stommel for years, and continued by several individuals for years thereafter, and were staged from the Bermuda Biological Station for Research. This effort stimulated other measurements to assess time-variant fluxes of particles to the deep ocean and was a key to initiation of the present time-series measurements in the Joint Global Ocean Flux Study (JGOFS) program.

Dissolved Trace Metals, Biological Processes, and Paleoceanography

While the large particles were being captured and analyzed, significant efforts were underway to measure dissolved trace metals in seawater using new, improved "clean" techniques largely provoked by the work of Patterson and coworkers on measuring lead in seawater (Martin, 1991). As Pilson (1998, p. 209) describes the situation, "The first real breakthrough in attempts to learn the true concentrations of these metals in seawater came in 1975 with the publication by Boyle and Edmond of a paper showing that their data from measurements of copper in surface waters south of New Zealand made sense when plotted against another oceanographic variable, in this case nitrate" (Boyle and Edmond, 1975). Boyle continued this line of research with other examples such as relationships between cadmium and phosphate. Bruland and coworkers and others added several more examples of dissolved trace-metal depth profiles (e.g., see review by Donat and Bruland, 1995). Boyle took the connection of selected trace metal and nutrient cycles and depth profiles a step further in the significant finding that cadmium could be used as a paleoceanographic tracer (Boyle, 1988).

Progress in analytical chemistry has been crucial to many of the advances in our knowledge of trace-metal biogeochemistry, and other biogeochemical processes in the oceans, as it was in the early days of chemical oceanography-marine geochemistry (Johnson et al., 1992). Figure 4, taken from the Johnson et al. (1992) paper, provides an impressive compilation of the 15 orders of magnitude range of concentrations of seawater components now measured in studies of the oceans.

The Iron Hypothesis and a Return to One "Root" of Modern Chemical Oceanography

Nearly simultaneous with the sediment trap research of Honjo and Deuser and their colleagues, the VERTEX Pro-

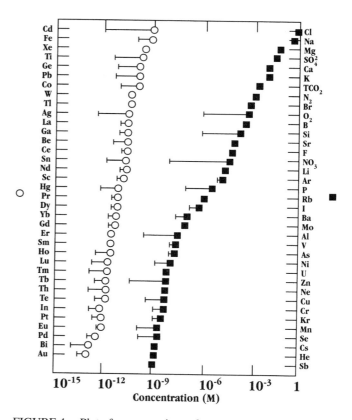

FIGURE 4 Plot of concentrations of seawater components spanning 15 orders of magnitude in concentration. SOURCE: Figure 1 in Johnson et al. (1992). Reproduced with permission from *Analytical Chemistry*, volume 68, pp. 1065-1075. Copyright 1992 by the American Chemical Society.

gram, led by John H. Martin of Moss Landing Marine Laboratory, and Ken Bruland and Mary Silver of the University of California-Santa Cruz, undertook efforts to study the fluxes of particles in the upper ocean and midwater regions and to couple these with both chemical and biological processes (Martin et al., 1983). From these and other studies (e.g., Martin and Fitzwater, 1988), Martin and his coworkers obtained results that led them to an important and stimulating hypothesis that iron was limiting productivity in many areas of the open ocean (Martin, 1991). This hypothesis involves atmospheric transport of dust and associated iron to the iron-limited areas of the oceans where the iron, as an essential limiting factor, stimulates biological primary production. There is even a link to carbon dioxide and climate; Martin suggested that during glacial times, atmospherically transported dust would increase in the southern ocean areas and cause higher productivity, thereby drawing down carbon dioxide levels in the atmosphere. Earlier in this volume, Dick Barber discusses this from the perspective of biological productivity.

This example from the work of Martin and coworkers returns us to one of the early and continuing themes in chemical oceanography noted in the beginning paragraphs of this paper, the connection between chemical oceanography and biological productivity. As reported earlier in this paper, Rakestraw had noted (see Shor, 1978):

> One of the most striking observations of marine biology is the fact that some parts of the ocean are fertile while other parts are quite barren. There must be chemical factors which determine fertility, and an explanation of this was perhaps the first serious question which oceanographers asked the chemist. In the year 1930 there were probably no more than a dozen professional chemists in the world who were actively interested in the ocean, and practically every one of them was trying to answer this question. (p. 231)

Have John Martin and coworkers answered the question at long last?

THE RELATIONSHIP BETWEEN CHEMISTRY, GEOLOGY, PHYSICS, AND CHEMICAL OCEANOGRAPHY/MARINE GEOCHEMISTRY

The 1950s and 1960s were periods of time when few graduate students actually formally received degrees in chemical oceanography or marine geochemistry. Instead, many who contributed to advances in this arena of research were formally educated for their graduate degrees in chemistry, geology, geochemistry, or physics. Examples from the efforts cited above are Max Blumer, Harmon Craig, Ed Goldberg, Bill Jenkins, John Hunt, Frank Millero, Claire Patterson, and Oliver Zafiriou, to name just a few. Beginning in the late 1960s, formal graduate education in chemical oceanography, marine geochemistry, and marine chemistry expanded, and now a majority of those conducting research in this arena have received formal degrees in chemical oceanography (or marine geochemistry, marine chemistry). However, it is important that research and graduate education in chemical oceanography and marine geochemistry maintain connectivity to the advances in the various areas of chemistry, physics, and geology.

I use personal experience to illustrate the point. My Ph.D. graduate education and thesis research in chemical oceanography was directed by Professor James G. Quinn. Jim was attracted to an assistant professorship position in oceanography at the Graduate School of Oceanography, University of Rhode Island, in the late 1960s because of the emergence of Sea Grant—at that time an NSF effort. He was a biochemist with no training or formal education in oceanography. I recall one of our first meetings to discuss what I would do as part of my Sea Grant-funded graduate research assistantship in the fall of 1968. Jim stated that he did not know very much about oceanography, but that he was knowledgeable about lipid biochemistry and thought that there were some exciting and important things to learn about the chemistry and biochemistry of lipids in the marine environment. He thought that perhaps we could learn about oceanography together. He was correct in both accounts! I

benefited greatly in my thesis research and throughout my career, as did others of his students and associates, from Jim Quinn's knowledge of lipid biochemistry.

Although I wholeheartedly support graduate education in chemical oceanography, marine chemistry, or geochemistry, I submit that we will be much poorer in the study of the chemistry of the sea and marine sediments unless we continue to attract people such as Jim Quinn to these studies from other arenas of chemistry and biochemistry.

While on the subject of graduate education, I would be remiss if I did not acknowledge the wonderful practice initiated in 1978 by Neil Anderson and Rodger Baier of NSF and Ed Green of ONR to gather together every two years a cross section of senior graduate students (a year away from their Ph.D.) or recent Ph.D.s in chemical oceanography, geochemistry, and aquatic chemistry in a symposium to share their thesis research and ideas. These "Dissertations in Chemical Oceanography" (DISCO) symposia have enriched early careers to the betterment of chemical oceanography and marine geochemistry.

THE SUPPORT AND ENABLING PEOPLE

National Science Foundation, Office of Naval Research, Atomic Energy Commission, Department of Energy, and other agency program managers and staff people support and enable the acceptance, review, and funding of proposals submitted by scientists. More than this, they work—often tirelessly—behind the scenes to bring the community of chemical oceanographers, marine geochemists, and others together. They have the thankless task of trying to stretch too often inadequate budgets to the maximum benefit of the science. Since this is an NSF-related activity, I confine my citation to those "career" NSF program directors and managers in the Ocean Sciences and IDOE sections with whom I have been acquainted over the years in their support of chemical oceanography and marine geochemistry research— Neil Andersen, Roger Baier, and Michael Heeley. Many scientists who spent one or two years in a temporary rotating appointment assignment at NSF ably assisted them. Dr. Neil Andersen has been recognized formally by the ocean science community for his important and wide-ranging contributions with the 1994 Ocean Sciences Award from the American Geophysical Union.

In a similar vein, numerous people at various universities, marine laboratories, and oceanographic institutions have provided the administrative, logistical, and laboratory support to enable the research described above. Especially important among these are the officers and crews of the research vessels. I cannot pay tribute to all by name; thus, I will use just two of these many folks as examples from my personal experience. For many years, Emerson Hiller was master of the R/V *Atlantis II* and then of R/V *Knorr* at the Woods Hole Oceanographic Institution. Most chief scientists I know, at least it was true for me, felt that there was no doubt that the science had first priority and that you were in "good hands" when at sea with Captain Hiller. His knowledgeable suggestions about cruise tracks and cruise execution in the face of various unforeseen challenges, and his ship handling, are legendary. Another example is Jerry Cotter the bo's'n of the R/V *Knorr*. There is no doubt that the Atlantic GEOSECS expedition (and many other expeditions and cruises) benefited enormously from the combined expertise and talents of these two individuals and their fellow officers and crew. There is no greater tribute to Jerry than to quote Hank Stommel about the bo's'n: "Lest we think that science is all instrumentation, data, and theory, some time when the weather is making up, and our gear is hopelessly fouled in the water over the side, and the wire has jumped the sheave, the sight of the bo's'n coming on deck is the most important of all." (Hogg and Huang, 1995, p. I-197).

CONCLUDING COMMENTS

The 1950s to the present have been years of significant advances in the application of chemistry to elucidate and quantify oceanic processes. A mix of individual investigator and larger group efforts involving innovative ideas and determined hard work has advanced the field. The development and application of sophisticated analytical methods of trace chemical measurements have been impressive. They have unlocked many secrets of natural and human-forced processes. The power of stable and radioactive isotope chemistry to elucidate and quantitatively unravel physical, chemical, and biological processes in the oceans and underlying sediments moved from concept to reality during the past 50 years and is still evolving rapidly. Mass spectrometers of all sorts have replaced titration burettes as common analytical equipment in the laboratories of chemical oceanography and marine geochemistry.

Data sets of unprecedented size and complexity are being interpreted more routinely. Both equilibrium and nonequilibrium approaches are used commonly to model data collected from the field and laboratory experiments. A rich mix of theory, experimentation, and observation has been at the heart of advances in chemical oceanography and marine geochemistry. As FOCUS (1998) notes, much more exciting and important science is already over the horizon and confronting us today. There are crucial societal needs in the global, regional, and local arenas to be served by improved knowledge in chemical oceanography-marine geochemistry. For this reason and because of the intrinsic excitement of unraveling the beauty and secrets of natural processes, let us hope that the efforts of the next 50 years will at least meet the impressive standard set by the past 50 years!

DEDICATION

In the spirit of this presentation in choosing examples to illustrate the contributions of many, I dedicate this paper to a person whom I had the privilege of collaborating with in two major efforts, the U.S. Mussel Watch Program and the VERTEX Program; I am speaking of John Holland Martin. In late 1993, at the invitation of Professor Margaret Leinen, Dean of the Graduate School of Oceanography (GSO) at the University of Rhode Island (URI), I was approached by the editors of the *Bulletin of the Graduate School of Oceanography* and asked as an alumnus to write a short *in memoriam* dedication of the bulletin for 1994-1995 to John H. Martin. John had earned his Ph.D. there in 1966. This was a great honor for me. However, I knew of a person who should aid in this venture, Dr. Donald K. Phelps, a long-standing friend and colleague of John's, who earned his Ph.D. from the GSO in 1964 and had recently retired from the U.S. EPA laboratory in Narragansett. I first met Don when he was a member of my Ph.D. thesis committee. Don offered a poem as his contribution to our effort.

Don's poem is a moving tribute to a friend and colleague who wanted a better world for all of us:

JOHN HOLLAND MARTIN

John Holland Martin, indomitable spirit.
Challenges barriers as a scythe to an emery wheel -
   well honed for the encounter.

Fallen from the playing field.
Incubated in an iron lung.
Emerges: wheelchair.
Braced for halting steps that bind his movement,
   he travels around the world
Where many have not.

Ideas.
Data.
Image in electronic transport speeds like light to spark
   controversy and awaken sleeping minds and
   tradition bound visions.

Ideas spark as iron against stone.
The iron limitation hypothesis.
Now tested further.
Now closer to crystallizing that vision.

Loves much.
Family, friends, this country, its story-tellers,
The mysteries of the ocean, his students and a good time.

His legacy:
When the Fates give you a bad deal, pick up the cards
   and play the game.

DONALD K. PHELPS, 1993
URI-GSO Bulletin for 1994-1995.

ACKNOWLEDGMENTS

I have many people to thank for an enjoyable career in chemical oceanography and environmental chemistry, and I hope that I am not yet finished with this career. First and foremost my family was very supportive (my wife Shirley, daughter Karen, and son Jeff) and endured countless late nights at the laboratory and weeks to months while I was away at sea—including a Thanksgiving, a Christmas, and a New Year. I wish to acknowledge my Ph.D. supervisor Professor James G. Quinn of the GSO-URI and to thank him once again for his unselfish devotion to his students and their careers. I also extend my appreciation to Professor Michael E.Q. Pilson of the GSO-URI for introducing me to chemical oceanography in his graduate class on the subject.

I am grateful to colleagues and coworkers, students and postdocs at the GSO-URI, the University of Massachusetts-Boston, and especially the Woods Hole Oceanographic Institution (where I have been in some category of appointment for twenty-six years) who enriched my life throughout my career. Professor Edward D. Goldberg of the Scripps Institution of Oceanography has provided guidance and advice for many aspects of my research and included me in many of his national and international ventures since we first met in 1972. I hope that other valued colleagues will understand my not mentioning them each by name. The list is extensive! Contribution Number 10083, Woods Hole Oceanographic Institution.

REFERENCES

Andersen, N. (ed.). 1973. *Chemical Oceanographic Research: Present Status and Future Direction*. Deliberations of a workshop held at Naval Postgraduate School, Monterey, California, December 11-15, 1972. Office of Naval Research, Arlington, Virginia.

Andersen, N. (ed.). 1977. Concepts in marine organic chemistry. *Marine Chemistry* 5:303-638.

Arons, A.B., and H. Stommel. 1967. On the abyssal circulation of the world ocean. — III. An advection-lateral mixing model of the distribution of a tracer property in an ocean basin. *Deep-Sea Res.* 14:441-457.

Barth, T.W. 1952. *Theoretical Petrology*. Wiley, New York.

Berg, W.W., Jr., and J.W. Winchester. 1978. Aerosol chemistry of the marine atmosphere. Chapter 38. Pp. 173-231 in J.P. Riley and R. Chester (eds.), *Chemical Oceanography* (2nd Edition) Volume 7. Academic Press, London.

Berner, R.A. 1963. Electrode studies of hydrogen sulfide in marine sediments. *Geochim. Cosmochim. Acta* 27:563-575.

Berner, R.A. 1964. An idealized model of sulfate distribution in recent sediments. *Geochim. Cosmochim. Acta* 28:1497-1503.

Berner, R.A. 1980. *Early Diagenesis: A Theoretical Approach*. Princeton University Press, Princeton, New Jersey. 241 pp.

Blumer, M., and J. Sass. 1972a. Oil pollution: Persistence and degradation of spilled fuel oil. *Science* 176:1120-1122.

Blumer, M., and J. Sass. 1972b. Indigenous and petroleum derived hydrocarbons in an oil polluted sediment. *Mar. Pollut. Bull.* 3:92-92.

Blumer, M., G. Souza, and J. Sass. 1970. Hydrocarbon pollution of edible shellfish by an oil spill. *Mar. Biol.* 5:195-202.

Bolin, B., and H. Stommel. 1961. On the abyssal circulation of the world ocean. IV. Origin and rate of circulation of deep ocean water as determined with the aid of tracers. *Deep-Sea Res.* 8:95-110.

Bowen, V.T., and T.T. Sugihara. 1957. Strontium 90 in the North Atlantic surface water. *Proc. National Acad. Sci. U.S.A.* 43:576-580.

Boyle, E. 1988. Cadmium: Chemical tracer of deepwater paleoceanography. *Paleoceanography* 3:471-489.

Boyle, E., and J.M. Edmond. 1975. Copper in surface seawaters south of New Zealand. *Nature* 253:107-109.

Brewer, P.G., and D.M. Glover. 1987. Ocean chemical fluxes 1983-1986. *Reviews of Geophysics* 25(6):1376-1386.

Brewer, P.G., N. Yoshiyuki, D.W. Spencer, and A.P. Fleer. 1980. Sediment trap experiments in the deep North Atlantic: Isotopic and elemental fluxes. *Journal of Marine Research* 38(4):703-728.

Broecker, W.S. 1957. *Application of Radiocarbon to Oceanography and Climate Chronology*. Ph.D. Thesis. Columbia University, New York.

Broecker, W.S. 1974. *Chemical Oceanography*. Harcourt Brace Jovanovich, New York. 214 pp.

Broecker, W.S. 1985. *How to Build a Habitable Planet*. ELDIGIO Press, Columbia University, Palisades, New York. 291 pp.

Broecker, W.S. 1992. Statement. P. 73 in A Tribute to Henry Stommel. *Oceanus* 35 (special issue), Woods Hole Oceanographic Institution, Woods Hole, Massachusetts.

Broecker, W.S., R.D. Gerard, M. Ewing, and B.C. Heezen. 1960. Natural radiocarbon in the Atlantic Ocean. *J. Geophys. Res.* 65:2903-2931.

Broecker, W.S., R.D. Gerard, M. Ewing, and B.C. Heezen. 1961. Geochemistry and physics of ccean circulation. Pp. 301-322 in M. Sears (ed.), *Oceanography*. American Association for the Advancement of Sciences, Washington, D.C.

Broecker, W.S., and T.H. Peng. 1982. *Tracers in the Sea*. Columbia University, Palisades, New York. 690 pp.

Broecker, W.S., K.K. Turekian, and B.C. Heezen. 1958. The relationship of deep sea (Atlantic Ocean) sedimentation rates to variations in climate. *American Journal of Science* 256:503-517.

Burns, K.A. 1976. Hydrocarbon metabolism in the intertidal fidler crab, *Uca pugnax*. *Mar. Biol.* 36:5-11.

Burns, K.A., and J.M. Teal. 1971. Hydrocarbon incorporation into the salt marsh ecosystem from the West Falmouth oil spill. Woods Hole Oceanographic Institution Technical Report 71-69.

Carson, R. 1962. *Silent Spring*. Houghton Mifflin, Boston, Massachusetts.

Chow, T.J., and C.C. Patterson. 1966. Concentration profile of barium and lead in Atlantic waters of Bermuda. *Earth Planet. Sci. Lett.* 1:397-400.

Clarke, W.B., W.J. Jenkins, and Z. Top. 1976. Determination of tritium by spectrometric measurement of ^{3}He. *International Journal of Applied Radioisotopes* 27:515.

Corliss, J.B., J. Dymond, L.I. Gordon, J.M. Edmund, R.P. von Herzen, R.D. Ballard, K. Green, D. Williams, A. Bainbridge, K. Crane, T.H. van Andel. 1979. Submarine thermal springs on the Galapagos Rift. *Science* 203:1073-1083.

Craig, H. 1953. The geochemistry of stable carbon isotopes. *Geochim. Cosmochim. Acta* 3:53-92.

Craig, H. 1972. The GEOSECS Program: 1970-1971. *Earth Planet. Sci. Lett.* 16:47-49.

Craig, H. 1974. The GEOSECS Program: 1972-1973. *Earth Planet. Sci. Lett.* 23:63-64.

Craig, H. 1992. Letter to the President's Committee on the National Medal of Science in support of the nomination of Henry Stommel for the National Medal of Science. Pp. 128-129 in *Oceanus* 35 (special issue).

Craig, H., and K.K. Turekian. 1980. The GEOSECS Program: 1976-1979. *Earth Planet. Sci. Lett.* 49:263-265.

Davin, E.M., and M.G. Gross. 1980. Assessing the seabed. *Oceanus* 23:20-32.

Deuser, W.G., and E.R. Ross. 1980. Seasonal changes in the flux of organic carbon to the deep Sargasso Sea. *Nature* 283:364-365.

Deuser, W.G., E.R. Ross, and R.F. Anderson. 1981. Seasonality in the supply of sediment to the deep Sargasso Sea and the implications for the rapid transfer of matter to the deep ocean. *Deep-Sea Res. Part A: Oceanographic Research Papers* 28:495-505.

Donat, J.R., and K.W. Bruland. 1995. Trace elements in the oceans. Pp. 247-281 in B. Salbu and E. Steinnes (eds.), *Trace Elements in Natural Waters*. CRC Press, Boca Raton, Florida.

Dow, W.G. 1992. Reflections on the career and times of John M. Hunt. Pp. 9-19 in J. Whelan and J. Farrington (eds.), *Organic Matter. Productivity, Accumulation, and Preservation in Recent and Ancient Sediments*. Columbia University Press, New York

Duce, R.A. (guest ed.). 1989. SEAREX: The Sea/Air Exchange Program. Volume 10 of J.P.Riley and R. Chester (eds.). *Chemical Oceanography*. Academic Press, Ltd., London.

Duursma, E.K. 1965. The dissolved organic constituents of sea water. Pp. 433-475 in J.P. Riley and G. Skirrow (eds.), Chapter 11 of *Chemical Oceanography*. Academic Press, New York.

Edmund, J.M. 1980. GEOSECS. *Oceanus* 23(1):33-39.

Edmund, J.M. 1982. Ocean hot springs: A status report. *Oceanus* 25:22-27.

Epstein, S. 1981. Introduction of Claire C. Patterson for the V.M. Goldschmidt Medal 1980. *Geochim. Cosmochim. Acta* 45:1383-1384.

Farrington, J.W. 1978. In memoriam, Dr. Max Blumer, organic geochemist and environmental scientist, 1923-77. *J. Fish. Res. Board Can.* 35:501-504.

Farrington, J.W. 1987. Review of marine organic geochemistry. *Rev. Geophys. Space Phys.* 25:1395-1416.

Farrington, J.W. (ed.). 1992. Marine organic geochemistry: Review and challenges for the future. *Marine Chemistry* 39(1-3):1-242.

Farrington, J.W., and J.G. Quinn. 1973. Petroleum hydrocarbons in Narragansett Bay. I. Survey of hydrocarbons in sediments and clams, *Mercenaria mercenaria*. *Est. Coast. Mar. Sci.* 1:71-79.

FOCUS (1998). FOCUS Workshop Report. Http://www.joss.ucar.edu/joss_psg/project/oce_workshop/focus/report/html.

Gagosian, R.B. 1983. Review of marine organic chemistry. *Rev. Geophys. Space Phys.* 21:1245-1258.

Gardner, W. 1977. Fluxes, Dynamics and Chemistry of Particulates in the Ocean. Ph.D. thesis, Massachusetts Institute of Technology-Woods Hole Oceanographic Institution Joint Program. 394 pp.

Garrels, R.G., and F.T. MacKensie. 1971. *Evolution of Sedimentary Rocks*. W.W. Norton, New York.

Garrels, R.G., and M.E. Thompson. 1962. A chemical model for sea water at 250°C and one atmosphere total pressure. *American Journal of Science* 260:57-66.

Gieskes, J.M. 1974. The alkalinity-total carbon dioxide system in seawater. Pp. 123-151 in E.D. Goldberg (ed.), *Marine Chemistry*. Volume 5 of *The Sea*. John Wiley and Sons, New York.

Goldberg, E.D. 1954. Marine geochemistry. 1. Chemical scavenging of the sea. *Journal of Geology* 62:249-265.

Goldberg, E.D. 1965. Minor elements in sea water. Chapt. 5. Pp 163-196 in J.P. Riley and G. Skirrow (eds.)., *Chemical Oceanography*. Academic Press, New York.

Goldberg, E.D. (convener). 1972. *Baseline Studies of Heavy Metal, Halogenated Hydrocarbon, and Petroleum Hydrocarbon Pollutants in the Marine Environment and Research Recommendations*. Deliberations of the International Decade of Ocean Exploration (IDOE) Baseline Conference, May 24-26. National Science Foundation, Washington, D.C.

Goldberg, E.D. (ed.). 1974. *Marine Chemistry*. Volume 5 of *The Sea*. John Wiley and Sons, New York. 895 pp.

Goldberg, E.D. 1975a. *The Health of the Oceans*. UNESCO, Paris.

Goldberg, E.D. 1975b. The Mussel Watch: A first step in global marine monitoring. *Mar. Pollut. Bull* 6(5):111.

Goldberg, E.D., and G.O.S. Ahrrenius. 1958. *Geochim. Cosmochim. Acta* 13:153.

Goldberg, E.D., and M. Koide. 1958. Io-Th chronology in deep-sea sediments of the Pacific. *Science* 128:1003.

Goldberg, E.D., V.T. Bowen, J.W. Farrington, G.R. Harvey, J.H. Martin, P.L. Parker, R.W. Risebrough, W. Robertson, E. Schneider, and E. Gamble. 1978. The Mussel Watch. *Envir. Conservation*. 5:101-125.

Goldschmidt, V.M.. 1933. Grundlagen dre quantitativen Geochemie. *Fortschr. Miner.* 17:112-156.

Goldschmidt, V.M. 1937. The principles of distribution of chemical elements in minerals and rocks. *J. Chem. Soc.* Part 1:655-673.

Goldschmidt, V.M. 1954. *Geochemistry*. Oxford University Press, London.

Gordon, D.C., Jr., and P. Michalik. 1971. Concentration of bunker C fuel oils in waters of Chebabucto Bay, April, 1971. *J. Fish. Res. Board Can.* 28:1912-1914.

Grice, G.D., and M.R. Reeve (eds.). 1982. *Marine Mesocosms: Biological and Chemical Research on Experimental Ecosystems*. Springer-Verlag, New York.

Harvey, H.W. 1928. *Biological Chemistry and Physics of Seawater*. Cambridge at the University Press, London.

Hazzigna, M.S. 1998. How to change the university. *Science* 282:237.

Hogg, N.G., and R.X. Huang (eds.). 1995. Pp. 1-197 in *Collected Works of Henry M. Stommel*. American Meteorological Society, Boston, Massachusetts.

Honjo, S. 1978. Sedimentation of materials in the Sargasso Sea at a 5,367 m deep station. *J. Mar. Res.* 36:469-492.

Hutchinson, G.E. 1947. The problems of oceanic geochemistry. *Ecological Monographs* 17:299-315.

Jenkins, W.J., M.A. Beg, W.B. Clarke, P.J. Wangersky, and H. Craig. 1972. Excess He in the Atlantic. *Earth Planet. Sci. Lett.* 16:122.

Jennings, F.D. and L.R. King. 1980. Bureaucracy and science: The IDOE in the National Science Foundation. *Oceanus* 23:12-17.

Johnson, K.S., K.H. Coale, and H.W. Jannasch. 1992. Analytical chemistry in oceanography. *Analytical Chemistry* 64:1065-1075.

Keeling, C.D. 1958. The concentration and isotopic abundance of atmospheric carbon dioxide in rural areas. *Geochim. Cosmochim. Acta.* 13:322-334.

Keeling, C.D. 1968. Carbon dioxide in surface ocean waters 4: Global distributions. *J. Geophys. Res.* 73:4543-4553.

Keeling, C.D., J.A. Adams, C.A. Ekdahl, and P.R. Guenther. 1976b. Atmospheric carbon dioxide variations at the South Pole. *Tellus* 28:552-564.

Keeling, C.D., R.B. Bascastow, A.E. Bainbridge, C.A. Ekdahl, P.R. Guenther, L.S. Waterman, and J.F.S. Chin. 1976a. Atmospheric carbon dioxide variations at Mauna Loa Observatory, Hawaii. *Tellus* 28:538-551.

Ketchum, B.H. (ed.). 1972. *The Water's Edge*. The Massachusetts Institute of Technology Press, Cambridge, Massachusetts.

Knauer, G.A., J. H. Martin, and K.W. Bruland. 1979. Fluxes of particulate carbon, nitrogen and phosphorous in the upper water column of the northeast Pacific. *Deep-Sea Res.* 26:97-108.

Knecht, R.W. 1982. Introduction: Deep ocean mining. *Oceanus* 25:3-11.

Koczy, F.F. 1956. Geochemistry of radioactive elements in the ocean. *Deep-Sea Res.* 3:93-103.

Kvenvolden, K.A. (ed.). 1980. *Geochemistry of Organic Molecules*. Benchmark Papers in Geology Vol. 52. Dowden, Hutchinson and Ross. Stroudsburg, Pennsylvania.

Lee, C., and S.G. Wakeham. 1989. Organic matter in seawater: Biogeochemical processes. Pp 1-51 in J.P. Riley (ed.), *Chemical Oceanography*, vol. 9. Academic Press, New York..

MacIntyre, F.G. 1974. Chemical fractionation and sea-surface microlayer processes. Chapter 8. Pp. 245-299 in E.D. Goldberg (ed.), *The Sea*. John Wiley and Sons, New York.

MacLeisch, W.H. 1982-1983. Roger Revelle: Senior Senator of Science. *Oceanus* 25(4):67-70.

Marsh, E.R. 1973. Alfred C. Redfield, naturalist: His scientific career. Transcript of a taped interview with her father, by Elizabeth R. Marsh, Woods Hole, Massachusetts.

Martin, J.H. 1991. Iron. Leibig's Law and the Greenhouse. *Oceanography* 4: 52-55.

Martin, J.M., J.D. Burton, and D. Eisma. 1981. River Inputs to Ocean Systems. Proceedings of a SCOR/ACMR/ECOR/IAHS/CNG/IABO/IAPSO Review and Workshop. UNEP and UNESCO, Switzerland.

Martin, J.H., and S.E. Fitzwater. 1988. Iron deficiency limits growth in the north-east Pacific subarctic. *Nature* 331:341-343.

Martin, J.H., G.A. Knauer, W.W. Broenkow, K.W. Bruland, D.M. Karl, L.F. Small, M.W. Silver, and M.M. Gowing. 1983. Vertical transport and exchange of materials in the upper waters of the oceans (VERTEX): Introduction to the program, hydrographic conditions and major component fluxes during VERTEX I. *Moss Landing Marine Laboratory Technical Publication 83-2*, 40 pp.

McElway, B. 1983. Karl K. Turekian: "Academic Gladiator." *Oceanus* 26(2):61-66.

Menzel, D.W. 1974. Primary productivity, dissolved and particulate organic matter, and the sites of oxidation of organic matter. Pp 659-678 in E.D. Goldberg (ed.), *Marine Chemistry*. Volume 5 of *The Sea*. John Wiley and Sons, New York.

Millero, F. 1974. Seawater as a multicomponent electrolyte solution. Pp. 3-80 in E.D. Goldberg (ed.), *Marine Chemistry*. Volume 5 of *The Sea*. John Wiley and Sons, New York.

Milliman, J.D., and R.H. Meade. 1983. World-wide delivery of river sediment to the oceans. *J. Geol.* 91:1-21.

National Academy of Sciences-National Research Council (NAS/NRC). 1957. *The Effects of Atomic Radiation on Oceanography and Fisheries*. NAS-NRC Publ. 551, Washington, D.C. 137 pp.

National Academy of Sciences (NAS). 1959. *Conference of Physical and Chemical Properties of Sea Water*. Committee on Oceanography. Washington, D.C. 202 pp.

National Academy of Sciences (NAS). 1971a. *Marine Chemistry: A Report of the Marine Chemistry Panel of the Committee on Oceanography*. Washington, D.C. 60 pp.

National Academy of Sciences (NAS). 1971b. *Radioactivity in the Marine Environment*. Report of the Panel on Radioactivity in the Marine Environment of the Committee on Oceanography Washington, D.C. 272 pp.

National Academy of Sciences (NAS). 1971c. *Marine Environmental Quality*. Washington, D.C.

Patterson, C.C. 1981. Acceptance speech for the V.M. Goldschmidt Medal. *Geochim. Cosmochim. Acta* 45:1385-1388.

Patterson, C., D. Settle, B. Schaule, and M. Burnett. 1976. Transport of pollutant lead to the oceans and within ocean ecosystems. Pp. 23-28 in H.L. Windom and R.A. Duce (eds.), *Marine Pollutant Transfer*. Chapter 2. Lexington Books, D.C. Heath and Company, Lexington, Massachusetts.

Pilson, M.E.Q. 1998. *An Introduction to the Chemistry of the Sea*. Prentice Hall, Upper Saddle River, New Jersey. 431 pp.

Redfield, A.C. 1958. The biological control of chemical factors in the environment. *Am. Scientist* 46:205-222.

Redfield, A.C., B.W. Ketchum, and F.A. Richards. 1963. The influence of organisms on the composition of seawater. Pp. 26-87 in M.N. Hill (ed.), *The Sea*. Volume 2. Interscience Publishers, New York.

Reeve, M. 1998. Assessment of NSF Grants in Geosciences in the 1950s (personal communication). Ocean Sciences Division, Geosciences Directorate, National Science Foundation, Arlington, Virginia.

Revelle, R., and H.E. Suess. 1957. Carbon dioxide exchange between the atmosphere and ocean and the question of an increasing atmospheric CO_2 during past decades. *Tellus* 9:18-27.

Richards, F.A. 1965. Anoxic basins and fjords. Pp. 611-645 in J.P. Riley and G. Skirrow (eds.), *Chemical Oceanography*. Chapter 13. Academic Press, New York.

Riley, J.P. 1965. Historical introduction. Pp. 1-41 in J.P. Riley and G. Skirrow (eds.), *Chemical Oceanography*. Chapter 1. Academic Press, New York.

Riley, J.P., and G. Skirrow (eds.). 1965. *Chemical Oceanography*. Academic Press, New York.

Rubey, W.W. 1951. Geologic history of seawater: An attempt to state the problem. *Geological Society of America Bulletin* 62:1111-1147.

SCEP. 1970. *Man's Impact on the Global Environment*. Report of the Study of Critical Environmental Problems. Massachusetts Institute of Technology, Cambridge, Massachusetts.

Sanders, H.L. 1978. *Florida* oil spill impact on Buzzards Bay benthic fauna: West Falmouth. *J. Fish. Res. Board Can.* 35:717-730.

Sanders, H.L., J.F. Grassle, G.R. Hampson, L.S. Morse, S. Price-Gartner, and C.C. Jones. 1980. Anatomy of an oil spill: Long term effects from the grounding of the barge *Florida* off West Falmouth, Massachusetts. *J. Mar. Res.* 38:265-380.

Sayles, F.L. 1979. The composition and diagensesis of interstitial solutions. I. Fluxes across the sediment-water interface in the Atlantic Ocean. *Geochim. Cosmochim. Acta* 43:527-545.

Sayles, F.L. 1981. The composition and diagensesis of interstitial solutions. II. Fluxes and diagenesis at the sediment-water interface in high latitude North and South Atlantic. *Geochim. Cosmochim. Acta* 45:1061-1086.

Sayles, F.L., P.C. Mangelsdorf, T.R.S. Wilson, and D.N. Hume. 1976. A sampler for the in situ collection of marine sedimentary pore waters. *Deep-Sea Res.* 23:259-264.

Sholkovitz, E.R. 1976. Flocculation of dissolved organic and inorganic matter during the mixing of river water and seawater. *Geochim. Cosmochim. Acta* 40:831-845.

Shor, E.N. 1978. *Scripps Institution of Oceanography: Probing the Oceans 1936-1976*. Tofua Press, San Diego, California.

Siever, R., R.M. Garrels, J. Kanwisher, J. Willis, R.A. Berner. 1961. Interstitial waters of Recent muds off Cape Cod. *Science* 134:1071-1072.

Sillen, L.G. 1961. The physical chemistry of seawater. Pp. 549-581 in M. Sears (ed.), *Oceanography*. American Association for the Advancement of Sciences, Washington, D.C.

Spencer, D.W., P.G. Brewer, A. Fleer, S. Honjo, S. Krishnaswami, and Y. Nozaki. 1978. Chemical fluxes from a sediment trap experiment in the deep Sargasso Sea. *J. Mar. Res.* 36:493-523.

Staresinic, N., G.T. Rowe, D. Shaughnessey, and A.J. Williams III. 1978. Measurement of the vertical flux of particulate organic matter with a free-drifting sediment trap. *Limnol. Oceanogr.* 23(3):559-563.

Stommel, H. 1957. The abyssal circulation of the ocean. *Nature* 180:733-734.

Stommel, H. 1958. The abyssal circulation. *Deep-Sea Res.* 5:80-82.

Stommel, H., and A.B. Arons. 1960a. On the abyssal circulation of the world ocean-I. Stationary planetary flow patterns on a sphere. *Deep-Sea Res.* 6:140-154.

Stommel, H., and A.B. Arons. 1960b. On the abyssal circulation of the world ocean-II. An idealized model of the circulation pattern and amplitude in oceanic basins. *Deep-Sea Res.* 6:217-233.

Strickland, J.D.H., and T.R. Parsons. 1965. Manual of sea water analysis with special reference to more common micronutrients and particulate organic material. Fisheries Research Board of Canada. Ottawa, Canada.

Strickland, J.D.H., and T.R. Parsons. 1972. A Practical Handbook for Seawater Analysis. 2nd edition. Fisheries Research Board of Canada, Ottawa.

Stuum, W., and J.J. Morgan. 1970. *Aquatic Chemistry: An Introduction Emphasizing Chemical Equilibria in Natural Waters*. Wiley-Interscience, New York.

Stumm, W., and J.J. Morgan. 1981. *Aquatic Chemistry: An Introduction Emphasizing Chemical Equilibria in Natural Waters*. 2nd edition John Wiley and Sons, New York.

Stuum, W., and J.J. Morgan. 1995. *Aquatic Chemistry: Chemical Equilibria and Rates in Natural Waters*. 3rd edition. John Wiley and Sons, New York.

Sverdrup, H.U., M.W. Johnson, and R.H. Fleming. 1942. *The Oceans: Their Physics, Chemistry, and General Biology*. Prentice-Hall, Inc.

Turekian, K.K. 1955. Paleoecological significance of the strontium-calcium ratio in fossils and sediments. *Geological Society of America Bulletin* 66(1):155-158.

Turekian, K.K. 1957. The significance of variations of strontium content in deep sea cores. *Limnol. Oceanogr.* 2:309-314.

Turekian, K.K. 1958. Rate of accumulation of nickel in Atlantic equatorial deep sea sediments and its bearing on possible extra-terrestrial sources. *Nature (London)* 182:1728-1729.

Turekian, K.K. 1968. *Oceans.* Prentice Hall, Engelwood Cliffs, New Jersey.

Turekian, K.K., and J.L. Kulp. 1956. The geochemistry of strontium. *Geochem. Cosmochim. Acta* 10:245-296.

Urey, H.C., H.A. Lowenstam, S. Epstein, and C.R. McKinney. 1951. Measurement of paleotemperatures and temperature of the Upper Cre-taceous of England, Denmark, and the Southeastern United States. *Geological Society of America Bulletin* 62:399-416.

Wallace, W.J. 1974. *The Development of the Chlorinity/Salinity Concept in Oceanography.* Elsevier Scientific Publishing Company, New York. 227 pp.

Windom, H.L., and R.A Duce (eds.). 1976. *Marine Pollutant Transfer.* Lexington Books, D.C. Heath and Company, Lexington, Massachusetts.

Zafiriou, O.C. 1977. Marine organic photochemistry previewed. *Mar. Chem.* 5:497-522.

Zika, R.G. 1987. Advances in marine photochemistry 1983-1987. *Reviews of Geophysics* 25:1390-1394.

Achievements in Physical Oceanography

Walter Munk

Scripps Institution of Oceanography, University of California, San Diego

ABSTRACT

The last 50 years have seen a revolution in our understanding of ocean processes. I consider the major developments in three eras: (1) ending roughly 1970, observations were generally interpreted in terms of steady circulation models of large scale, with variability regarded as "noise"; (2) following 1970, emphasis was on mesoscale variability (which was found to contain 99 percent of the oceanic kinetic energy), internal waves, edge waves, mixing events, and other time-dependent processes; (3) the "now" era returns to some of the large-scale problems of the first era, but with allowance for the decisive role played by the time-dependent processes and a growing appreciation that the large-scale features are themselves subject to slow climate-connected changes. Technological developments generally led (rather than followed) new ideas. Underlying all these developments is a half-century transition from grossly inadequate sampling to an appreciation of a rational sampling strategy.

I am to speak on "Landmark Achievements in Physical Oceanography." Why not call it "Seamarks"? I was in Lisbon in August at a meeting on satellite oceanography. Following the welcome by Mr. M. Gago, Minister of Science and Technology, I was assigned a generous 5 minutes to cover the subject of "Oceanography Before Satellites." For illustration I used Plate 5, showing how Moses was saved by a tsunami with high nonlinear distortion, what we now call a soliton of depression. Going back to the early 1950s is like going back to Exodus.

What are the seamarks that led us from the Exodus stage to our present theology? The following choice is highly subjective. I am an ocean adventurer, not an historian of science.[1] I have paid little attention to the extent of National Science Foundation (NSF) support as compared to the Office of Naval Research (ONR) and other support; Michael Reeve, Richard Lambert, and others discuss this in other papers in this volume.

THEOLOGY OF A "STEADY" OCEAN

I will remind you that the field of oceanography immediately before NSF was founded was just coming to terms with Sverdrup's 1947 solution for the mid-ocean circulation in response to wind torquing and with Stommel's 1948 explanation of the intensification of currents along western boundaries (e.g., Gulf Stream).

Equatorial Undercurrent

In 1952, NSF awarded $6,100 for each of two years to Ray Montgomery for "Analysis of Serial Data." What Ray was really working on was the Equatorial Undercurrent, the last major current system missing from the lexicon of oceanography.[2] T. Cromwell and J. Knauss were the major actors.

---

[1] I am certain to have neglected to report some vital contributions, and I apologize for this.

[2] The undercurrent was actually discovered by Buchanan in 1885 and then forgotten.

Abyssal Circulation

The basic elements of the deep (thermohaline) ocean circulation were known in Sverdrup's time. (He mapped global volume fluxes in units of million cubic meters per second, now known as *sverdrups*.) At the time it was believed that deep water known as Montgomery's "common water"[3] was formed in a few concentrated areas south of Greenland and along the Antarctic Shelf by top-to-bottom convection. But there is no top-to-bottom convection. The work of V. Worthington, J. Reid, A. Gordon, and D. Roemmich has since shown that the formation of Montgomery's common water requires a complex interplay of water masses. Starting in 1960, Stommel and Arons provided a dynamical (though highly idealized) framework, with deep water transported to lower latitudes along western boundaries and communication between the oceans basins accomplished via the Antarctic Circumpolar Current. A subsequent visualization called "the great global conveyor belt" has enjoyed popular support because of its vividness, and support by chemists because of its simplicity, but it is important to keep in mind that this subject is still under active development.

Hydrography

The fashion at the time was to map the measured scalar fields of temperature and salinity and to infer the current velocities by a joint application of the hydrostatic and geostrophic equations. Since the scalar fields were relatively smooth and steady, the inferred currents were relatively smooth and steady. We had so much confidence in the method that we issued current charts on pocket handkerchiefs to our World War II pilots in the Pacific so that they could navigate the "known" surface currents toward the nearest islands.

There are two shortcomings to the hydrographic method. First, smooth scalar distributions do not necessarily call for smooth, steady current systems, the scalar fields being space and time integrals of the motion field. One has found smooth scalar fields in the presence of extremely complex float trajectories. The downed flyers would have found the current charts useful only if they had been willing to integrate their drifting experience over a year or two.

The second shortcoming is that the hydrographic method gives only *relative* currents, and much effort has been expended to find the so-called depth of no motion. The problem was treated in the 1970s by Stommel (with Schott, Behringer, and Armi) in the work on the β-spiral, and similarly by Wunsch in his application of inverse methods. It is ironic that progress on the problem of the depth of no motion

came about just as it was becoming clear that ocean currents were seriously time dependent at all depths.

Ekman Spiral

All students of oceanography learn about the Ekman spiral, an elegant early-century mathematical solution to the wind-driven current profile. But it has been very difficult to extract a clear spiral signature from a noisy environment until the work of Price and Weller, and Niiler's recent statistical analysis of 50,000 float observations. In more general terms, "Ekman dynamics" has been observationally confirmed by Davis in the Mixed Layer Experiment (MILE), and by Rudnick in an acoustic Doppler current profiler (ADCP) transect across the Atlantic.

UNDERSTANDING VARIABILITY

Fifty years ago physical oceanographers were deploying around the ocean in a few vessels taking Nansen casts and bathythermographs (BTs). The underlying theology was that of a *steady* ocean circulation: differences between stations were attributed to the difference in station *position*, not the difference in station *time*.[4] We now know that more than 99 percent of the kinetic energy of ocean currents is associated with variable currents, the so-called mesoscale of roughly 100 km and 100 days. Incredible as it may seem, for one hundred years this dominant component of ocean circulation had slipped through the coarse grid of traditional sampling. Our concept of ocean currents has changed from something like 10 ± 1 cm/s to 1 ± 10 cm/s. This first century of oceanography, since the days of the *Challenger* expedition in the 1870s, came to an abrupt end in the 1970s.

The Mesoscale Revolution[5]

By 1950, the oceanographic community had become aware of the meandering of the Gulf Stream. If there was any doubt, the multiple ship Operation Cabot (the first of its kind), under the leadership of Fritz Fuglister, dramatically demonstrated the shedding of a cold-core eddy. At first it was thought that transients are confined to the regions of the western boundary currents. But the acoustic tracking of neu-

---

[3]Referring to 9 percent of global ocean volume within the narrow limits of 1.0–1.5°C and 34.7–34.8 ppt salinity.

[4]But Helland-Hansen and Nansen in their classical 1909 paper on the physical oceanography of the Norwegian Sea were aware of the mesoscale variability.

[5]By "revolution" I mean that an oceanographer totally familiar with the topic at the beginning of the period, but with no further learning experience, would flunk a freshman exam at the end of the period. Other topics have been remarkably stationary. (See a delightful review of Sverdrup's chapter in the "Ocean Bible" [Sverdrup, H.U., M.W. Johnson, and R.H. Fleming 1942. *The Oceans: Their Physics, Chemistry, and General Biology*, Prentice-Hall, Inc.] by Bruce Warren. 1992. Physical oceanography in *The Oceans. Oceanography* 5:157-159).

trally buoyant floats by Swallow (who credits Stommel for suggesting this idea) soon demonstrated that variability in space and time was the rule, not the exception (though more intense near the boundary currents). There was an urgent need for a systematic exploration of the ocean variability. The development of deep-ocean mooring technology provided such an opportunity, and the Mid-Ocean Dynamics Experiment (MODE) starting in 1973 under the leadership of Stommel and Robinson defined the parameters of variability. (Soviet oceanographer Brekhovskikh got there first, but failed to get definitive results because of a high failure rate of current meters.) We now think of this mesoscale variability as the ocean *weather* and the underlying circulation as the ocean *climate* (itself subject to slow variations that are discussed later). Climate came first, weather later—rather the opposite of what happened in meteorology.

The era coincides with a flowering of geophysical fluid dynamics (GFD). Nearly everyone in GFD had their initiation at the summer sessions in Walsh Cottage at Woods Hole first organized by W. Malkus. Mesoscale variability was incorporated into general circulation models (GCMs). We recall the excitement of seeing B. Holland's first spontaneously unsteady wind-driven circulation model.

I believe that the numerical modeling reached a plateau in later decades as a result of a dependence on semiempirical nonphysical parameterization. Ironically, modeling came to the rescue, but in the new form of process-oriented modeling (as opposed to simulations of actual conditions), leading to appreciation of the ventilation of deep layers, of constant potential vorticity pools, and so forth. A resurgence of theoretical thinking has evolved into an indispensable complement to big numerical models.

Internal Waves

On a smaller scale, internal waves (long recognized as a curiosity) became part of the oceanographic mainstream. At periods of less than a day, internal waves are the principal contributors to the velocity variance. This development owes a great deal to the application of power-spectral analysis, which in turn was made possible only by the computer revolution. Fifty years ago no oceanographer knew how to handle the wiggly records associated with random-phase wide-band processes. (Yet acousticians and opticians had done so for many years.) We could manage the discrete tidal line spectrum, and get away with the analysis of narrow-band processes such as distant swell, but we failed miserably in the analysis of storm waves or internal waves. Most ocean processes are wide-band!

In 1931, Ekman took some current measurements with a string of Ekman meters suspended on a vertical mooring. When I met Ekman in Oslo in 1949, he expressed disbelief that currents separated vertically by as little as 100 m could bear so little resemblance, and he delayed publishing an analysis until shortly before his death. But there is nothing

mysterious in the result; processes with vertical bandwidth $\Delta\kappa$ are incoherent at separations exceeding $\Delta z = \Delta\kappa^{-1}$!

Among the seamark achievements are the recognition of an astounding spectral universality (within a factor 2) under a wide variety of conditions (still not understood) and of the role played by internal waves in ocean mixing processes. The transformation to internal solitons (solibores) in near-shore regions (first recognized on satellite images) is becoming an important component in coastal studies.

Edge Waves

There exists a class of wave motion that is coastally trapped. Wave crests and troughs extend *perpendicular* to shore and diminish exponentially with distance from shore. Propagation is in a direction parallel to shore. There are two scales: the rotationally trapped Kelvin edge waves, and the gravitationally trapped Stokes edge waves. Both were discovered in the nineteenth century and considered curiosities. Referring to the latter, Lamb writes: "it does not appear that the type of motion here referred to is very important." In fact, these curiosities are the very centerpiece of a rapidly developing coastal dynamics—one that is amazingly different and almost isolated from the deep ocean dynamics. It has turned out that the linear edge waves provide a linear core to the highly nonlinear coastal and littoral dynamics.

Gravitational edge waves are excited by incoming surface waves depending in a complex (but predictable) way on the character of the wave system. The edge waves, in turn, determine the littoral dynamics, the bar formation and cusps in the beach profile, and the spacing of rip currents. For a given medium size of sand grain and representative values of wave height, period, and direction, it is now possible to predict an equilibrium beach profile. Crucial elements in this development were the radioactive and fluorescent tagging of sand grains and the spectral representation of the incoming wave system. In a larger sense the underlying parameter space depends on the type of coast as determined by plate tectonics, and a mass balance determined by river discharge, cliff erosion, and the presence of submarine canyons.

Surface Waves

This is another old subject that was revived by modern spectral analysis. In 1957, Miles and Phillips in two seamark papers[6,7] discussed the generation of waves by wind, and a year later Phillips[8] introduced the famous k^{-4}

[6]Phillips, O.M. 1957. On the generation of waves by turbulent wind. *J. Fluid Mech.* 2:417-445.

[7]Miles, J.W. 1957. On the generation of surface waves by shear flows. *J. Fluid Mech.* 3:185-204.

[8]Phillips, O.M. 1958. The equilibrium range in the spectrum of wind-generated waves. *J. Fluid Mech.* 4:426-434.

equilibrium spectrum. In 1963, Hasselmann first pointed out the crucial role played by the nonlinear energy transfer from the short and long components to the energetic central spectrum. The subject has now advanced to a point where wave prediction based on a given (past and future) wind field is routinely used in a wide range of human activities. Now that the wave field can be measured by synthetic aperture radar (SAR) satellites, I predict that the deconvolution of the wave field to provide wind data will become an important future application.

Tides

Tides are the earliest application of oceanography to human activities[9] and were a favorite subject of Victorian mathematicians. This field, too, has been revived by the computer revolution. In 1969, Pekeris and Accad solved the Laplace tide equation over a world ocean with realistic topography, using the new GOLEM computer built at the Weizmann Institute.[10] There was a need to compare the global computations with measurements in the open sea. Coastal tide gauges have been around for centuries, but the ability to measure deep sea tides did not come until the early 1960s when pressure gauges could be dropped freely to the deep seafloor and subsequently recalled acoustically; about 350 pelagic stations have been occupied (mostly by Cartwright) in the 30-year window before satellite altimetry provided the means of truly global measurements. Tidal dissipation from the principal lunar tide is $2.50 \pm .05$ Terawatts (TW), very accurately derived from the measured rate of 3.82 cm/s at which the Moon moves away from the Earth. Tidal dissipation may have important implications to ocean mixing (as discussed below).

There are other achievements. We have learned the importance in tidal modeling of allowing for the elastic yield of the solid Earth. A seamark achievement is G. Platzman's expansion into global ocean normal modes. Tidal studies have not been in the oceanographic mainstream; I am one of the very few people who think that lunar studies will become fashionable once more (there is a name for such people).

The Microscale Revolution

At the opposite end of the general circulation scale is the micro- (or dissipation) scale where energy is irreversibly converted into heat. We are talking about millimeters to centimeters, but just because the process scales are small does not mean their importance is small.

It was not always clear that the deep ocean was cold. In the seventeenth century, Boyle argued that the temperature

must increase with pressure according to his law PV = NRT (as it does in the Mindanao Deep, from 1.7°C at 5 km to 2.5°C at 10 km). While passing through the tropics on a voyage to the East Indies, Boyle noticed that the cook was lowering some bottles of white wine over the side. "And why should you be doing this?" he asked, to which the cook replied, "Every gentleman knows that white wine must be chilled before serving." Surely this was one of the most decisive oceanographic experiments of all time.[11]

At the rate of 25 sverdrups of bottom water formation, the oceans would fill up with ice cold water in 3,000 years, forming a 1-m-thick thermal surface boundary layer controlled by molecular conductivity. Why is it you do not freeze your toes every time you go swimming?

The answer is that turbulent mixing brings warm water downward. A scale depth of 1000 m (roughly as observed) requires 1000 times the molecular diffusivity, or about 10^{-4} m²/s. Is this in accordance with fact? It has taken 30 years to find out that it is not. Cox, Gregg, and Osborn, among others, have developed the instrumentation with the required vertical resolution and found typical pelagic values of 10^{-5} m²/s. Ledwell confirmed these values by *in situ* measurements of the diffusion of a dye patch. Although a discrepancy by a factor of 10 is not large in this context, it appears to be real. A possible interpretation is that most of the ocean mixing takes place in a few regions of rough topography and very high turbulence. Far higher diffusivities have in fact been measured by Schmitt, Toole, and Polzin near rough bottom topography in the South Atlantic Ocean. An ambitious experiment along the Hawaiian ridge is being planned.

Mixing associated with 10^{-4} m²/s required 2 TW, the pelagic mixing rate of 10^{-5} m²/s requires 0.2 TW globally. Where does the energy for the mixing come from? Wind is an obvious candidate, tidal dissipation is another (2.5 TW are dissipated by the M_2 tide alone, but nearly all of this has been claimed for dissipation in marginal seas).

Getting the mixing right is vital to any realistic modeling of ocean circulation and heat transport. In this connection we need to mention two other important developments. In 1956, Stommel (with Arons and Blanchard) published a paper: "An Oceanographical Curiosity: The Perpetual Salt Fountain."[12] In a temperature-stable and salt-unstable stratification, a vertical hose, once primed, will pump up cold, salty (and nutrient-rich) deep water forever. Stern realized that this was associated with a fundamental instability (hose or no hose), and Turner developed this into the discipline of double-diffusive mixing.

The MEDOC (Mediterranean Deep Ocean Convection)

---

[9]Cartwright, D.E. 1999. *Tides: A Scientific History*. Cambridge University Press.

[10]Supported by the first overseas grant from NSF.

[11]For a more accurate account, see page 6 in McConnell, A. 1982. *No Sea Too Deep*. Adam Hilger, Ltd., Bristol.

[12]Stommel, H., A.B. Arons, and D. Blanchard. 1956. An oceanographical curiosity: The perpetual salt fountain. *Deep-Sea Research* 3:152-153.

experiment in 1969 (another Stommel brainchild) provided direct measurements of convective overturning. Prior to MEDOC there had been very little direct observational evidence for deep water formation.

THE CLIMATE REVOLUTION

In 1960, NSF awarded J. Bjerknes a grant of $30,000 per year for three years to study "Sea Surface Temperature and Atmospheric Circulation." This was the beginning of ENSO, a combined ocean (El Niño) and atmosphere (Southern Oscillation) phenomenon.

Milankovitch long ago computed long-term variations in the orbital parameters of the Earth-Sun-Moon system with periods from 20,000 to 100,000 years. In a remarkable development pioneered by Imbrie, the terms have now been detected in the ocean sediment record, and they provide important information concerning the atmosphere-ocean response to harmonic forcing.

Hasselmann pioneered an approach that in some sense is opposite to that of Milankovitch. He suggested a "random walk" of the climate state in response to random pulses associated with short-term "weather." The character of such random walks is that they lead to large long-time departures from the mean. It has been demonstrated that the random-walk excitation accounts for the dominant part of the observed climate variance.

The coupled ocean-atmosphere system is capable of complex feedback systems. A number of these have been identified: ENSO, the Pacific "decadal variation," and the North Atlantic Oscillation. It would appear that the three phenomena can account for a significant fraction of the ambient variance. El Niño has a recognizable linear component in a highly nonlinear equatorial dynamics: an equatorially trapped wave moving eastward at a rate of order 0.1 m/s (playing a role somewhat similar to the edge waves in highly nonlinear coastal and littoral dynamics). There has been significant progress in ENSO prediction.

Greenhouse warming has occupied center stage, largely because mankind can do something about this component of climate variability. Model predictions now have error bars of the same order as the predicted mean change. There is urgent need for observational testing. The inevitable result will be an improved modeling and an increased understanding of ocean processes.

In all of the climate problems, a first-order consideration is the oceanic and atmospheric equator-to-pole heat flux (3.7×10^{15} W across 24°N) required to maintain the global heat balance. In 1955, Sverdrup estimated that the ocean contributed 1.4×10^{15} W, and this was mostly in the wind-driven circulation. We now estimate that the ocean carries more than half the total load, with comparable contributions from the wind-driven and thermo-haline circulations. Quite a change!

THE TECHNOLOGY REVOLUTION

We all agree that there has been a technology revolution in ocean sciences; Larry Clark's paper, later in the volume, presents highlights of this revolution. It probably would have made more sense if I had organized this review along those lines; more often than not, new ideas have come out of new technology, rather than the other way around.

High-speed computers led to an explosion in the 1950s in every branch of physical oceanography (I have already listed a few examples). Readily available analysis of noisy records led at last to a sensible and reproducible description of surface waves and internal waves. It opened the door to objective analyses of extensive and diverse data sets, matched field processing of ocean acoustic transmissions, and the application of inverse theory to ocean measurements for an objective approach to estimating the validity of a given set of assumptions. Sadly, oceanographers had long found support for their favorite theory without such an objective assessment. In reviewing some past experiments designed to answer certain questions, one finds that the proposed measurements could not possibly have decided the issue with any reasonable degree of probability even if all measurements had worked (which is not always the case).

We have already referred to the revolution associated with the development of a deep-sea mooring technology. A similar case can be made for drifters, particularly those with a programmed depth strategy $z(t)$, which have spearheaded a Lagrangian renaissance led by T. Rossby, D. Webb, and R. Davis. The oceans are a remarkably good propagator of sound (but not of electromagnetic energy), and this has played a profound role in ocean exploration starting with the acoustically navigated Swallow floats. The application of inverse methods has made possible the interpretation of the entire recorded field of an acoustic transmission in terms of the properties of the intervening water.

We must not overlook low-tech developments. A U.S. patent for the O-ring was awarded to Niels Christensen in 1939 (so Rita cannot claim credit for this seamark). Until the mid-1960s we used to load our gear into numerous boxes and carry them aboard the vessels, only to find that a crucial item had been left ashore. I think Frank Snodgrass was the first to build portable laboratories with the equipment assembled and pretested. The portable laboratory (Figure 1) is then brought aboard, ready for action. Decks of all oceanographic vessels now provide bolt-downs 2 feet on center for securing the portable laboratories. In about the same period we learned how to drop unattached instruments to the relatively benign environment of the deep seafloor, later to be recalled acoustically. There was a psychological block to overcome; it is not easy to let go of a line from which you have a year's budget of equipment hanging.

Satellites constitute the most important technology innovation in modern times. Oceanographers are a conserva-

FIGURE 1   Frank Snodgrass and Mark Wimbush outside the portable laboratory preparing a deep-sea tide capsule for a freefall to the ocean floor (about 1964).

FIGURE 2   Henry Stommel came to Woods Hole in 1944 and died there in 1992. He is the dominant figure in the period reviewed here.

tive lot; they did not welcome satellites with open arms. Apel came to Scripps and Woods Hole in 1970 to look for advice and support in planning SEASAT (Earth Satellite dedicated to Oceanographic Applications). He got neither. When mentioning that satellite altimeters would measure dynamic height, a well-known oceanographer replied: "If you gave it to me, I would not know what to do with it." With regard to climate, given the reluctance to employ new technologies, given that some of the underlying processes are not yet understood, given the slow rate (as demonstrated over the last 50 years) at which new concepts are adapted, and given the requirement of long time series for testing models, given that long time series take long times, we cannot expect to "solve" the climate problem in the next decade.

CLOSING REMARKS

One final attempt at generalization. The key change between the century of the *Challenger* and the last 50 years is adequate sampling. The key product of the Technology Revolution is sampling. The key contribution of the conductivity-temperature-depth (CTD) profiler to vertical profiling was not more precision than the Nansen bottle (in fact it was less), but continuity in vertical sampling. The most important satellite contribution, I think, is not the instrument packages (remarkable as they are), but the ability to sample the global ocean and sample adequately. A key contribution of computers and the associated move from analog to digital discrete recording was that a new generation of oceanographers understood what the previous generation had not: the requirements of the sampling theorem. Beware of ignoring the theorem; it is unforgiving. Even the uncanny intuition of

a Fritz Fuglister for the behavior of the Gulf Stream was not able to overcome the inadequate sampling of his time.

One person easily stands out in this brief account: Henry Stommel (Figure 2). Stommel joined Woods Hole in 1944 and died there in 1992. In his later years, NSF provided him substantial and continuous support. Stommel was the first to develop an intuition about the conservation of potential vorticity, with far-reaching consequences. In 1954, he privately printed a pamphlet entitled: "Why Do Our Ideas About Ocean Circulation Have Such a Dream-Like Quality?" Dreamlike, indeed. An ocean with currents of 10 ± 1 cm/s (as we then thought) is an ocean far different from one with 1 ± 10 cm/s. My teacher Harald Sverdrup considered it one of the chief functions of physical oceanographers to provide biologists the background information for studying life in the sea. I am afraid that our concepts were too dreamlike to provide useful guidance. Today we can provide information that is useful. Surely this is a revolutionary change! Stommel has led this 50-year transition from a dreamlike to an (almost) realistic ocean.

John Knauss discusses the transition from ONR to NSF dominance earlier in this volume. The changes are profound, nothing short of another revolution (Plate 6). (I must confess to a certain nostalgia for the old ONR days). This is an inevitable result of going from adventure to public service. We enter the prediction arena at a high price; our failures (and there have been many) will now be publicly vented. My plea to NSF for the next 50 years is to support a few "curiosities" and other high-risk ventures and to retain a tolerance for failure.

ACKNOWLEDGMENTS

I thank L. Armi, M. Hendershott, R. Knox, J. Layton, P. Niiler, J. Reid, and C. Wunsch for discussions, but the (very limited) selection of achievements reported here is entirely mine.

Achievements in Marine Geology and Geophysics

Marcia K. McNutt

Monterey Bay Aquarium Research Institute

ABSTRACT

If any prominent researcher were to list the crowning achievements in science over the past 50 years, two discoveries in marine geology and geophysics would surely make everyone's top ten: the development of the theory of plate tectonics and the unraveling of Earth's paleoclimate history through the use of the deep-sea sediment record. The former has become the archetypal example of a scientific revolution, whereas the latter now provides the essential observational evidence for the magnitude, rates, and fundamental causes of climate change. These major discoveries cannot be completely understood apart from the people, the institutions, the organizations, and the funding agencies that led to their advancement. In particular, the establishment of the National Science Foundation at the dawn of the modern era of ocean exploration and on the eve of the discovery of plate tectonics fueled the rapid rise of marine geology and geophysics to become one of the most fundamentally exciting and societally relevant disciplines in all of science. This essay is a personal attempt to describe the context within which this rapid advancement took place and to relate how the institutions and funding structures adapted as the century aged.

INTRODUCTION

It is a formidable task to review all of the achievements sponsored by the National Science Foundation (NSF) in the area of marine geology and geophysics (MG&G). To begin with, this one program within the Division of Ocean Sciences at NSF sponsors research programs nearly as broad in scope as NSF's entire Division of Earth Sciences: marine petrology, tectonics, geomorphology, paleontology, geochemistry, sedimentology, stratigraphy, and geophysics. Personally, an even more immediate problem was the fact that the first NSF grant in MG&G was awarded the month I was born. It was not until some 20 years later that I even became aware that the National Science Foundation existed. Therefore, in order to focus this presentation, I made the conscious effort to concentrate on what I consider to be the two greatest achievements in this field over the past 50 years: the development of the theory of plate tectonics and the deciphering of Earth's paleoclimate record from deep-sea sediments. The fact that any one field could lay claim to two such fundamental paradigms in a single half-century is indeed remarkable, and it is only in the enormity of their im-

pacts that one can justify overlooking the myriad of other important, although more isolated, discoveries in MG&G.

Despite the very different natures of these two major discoveries, there are several parallels in their respective developments. Neither theory could have been advanced had it not been possible to establish a global chronology that allowed observations from different oceans to be intercompared. The plate tectonic "clock" is the history of reversals of Earth's magnetic field, as calibrated by radiometric dates of igneous rocks and biostratigraphy in sediments. In the case of Earth's climate record, the relevant clock arises from the variations in Earth's orbital parameters, such as eccentricity, obliquity, and precession. For both theories it was relatively simple to reconstruct what happened (e.g., how fast the plates moved in what direction or when Earth experienced relatively warm or cold periods), but it has been a far thornier task to understand why. Exchange of ideas within an international community of scientists was essential for both the plate tectonic theory and paleoclimate proxies, and advances in both seagoing and laboratory analytical instrumentation led to important breakthroughs. In looking over the histories of both revolutions, one discerns the influence of NSF in enabling the collection of the fundamental data sets, promoting international collaborations, providing access to technology, and stressing the importance of relating the observations to basic physical, chemical, and mathematical models in order to gain real understanding.

MG&G IN THE PRE-NSF ERA

The Challenger Expedition

Marine geology and geophysics as a field dates back at least to the HMS *Challenger* expedition in 1872-1876. The *Challenger* was a sailing ship of 2,300 tons with auxiliary steam power. With funding from the British Royal Society, that expedition systematically collected observations of the oceans stopping every 200 miles. At each station, depth to the seafloor and temperature at various depths were measured by lowering a sounding rope over the side. Water samples were collected, and the bottom was dredged for rocks and deep-sea marine life. The *Challenger* expedition set the pattern for all expeditions for the next 50 years. The results from the expedition were staggering and filled 50 volumes. Surprisingly, oceans were not the deepest in the middle—the first hint of the vast midocean ridge system that was so central to the seafloor spreading concepts to be proposed later. Although 715 new genera and 4,417 new species were identified, unexpectedly, none turned out to be the living fossil equivalents to the trilobites and other ancient marine creatures found in terrestrial strata. The types of sediments on the seafloor were unusually lacking in diversity compared with terrestrial equivalents and were categorized by Sir John Murray as being one of only two types: chemical precipitates or accumulations of organic remains. Despite the great improvements in sampling technology that have been achieved since the days of the *Challenger*, some things never change. The dredge is still a mainstay for bringing up samples of submarine rocks, and it can still be expected to return to the surface right at the dinner hour.

Between the Two Wars

The modern era of ocean sciences began in the years preceding the Second World War. It was in these years that Scripps Institution of Oceanography (SIO) grew from a coastal marine station to an oceanographic research laboratory. Founded as a coastal marine station in 1903 by William Ritter, chairman of zoology at the University of California, Berkeley, Scripps grew in national and international stature under its second director, T. Wayland Vaughn, but lacked a ship to truly explore the Pacific Ocean. In 1936, Harold Sverdrup took over and obtained $50,000 from a long-time benefactor of the institution, Robert P. Scripps. The funds were used to purchase the *E.W. Scripps*, a 100-foot sailing vessel. Scripps as an institution was now a viable deep-sea research institute. Of course the realities of

the endurance of a 100-foot vessel still meant that the institution was hardly global in scope. The great marine geologist Francis Shepard was the first to use the ship to take bottom cores and measure currents near the bottom of the ocean.

Woods Hole Oceanographic Institution (WHOI) was established in 1930, by direct intervention of the National Academy of Sciences (NAS, 1929). The U.S. Navy and other government officials saw the need to establish an East Coast equivalent to Scripps to concentrate on the Atlantic Ocean. Although a number of sites along the East Coast could have suited the purpose, the fact that the Marine Biological Laboratory (MBL) was already established in Woods Hole, Massachusetts, was a deciding factor (along with access by rail and an "equitable" climate year around). At one point, MBL was approached to ascertain whether the institute was interested in expanding its scope to be an interdisciplinary oceanographic center. MBL declined the offer, but helped to establish Woods Hole Oceanographic Institution and, to this day, retains close ties with its research neighbor.

The World War II Effort

In 1940, the threat of submarine warfare provided the national imperative to understand the marine environment. At the time, there were two differing views as to how to detect submarines. As recalled by Roger Revelle,

> [Ernest] Lawrence and his friends, reasoning with some justification that oceanographers were bumbling amateurs, quickly decided that underwater sounds were a poor way to catch submarines and that optical methods should be used instead. They constructed an extremely powerful underwater searchlight and sewed together a huge black canvas cylinder which could be towed underwater to imitate a submarine. Unfortunately, it turned out that when the searchlight was directed on this object, it could be detected out to a range of about 100 feet. Shortly thereafter, the physicists disappeared. (Shor, 1978, p. 25)

and the "bumbling amateurs" took over. It was only after the war that oceanographers learned that the contribution of these physicists to the war effort was not entirely useless. They had been shuffled off to New Mexico to design and build an atom bomb.

The current format of oceanography, which involves an interdisciplinary grouping of marine physicists, biologists, engineers, chemists, and geologists, was largely an invention of the Navy to meet its specific needs. Although at first glance it might seem odd that investigations undertaken for the purpose of antisubmarine warfare might lead to plate tectonics or paleoclimate reconstructions, many observations relevant to Navy interests turned out to be key ingredients for these future revolutions. For example, detecting the presence of submarines acoustically required knowledge of the shape of the bottom of the ocean and the sediment type, magnetic detection required knowing the ambient background field, and so forth. This strategic alliance between scientists and the military counted in the victory, and after the war the nation's leaders recognized the need to maintain a cadre of scientists and engineers trained in oceanography. The initiative for ONR came from the wartime Office of the Coordinator of Research and Development under Admiral J. A. Furer. Columbus Iselin of WHOI and Lieutenant Mary Sears encouraged Roger Revelle, then Director of Scripps, to become involved with the new ONR's Geophysical Branch. The National Science Foundation was founded soon after, in 1950. From the very beginning, marine geology and geophysics was supported by these two agencies.

THE EARLY DAYS OF NSF

The First MG&G Research Supported

In the very first round of awards from the NSF presented in February 1952, MG&G was represented. The Earth Science Program awarded Bob Ginsburg at the University of Miami $4,700 for one year for a project entitled, "Geological Role of Certain Blue-Green Algae." Forty-six years later, Bob Ginsberg is still submitting proposals to NSF.

In 1953, a second research project in MG&G supported Kulp at Lamont for carbon-14 measurements of ocean sediments. In the following year, three more awards were made in MG&G, all again going to Lamont or to the Geology Department at Columbia. In 1955, there were two more new awards, again both going to Lamont, same pattern in 1956: two more new awards to Lamont. Not until 1957 was the Lamont monopoly broken. In that year, there were three more new awards to Lamont, but also one to K.O. Emery, who was at the time at the University of Southern California, for research on the deposition of sediments off southern California. One of the Lamont awards was to Maurice Ewing for "Reduction of Magnetic Data." Typical awards during these first five years included up to about $30,000 in funds and durations of one to three years.

In the postwar era, oceanographers were given extensive freedom to direct the course of their investigations to serve basic science. The Navy recognized that it would not be possible to precisely predict which question it would need the answer to next, and therefore it was essential to have broadly trained problem solvers. Unlike the situation in later years when one might reliably guess the source of support for a particular research project by its topic, in the 1950s and 1960s ONR and NSF were funding similar sorts of research. This strong overlap in research interests between ONR and NSF at first caused me some discomfort. I was concerned as to how I would distinguish, many years later, which of the great discoveries in MG&G should be attributed to the Navy and which to NSF. I soon realized that this was a non-issue. ONR and NSF did not seem to mind if their researchers were not careful about distinguishing who was supporting what, and they were delighted to take credit for jointly funded achievements. In thumbing through Lamont's annual collections of published papers, a common acknowledgement is of the form: "This research was funded by the Office of Naval Research and the National Science Foundation."

For those of us raised by the NSF of the latter quarter of the century, it is easy to envy the funding rate in the early years of NSF. In January 1961, J.D. Fautschy at Scripps issued to all division directors a list of all proposals to NSF that had been submitted in the prior two years. Every one was funded. The average time between submission of the proposal and the awarding of funds was five months. However, Fautschy did not circulate the list for the purpose of marveling at the largess and efficiency of the Foundation. Rather, he was complaining that despite NSF's encouragement of proposals for three to five years duration if "consistent with the nature and complexity of the proposed research," several of the proposals for longer durations had been cut back such that no grant was for more than three years. It is clear that even at that time, we were at the mercy of our peers, regardless of policies NSF might be trying to promote. More than 30 years later, requests for more than three years of funding on a single grant request are still having difficulty in peer review.

Even in its earliest years, funding from the Foundation was not limited to individual research grants. In June 1955, Raymond J. Seeger, then acting Assistant Director of the National Science Foundation, wrote a letter to Walter Munk at Scripps requesting suggestions for facilities that NSF might sponsor that would be relevant to the Earth sciences. The general policy, as articulated by the Divisional Committee for the Mathematical, Physical, and Engineering Sciences was:

> The NSF should recommend as a national policy the desirability of government support through NSF of large-scale basic scientific facilities when the need is clear, the merit endorsed by panels of experts, and funds are not readily available from other sources. (Letter from R.J. Seeger to W.H. Munk, June 1, 1955)

Astronomical observatories, radioastronomy facilities, computers, accelerators, and reactors were all given as examples of what NSF was looking for, and the committee made it clear that funds for these large-scale projects should not compete under the normal grants program. In his reply to Seeger's letter, written a mere seven days later (coast-to-coast mail service was clearly faster then than now), Munk suggested computer facilities and support for research vessels, the latter being the oceanographer's equivalent of an astronomical observatory. It is hard to imagine what marine geology and geophysics would be like today had NSF not eventually acted on both of Munk's suggestions.

The Beginning of the MG&G Program

After the MG&G program was established at NSF in the late 1960s, its first program managers (e.g., Bob Wall) came from ONR and set up styles of doing business that were very reminiscent of ONR methods of operation. The community was still small enough that NSF program managers could take a very personal interest in the investigators they supported and mentor career development. Compared with their academic equivalents, program managers were well paid in those days, and it was considered a very prestigious position.

The community of ocean scientists was small enough and centralized enough in the 1960s and 1970s that NSF program managers could visit their constituents on a regular basis. I recall as a graduate student at Scripps in the mid-1970s sharing a Mexican meal at an inexpensive Del Mar restaurant with an NSF program manager and a number of more senior graduate students. I doubt any of my own graduate students have had a similar experience. Although the NSF MG&G program managers did not conduct ONR-type site reviews, they began the tradition of tagging along on ONR's own site reviews. (One worldly investigator insists that NSF's main function at these meetings was to prevent double-dipping for travel to cruises jointly funded by NSF and ONR.)

Dick Von Herzen recalls sitting on Wall's MG&G panel in the late 1960s. At the time, NSF was funding blocks of related proposals from the oceanographic institutions that would provide the ship funds, travel money, et cetéra, to support a sequence of interrelated research legs. Research planning for these expeditions was a group effort conducted within each institution. Each leg of an expedition tended to be multidisciplinary, with physicists, chemists, geologists, and biologists sharing the same ship. Emphasis was on making every conceivable co-located measurement, since almost everything was being observed for the first time. Dick says it was not until well into the 1970s that he recalls writing his first "heat flow proposal," in which the expedition was devoted to heat flow observations and all ancillary data collection was justified on the basis of needing it to interpret the heat flow data.

THE THREE INSTITUTIONS

During the 1960s and early 1970s, the three major institutions conducting research in marine geology and geophysics (Lamont, Scripps, and Woods Hole) had very different characters, and these characters were reflected in the form of NSF support.

Lamont

Lamont could be best described as a dictatorship. Maurice Ewing exerted strong leadership, on the sorts of data to be collected, the route of the ships, and the research being addressed. Of the 600 papers published at Lamont between 1950 and 1965, Ewing was co-author on 150 of them and first author on 55. The block funding for institutional operations from ONR and NSF enabled a visionary

like Ewing to set a firm course for Lamont. He insisted that his ships run their precision depth recorders at all times; tow magnetometers; and stop every day for a core, a bottom temperature measurement, and in later years, a heat flow station. His strategy in cruise planning was to keep his ships circling the Earth ("like two moons"), collecting data where no oceanographic ship had gone before. Ewing established a cataloging system to keep an inventory of samples collected on each expedition and was regularly in touch by radio with his ships, sometimes even directing the sampling from shore. Ewing's system was such that it was clear if there was a gap in the record that a daily sample was missed. Jim Cochran recalls sailing as chief scientist on Lamont's ship, the *Vema*, right after obtaining his Ph.D. His surveying was going so well that he neglected to stop for the daily core. In less than 24 hours, Captain Kohler (gleefully) delivered to him a tersely worded message from Ewing expressing his displeasure with the young chief scientist's failure to follow orders.

The depth and magnetic anomaly data amassed by Lamont in its first two decades were key in establishing the validity of the plate tectonic theory (although Ewing was at first a vocal critic of the Vine and Matthews hypothesis). Lamont cores were instrumental in establishing global climate history. Underway data from Lamont's ships dominated the geophysical data banks. Lamont led the way in perfecting the use of geophysical surveying and sampling systems in the oceans, including the use of marine magnetometers, marine seismic reflection, precision depth recorders, piston cores, heat flow probes, and marine gravimeters (although others, including Sir Edward Bullard and Vening Meinesz, were the true pioneers). There had been much pessimism whether some of these methods, especially the seismic and gravity methods so central to terrestrial geophysical exploration, could ever be used at sea. Lamont scientists under Ewing's leadership demonstrated not only that these techniques could be used in the oceans, but that they gave even better data with cleaner signal in the marine environment. In retrospect, it might appear that Ewing's foresight was 20-20. Lamont-Doherty became a vast storehouse for marine data and samples just waiting to confirm the new theories after they were proposed. But in some respects Ewing was lucky as well. For example, Bill Curry tells me that many of the key deep-sea cores that figured so prominently in reconstructing Cenozoic climate were actually collected for the purpose of determining thermal conductivity for heat flow measurements. In fact, to this day, we are still scratching our heads trying to make sense of the widely scattered heat flow measurements acquired on Lamont vessels.

Sir Edward Bullard once asked Ewing why he kept taking so many cores. He answered:

I go on collecting because now I can get the money; in a few years it will not be there anymore, then I shall have the material to keep my people busy for years. (Menard, 1986, p. 269)

And he was right.

Ewing's style did not always endear him to his counterparts elsewhere in the national oceanographic community. Walter Munk recalls having been asked by Roger Revelle to sit in on a meeting at SIO just after the war. The Dutch pioneer in making pendulum gravity measurements from submarines, Vening Meinesz, was offering to give to the United States three of his instruments. The question was how to divide up the instruments. Columbus Iselin was there to represent Woods Hole, Ewing for Lamont, and Revelle for Scripps. (Three major institutions and three available instruments—the solution seems obvious.) Ewing's answer to the problem was that all three instruments should go to Lamont. He stated that making marine gravity measurements was Lamont's number one priority, and therefore for the good of the nation he should have all of the instruments. Lamont thus began a marine gravity program using U.S. Navy submarines.

Scripps

If Lamont was a dictatorship, then Scripps might have been best described as a fiefdom ruled by grand dukes. Revelle, who was director of Scripps in the early days of NSF funding, played a key role in attracting first-rate researchers to Scripps and in organizing the expeditions, but he did not oversee the daily science activities in the way that Ewing did. Marine geology and geophysics already had a rich history at SIO, thanks to the pioneering work on submarine canyons of Francis Shepard. By the 1950s, Scripps had built up a strong staff in MG&G, some of whom came from the Division of War Research that had been established on the eve of World War II at Point Loma. These researchers came with a storehouse of paper records of echograms acquired on Navy ships.

Bill Menard arrived at Scripps in 1955 with initial interests in turbidites. He developed the ability to read echosounder records faster and better than anyone else. He discovered and named the great Pacific fracture zones, mapped the East Pacific Rise, and later defined the geometry of the tectonic plates. At the same time, R.L. Fischer explored the Indian Ocean. The dredging efforts of Fisher, Menard, and others at Scripps resulted in a collection of abyssal basalts that was second to none. Joe Curray continued Shepard's legacy of understanding the sediments of continental margins. Doug Inman combined academic training in physics with hands-on learning under Shepard to pioneer the application of physics and fluid mechanics to the study of shore processes.

One of the more interesting early discoveries was made by Russ Raitt, who along with Ewing was applying seismic techniques to study the distribution of sediments in the oceans. Both Raitt and Ewing were getting similar results: sediment thickness was only about 300 m in the Pacific and 450 m in the Atlantic. These thicknesses were far less than

what would be predicted if the ocean basins were as old as the continents.

William Riedel joined Scripps in 1956 and began studying radiolarians. By the mid-1960s he had developed a precise chronology using radiolarians that allowed for geologic dating. These silica-shelled organisms were preserved even in the deep ocean and thus provided age estimates below the levels of dissolution of carbonate organisms. Jerry Winterer joined the institution in 1961, developing a reputation for deciphering ocean history from core stratigraphy.

The Scripps "grand dukes" shared Ewing's philosophy that ships should be required to collect every conceivable data type regardless of the objectives on an individual mission, although no one individual had the authority of Ewing to enforce quite such catholic sampling as was required on the Lamont ships. Nevertheless, Menard insisted that the echosounder always be running, while R.G. Mason and Vic Vaquier encouraged towing a magnetometer. Acceptance of the value of the soundings was more widespread than appreciation of the value of the bizarre variations in scalar magnetic field sensed by the magnetometers. In the 1950s, Mason encountered substantial resistance to the use of the magnetometer from both the Navy and the United States Geological Survey (USGS), so much so that Scripps nearly had to pass on the opportunity to mount its magnetometer on a U.S. Coast and Geodetic Survey ship, the *Pioneer*, that was conducting a detailed survey of seafloor off the Washington-Oregon coast. Menard managed to obtain support from Revelle's discretionary fund to allow use of the magnetometer. With line spacing of only 5 miles, the lineated nature of the magnetic anomalies was clear to Mason when he plotted the data. The pattern changed at the fracture zones and was repeated 80 km to the west as the ship passed south across the Murray fracture zone. It was not until a decade later that the symmetric anomaly patterns were found in the Indian Ocean, along the Reykjanes Ridge, and in the South Pacific that allowed geophysicists to correctly identify the cause of the *Pioneer* magnetic anomalies. Sometimes one needs to go an ocean away to understand something in one's own backyard.

Clearly two of the greatest legacies of the Scripps MG&G program in these early years were the decision to put computers on the ships (a radical notion in the 1960s before the days of computers in every lab, home, and toaster) and the establishment of the Geological Data Center. Bill Menard spearheaded the effort to install the computers, with an identical machine on shore to analyze the data after each expedition. IBM actually provided the computers (the "Red Baron," "Blue Max," and "Yellow Peril") and a computer operator. Stu Smith recalls that after the computers arrived, they had but one month to set them up in preparation for the Scan Expedition cruise on the *Argo* in 1969, using software borrowed from Manik Talwani.

The Geological Data Center (GDC) was instigated by George Shor, prompted by interest from the oil companies in the growing amount of seismic reflection data at Scripps. The funding to establish the center was raised from the Scripps Industrial Associates, and it opened for business in 1970. State of California funds supported Stu Smith as the curator of the facility. Prior to the establishment of the center, there had been no place to archive geophysical data. Observations collected at sea were considered the property of the principal investigator (PI), and exchanges between PIs were accomplished by a sort of bartering system. In order to convince the PIs that they should place their data in the archive, George Shor came up with the policy of a two-year proprietary hold on the data before distribution to other investigators. Scripps' Geological Data Center and the National Geophysical Data Center, which was established under National Oceanic and Atmospheric Administration (NOAA) sponsorship at about the same time, changed the way marine geology and geophysics could be accomplished. Data could now be used by a much broader array of researchers to answer questions not yet posed at the time that the data were collected.

Although neither of these two great legacies can be directly attributed to NSF, the Foundation was quick to seize the advantage of geophysical archives and shipboard computers. NSF promoted the archives by insisting that all NSF-funded PIs place their data in an archive facility where it would be in the public domain. NSF funding has allowed both the shipboard computers and the Geological Data Center to continue by allowing some of the costs for these facilities to be included in the day rates for data collection in NSF-funded ship time. And most importantly, NSF provided the funds for countless peer-reviewed grants to use data collected by shipboard computers and archived in the GDC for outstanding science.

Woods Hole

Woods Hole was even less centralized than Scripps and had a smaller staff in MG&G, compared with either Lamont or Scripps. Harold Stetson founded the WHOI MG&G group about the same time that Shepard was building the Scripps department. Research at WHOI was not instituted from the top down, although Brackett Hersey, the MG&G department chair immediately after World War II, had a lot of influence. As at Scripps and Lamont, most of the collaborations were forged internally, with liberal use of WHOI adjunct positions as a means of inviting selected outsiders to use WHOI ships. Doc Ewing himself was an example of an outsider who benefited from access to the *Atlantis* before Lamont purchased the *Vema*.

In the 1960s, ship time at WHOI was funded apart from individual proposals by ONR and NSF. Department chairs had the ability to assign ship time to staff members, who would then write proposals to cover incidental expenses af-

ter ship time was awarded. Charlie Hollister recalls arriving at Woods Hole from Lamont in 1967. The first thing his department chair urged him to do was to pay a visit to the manager of the new NSF MG&G program in Washington in order to establish a rapport and let him know what sort of NSF support Charlie would need for his science. Charlie recalls how radical this sounded to him at the time. Back at Lamont, Doc Ewing would have considered it high treason for a junior staff member to cultivate his own personal relationships with funding managers.

Despite the advantages to the young PIs of having a very decentralized research system at Woods Hole, there was a downside. No one investigator had the ability to mandate the routine collection of data sets on the Woods Hole ships, and thus Woods Hole did not early on amass the samples and data series that fueled the plate tectonic revolution.

Woods Hole began to step to the forefront sometime later than Lamont and Scripps in the area of MG&G. The development of *Alvin* gave Woods Hole an asset that was nowhere else duplicated in the academic research community. Project FAMOUS (French-American Mid-Ocean Undersea Survey) in the mid-1970s defined a new way of doing marine science (see Ballard paper later in this volume). Whereas much of the work prior to this time had been reconnaissance in nature, FAMOUS concentrated on a small area of the Mid-Atlantic Ridge using the submersibles *Alvin* and *Cyana*. The data amassed during the FAMOUS expedition led to an examination of the details of accretionary plate boundaries at scales smaller than what the plate tectonic paradigm could predict.

THE TWO REVOLUTIONS

It is interesting to consider how the MG&G community was so uniquely able to capitalize on the ability to make key observations even on non-MG&G cruises and to rapidly store the information in computer-aided archives. Whereas marine bathymetry, magnetics, and gravity could be collected while underway without interfering in whatever other science was to be accomplished on the trip, marine chemists, biologists, and physical oceanographers needed to stop to lower their instruments and collect their samples. Whereas the pertinent information on depth, magnetic field, and gravity field could be reduced to a simple series of numbers, this was not the case for water and biological samples. Even sediment cores collected by oceanographic institutions or by the drilling program were carefully cataloged, subsampled, and archived in a systematic way unduplicated for samples of interest in the other oceanographic disciplines. The ease with which key measurements could be acquired and shared, with help from NSF funding, helped propel U.S. researchers in MG&G to the forefront in two of the most important revolutions in science.

Plate Tectonics

The saga of the plate tectonic revolution has been so oft cited that I will not take the time to repeat it here. It is the archetypal scientific revolution that had its roots back in Wegner's theory of continental drift in the 1920s. But plate tectonics was a concept that was poorly represented on the continents, and therefore there was little hope of getting the story straight before the post-World War II era of ocean exploration.

The decade of the 1950s was marked by a total lack of consensus on Earth history. Was the Earth expanding? Contracting? Did continents drift? Remain fixed? In 1959, Americans Harry Hess (from Princeton), Bill Menard, and Maurice Ewing were joined by the Canadian Tuzo Wilson and the British Sir Edward Bullard at an international oceanographic congress in New York City right after the end of the International Geophysical Year. All believed that the midocean ridges were the source of some wholesale motion of Earth's crust in a manner not compatible with continental drift. The data collected by Ewing and others showed that the midocean ridges were clearly the youngest part of the seafloor. Wilson thought that Earth was expanding along the midocean ridge system, whereas Ewing, Bullard, and Hess believed the ridges to be the rising limbs of thermal convection cells. Hess balanced the expansion with contraction at the trenches and mountain belts. Menard kept the continents in place while the seafloor recycled. After the congress, Hess and Robert Dietz wrote papers revising the notion of continental drift to include spreading seafloor. Most others were skeptical, citing the inability of rising and descending limbs of thermal convection to explain the fact that Antarctica is nearly entirely circled with midocean ridges.

In 1963 came the breakthrough that would allow the concept of seafloor spreading to take a firm hold. Fred Vine and Drummond Matthews of Cambridge University became the first to publish the hypothesis that the puzzling magnetic anomalies in the ocean basins were the result of seafloor spreading combined with aperiodic reversals of Earth's magnetic field. In reaching this conclusion, they relied heavily on evidence just published by Allen Cox, Richard Doell, and Brent Dalrymple (Cox et al., 1964) for reversals of Earth's magnetic field globally recorded in volcanic rocks. This is one clear example of how advances in terrestrial Earth science research helped fuel a great discovery in MG&G. For the most part, however, it was an advantage not to have been too indoctrinated by the theories of terrestrial geologists in order to embrace the new paradigm.

Despite the attractiveness of the Vine-Matthews hypothesis, most Americans were still skeptical. George Backus published a paper in *Nature* in 1964 that proposed an elegant test of the Vine-Matthews hypothesis. He reasoned that the rate of seafloor spreading should increase from north to south in the Atlantic as a consequence of plate motion on a sphere.

It should be simple enough to determine whether the pattern of magnetic stripes in the South Atlantic repeated that already found off Iceland, except with greater thickness to the stripes. His NSF proposal to fund just such an expedition was declined by a panel of his peers as being "too speculative." NSF would soon prove the validity of the plate tectonic hypothesis, but not through deliberate forethought.

In 1965, J. Tuzo Wilson published a new explanation for the offset of the magnetic lineations across fracture zones. The lineations were offset because the ridge itself was offset (Figure 1). Earthquakes occurred only along the segment of the fracture zone between the two ridges where he predicted, based on seafloor spreading, that crust was moving in opposite directions. Later, Lynn Sykes at Lamont would go on to prove Wilson's hypothesis by showing that the first motions of earthquakes were consistent with this theory.

The tide turned in favor of the acceptance of seafloor spreading with the publication of the *Eltanin*-19 profile (Figure 2). The *Eltanin* was a southern ocean research ship owned by the National Science Foundation and operated by Lamont until she was retired in 1973. Walter Pitman, a student at Lamont, was the first, in December 1965, to take a careful look at that profile across the South Pacific and note the nearly perfect symmetry in the magnetic lineations. *Eltanin*-19 was fortuitous; it was collected in the Southern Ocean near the magnetic pole such that the magnetic anomalies were large and barely skewed. The seafloor spreading history had been steady to first order, with no major plate reorganizations back to 80 million years. Pitman began numbering the magnetic anomalies on a paper record, beginning at the left edge. By the time he got to the midocean ridge, the numbers were large. He quickly realized that this would not do, erased his numbers, and began counting anew from the ridge outward. This original profile now hangs on the wall in John Mutter's office at Lamont. By this time, Cox et al. (1964) had firmed up the magnetic reversal time scale for the first few million years, and the correspondence with the spacing of the anomalies on the *Eltanin* profile was staggering. By February 1966, Pitman's colleagues at Lamont quickly embraced Vine-Matthews and the other tenets of the new theory. The institution with more than half of the existing magnetic and profiler records from the oceans and 80 percent of the deep-sea cores would from then on be working to help establish the evidence for seafloor spreading.

The conversion of Lamont came just before a National Aeronautics and Space Administration (NASA) conference at Columbia in 1966 on the "History of the Earth's Crust." The papers ultimately presented there bore in many cases little resemblance to the abstracts submitted months earlier. The field was moving too fast. At this meeting, Heirtzler presented the results of the *Eltanin* surveys. After his talk, Pitman recalls:

Menard from Scripps, who had opposed [continental drift] sat and looked at *Eltanin*-19, didn't say anything, just looked

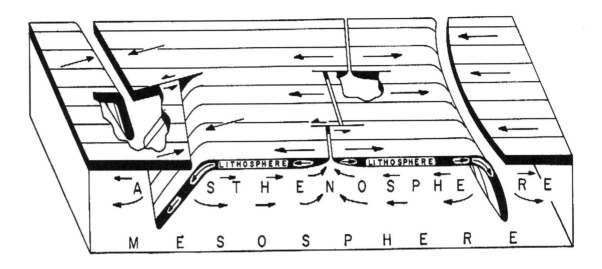

FIGURE 1   Lithospheric plate motion in three dimensions shows plate generation along the midocean ridges, transform motion associated with ridge offsets, and sinking of the plate at the ocean trenches. Reprinted from Isacks et al. (1968), with permission from the American Geophysical Union.

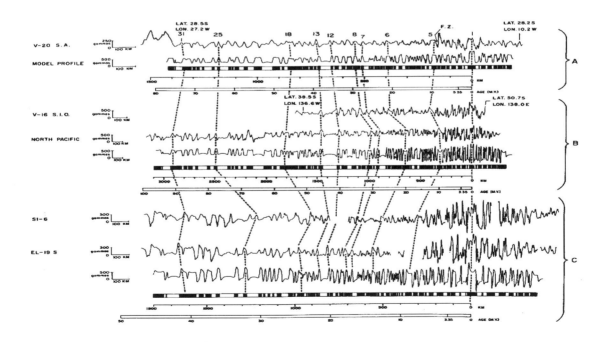

FIGURE 2   Comparison of magnetic anomaly profiles from the South Atlantic (A), and North Pacific (B) with the *Eltanin* profile (C) from the South Pacific. Correlations of individual anomalies are indicated with dashed lines. Shaded boxes are the magnetic reversal time scale. Reprinted from Heirtzler et al. (1968), with permission from the American Geophysical Union.

and looked and looked. Next, Lynn Sykes delivered the one–two punch by showing that earthquake focal mechanisms on transform faults were consistent with J. Tuzo Wilson's theory of ridge offset. Menard returned to Scripps a complete convert.

Although the battle for acceptance of plate tectonics was quickly waged and won in the mid-1960s, there were still a number of details to be filled in, much of which was done under the sponsorship of NSF. The present-day plate kinematics were to be sorted out using the azimuths of transforms and the width of the near-ridge magnetic anomalies. The history of plate motions and reorganizations needed to be worked out, a problem often requiring targeted expeditions funded by NSF to key areas where there were gaps or complexities in the magnetic records. Second-order effects, such as the existence of propagating ridges and microplates, were observed from detailed surveys and found to be important mechanisms for accommodating changes in the direction of relative plate motion.

The vertical motion of the seafloor was predicted from conductive cooling relations and compared with the depth data. The archives of heat flow observations were compared with what was predicted based on the thermal cooling model that fit the subsidence of the seafloor away from the ridges, but were found lacking. The conductive heat flow was less than predicted near the ridges and on the flanks, leading to the proposal that hydrothermal circulation was appreciable in young crust. Later expeditions funded by NSF, notably the RISE (RIvera Submersible Experiments) Expedition to the East Pacific Rise in 1979, found the "smoking gun" for hydrothermal circulation near midocean ridges in the form of hot vents and the completely unexpected chemosynthetic food chain associated with them. Thus, even the crowning achievement in the field of marine biology can be claimed by MG&G.

Hotspots, although not a natural component of the plate tectonic paradigm, proved to be a useful indicator of the direction and speed of absolute plate motion. Observations of the flexure of the lithosphere beneath the weight of the hotspot islands and seamounts, and seaward of subduction zones, were used to calibrate the strength of the oceanic plates. These studies, funded mostly by NSF, led to unprecedented abilities to predict the horizontal and vertical history of seafloor in all of the world's oceans.

I recall the first time I heard about the theory in 1972. I was an undergraduate at Colorado College majoring in physics, soon to graduate. One of my physics professors gave me an article from *Scientific American* written by John Dewey describing the new theory. After the geology courses I had taken that spoke of geosynclines deformed under unknown forces, plate tectonics seemed so simple and elegant. Soon after, J. Tuzo Wilson came to speak at the college. I was hooked. I had already applied to graduate school in physical oceanography, but quickly decided that geophysics was what I really wanted to study—nothing like getting in on the first

decade of a major paradigm shift. On my first oceanographic expedition, there was no one more senior than the graduate students, including the two co-chief scientists, Peter Lonsdale and Kim Klitgord. Everything had to be discovered anew and reinterpreted in terms of the new model, and who better to do it than the graduate students who had no stake in any previous ideas?

It is impossible to understate the importance of plate tectonics. It grandly explained the distribution of earthquakes and volcanic eruptions. It exactly predicted evolutionary patterns and distributions of related species. It predicted the history of possible pathways for ocean circulation, trends in ocean volume that controls sea level, and alteration of seawater chemistry via fluid circulation at ridges and trenches. In the chemosynthetic colonies in the hot vents, it might even explain the origin of life.

Reconstruction of Earth's Paleoclimates

The impact on society of the use of MG&G observations to reconstruct paleoclimates has been no less important and followed fast on the heels of the plate tectonic revolution. Whereas the time scales for plate tectonics are measured in millions of years, the deep sea record from sediment cores has taught us that Earth's climate vascillates on thousand-year time scales, and possibly much less. No great revolution sparked the acceptance of the climate proxies from the deep sea, as was the case in plate tectonics, but the impact on mankind could be much greater. We doubt that plate tectonics will render Earth uninhabitable for mankind on a human time scale, but there is every reason to believe that natural climate cycles enhanced by man's degradation of air, water, and land could result in an Earth unable to support the present population in a matter of centuries or less.

The climate story is also one of fortuitous gathering of samples, specifically the deep-sea cores, before their significance was established. A large number of researchers labored long and hard to work out the biostratigraphy of the cores using the carbonate and siliceous shells of microscopic marine animals. These cores demonstrated that the carbonate compensation depth in the oceans had varied over time, for not completely understood reasons, as had sea level. Furthermore, the microfossils indicated that there had been sudden swings of climate from warm-loving to cold-loving marine planktonic microfossils and back again at rates too fast to have been caused by plates drifting into different climate zones. But the resolution in the biostratigraphy was too poor to work out the rates of climate shift and to establish absolute global synchronicity. Here again the pioneering work of Cox et al. (1964) proved useful, in that the reversal of Earth's magnetic field at the beginning of the Bruhnes epoch, about 700,000 years ago, was often faithfully preserved in the paleomagnetic field of the core, such that it

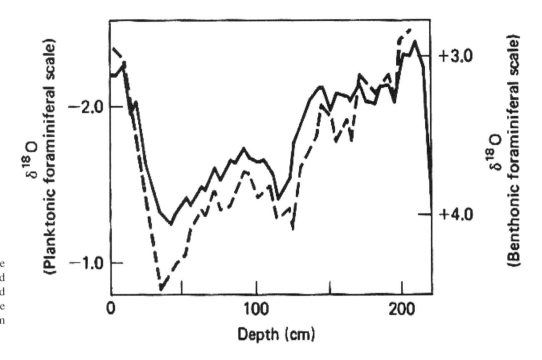

FIGURE 3 Oxygen isotope records of planktonic and benthic foraminifera. Reprinted from Shackleton and Opdyke (1973), with permission from Academic Press, Inc.

provided at least one absolute calibration point for estimating average rates of sediment accumulation.

Nick Shackleton, a British marine geologist, was the foremost figure in promoting another proxy for climate change, stable isotopes. Working in England, he used a high-resolution mass spectrometer to analyze the down-core oscillations in the ratio of the heavy oxygen isotope, ^{18}O, to the light oxygen isotope, ^{16}O. Based on the correlation with the biostratigraphy, these variations were clearly correlated with changing climate, but it was unclear whether the isotopic variations were caused by changes in ocean temperature or in terrestrial ice volume. With the encouragement of NSF, Shakleton became the first international corresponding member of NSF's CLIMAP program, which sought to decipher Earth's paleoclimate during the last glacial maximum. U.S. researchers were intrigued by Shakleton's stable isotope work, and Shakleton badly needed better samples on which to work. He had been using samples collected 100 years earlier by the HMS *Challenger*! Under CLIMAP sponsorship, Shakleton came to the United States and worked on core V28-238, a high-resolution core in the Lamont data bank collected by the *Vema* from the Ontong Java Plateau (Figure 3). This core contained well-preserved benthic and planktonic foraminifera, which showed the same oxygen isotopic signal. The argument was that whereas surface waters are very prone to temperature changes, the deep sea is roughly isothermal. Therefore, the fact that the signal was the same in the surface waters as the deep sea argued that the ultimate cause was climate-related changes in ice volume, not temperature directly.

The impact of the development of the stable isotope proxy on paleoceanography was substantial. On the assump-

tion that sedimentation rates were constant throughout the entire Bruhnes epoch, the oscillations in the stable isotopes became the paleoclimate equivalent of the magnetic reversals for plate tectonics. The pattern could be used for global correlation. But unlike the magnetic reversal signal, which defies prediction and is likely an excellent example of chaos, there was a pattern to the variations in the oxygen isotopes. In 1976, Hays at Lamont, working with Imbrie at Brown and Shackleton, applied spectral techniques to the signals from cores that were thought to be fairly well dated such that the isotopic signal as a function of depth could be accurately converted to a time series. The result was the identification of spectral peaks that matched the predictions of the Milankovitch hypothesis (Figure 4). According to this theory, variations in Earth's orbital parameters (eccentricity, tilt, and precession of the equinoxes) caused variations in solar insolation that resulted in changes in climate.

Although there was some cause to question how well core depth had been converted to time, the strength of the spectral peaks and the repeatability of the pattern won many converts—so much so that now cores with poor age control are assigned dates by assuming that the isotopic peaks and troughs should correspond in time to what is predicted by the Milankovitch hypothesis ("orbital tuning"). Not all is completely understood, however. For example, northern and southern hemispheres would be predicted to be out of phase for the precession period, but they are not. Overall, phase relationships demonstrate that regional insolation is not important. The net effect on the whole globe with its unequal distribution of continents and oceans must be taken into account. In addition, the strength of the spectral peaks is not consistent with the hypothesis that it is variations in solar

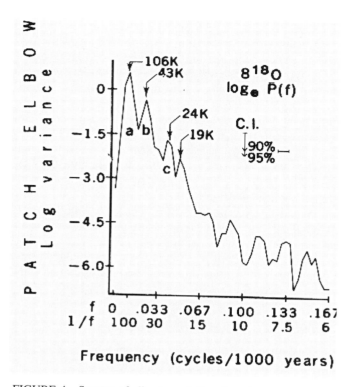

Frequency (cycles/1000 years)

FIGURE 4 Spectra of climate variations in sub-antarctic piston cores as inferred from variations in oxygen isotopes. Prominent spectral peaks, labeled a, b, and c, correspond to the predicted periods of eccentricity, obliquity, and precession of the Earth's orbit. Reprinted from Hays et al. (1976) in *Science*, Vol. 219 with permission from the American Association for the Advancement of Science.

insolation that leads to ice volume variations, and the spectral amplitudes are not stationary in time. Despite these remaining questions, the deep sea has provided a well-calibrated record of Earth's natural climate changes that can be used to help assess the future impact of man's activities.

The National Science Foundation was by far the greatest supporter of climate research, including the very successful CLIMAP project (Figure 5). A large amount of the paleoclimate work was supported and continues to be supported by NSF-MG&G. However, the Division of Atmospheric Sciences and the Ocean Drilling Program were also major players. MG&G has benefited greatly from broader NSF initiatives in global change that support paleoceanographic research beyond what the MG&G program could afford.

THE ASCENDANCE OF NSF SUPPORT

During the course of my interviews for this assignment, I asked a number of people when they recalled NSF taking over from ONR as the principal source of funding in MG&G. The universal answer was that the changeover occurred in the mid- to late-1970s. And yet the numbers from Lamont (Figure 6) and Scripps in no way support this impression. Even in the early 1970s (as far back as, it seems, anyone bothered to keep records), NSF was providing more dollars to the oceanographic institutions than ONR. Why was the impression just the opposite?

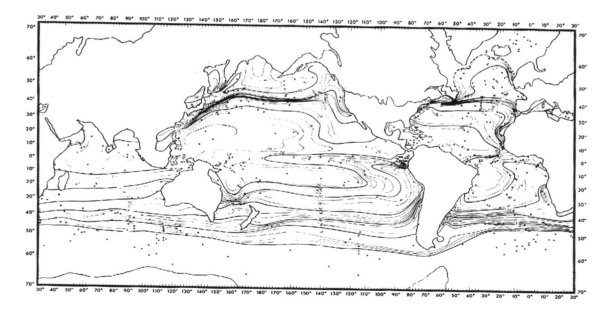

FIGURE 5 Sea-surface temperatures for northern hemisphere summer 18,000 years ago as determined by climate proxies mapped by the CLIMAP project. Contour intervals are 1°C for isotherms. Black dots show the locations of cores used to determine paleoclimate. Extent of continental glaciers is shown for the northern hemisphere, and coastlines reflect the corresponding lowering of sea level. Reprinted from CLIMAP (1976) with permission from the American Association for the Advancement of Science.

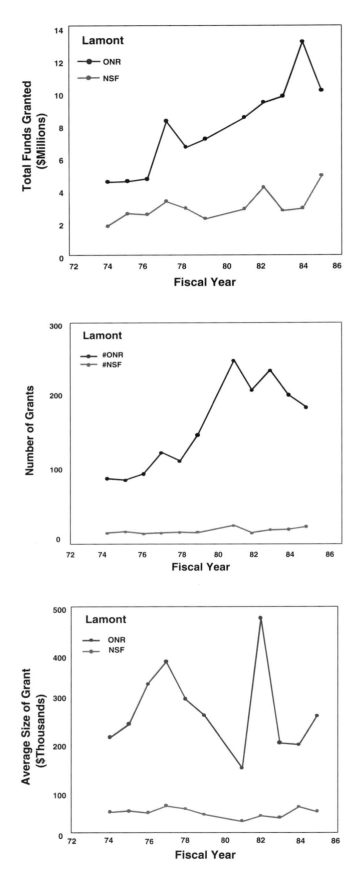

Deborah Day, the Scripps archivist, suggested a possible answer to this question. The prelude to many key MG&G experiments was the development of a new technology—Woods Hole's *Alvin*, Scripps' ocean bottom seismometers, Lamont's airguns, swath mapping systems, and so forth. The Navy tended to take the lead in instrument development in MG&G, but once the technology was proven, NSF would support the science programs that used the technology. In a few cases, successful science programs initiated by ONR would be continued by NSF. It is possible that into the 1970s, ONR was still getting credit for programs it had started but handed off to NSF.

With some exceptions, NSF's decision-making process of judgment by our peers has not been a good source of "venture capital" in MG&G. Rather, the community found this venture capital at ONR, from industry, and from the discretionary funds of institute directors. NSF was quick to support the successful venture, and make them pay off.

The Impact of International Programs

One place in which NSF clearly set a policy direction different from that of ONR was in the encouragement of international collaborations. Initially through the International Geophysical Year (IGY), and later via the International Decade of Ocean Exploration (IDOE), U.S. investigators were encouraged to invite foreign colleagues to the United States with travel support from NSF. This sort of attitude would have been uncharacteristic for an agency like ONR responsible for maintaining a competitive advantage in U.S. science for the sake of national security.

In the area of MG&G, the international program that has had the greatest impact has been the Deep Sea Drilling Program (DSDP) and its successors. Because this is the topic of another paper (see paper by Winterer, this volume), I mention here a few of the highlights. DSDP sampled the basal sediments in Leg 3 along a magnetic profile in the South Atlantic that established beyond a shadow of a doubt that the seafloor just beneath was indeed the age predicted by the Vine-Matthews hypothesis. The ocean drilling program developed the hydraulic piston corer that became the mainstay for sampling thick, continuous sequences in areas of high sedimentation rate in order to investigate climate change on orbital and suborbital time scales. DSDP and its successors established repositories for logging data and cores and thick volumes of results. It set the standard for interna-

FIGURE 6   Total funds granted (*top*), number of grants (*middle*), and average size of grant (*bottom*) for NSF versus ONR awards given to Lamont, 1974-1985. Similar trends are seen in data from Scripps, but Lamont numbers are used here since they can reasonably be expected to represent trends in MG&G as opposed to those in marine biology, chemistry, or physical oceanography.

tional scientific cooperation and became the vehicle for exporting American science and our scientific system to the rest of the world.

The "Democratization" of Ocean Science

The plate tectonic revolution led to an explosion in the number of young graduate students studying marine geology and geophysics. At first, in the late 1960s and early 1970s, many of the most promising researchers were retained by their Ph.D. institutions or one of the other oceanographic institutions in order to complete the data analysis for the revolution. But by the mid-1970s, the slots within the institutions were filled by a young cohort, and nonoceanographic institutions began hiring the MG&G students to teach plate tectonics to undergraduates and graduates. As these former students who found themselves at nonoceanographic institutions sought to develop their own research programs, they saw the lock that their former alma maters had on MG&G funds and ship time, and they cried, "Foul!" By the end of the 1970s, the democratization of MG&G was well underway, as perhaps best illustrated by the increase in non-Lamont chief scientists on her ships (Figure 7).

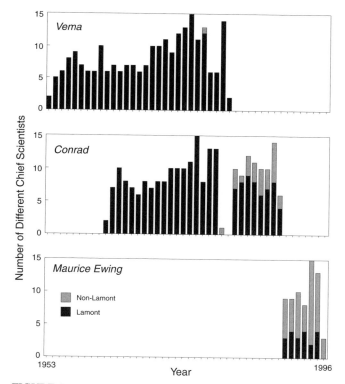

FIGURE 7   Comparison of the number of different chief scientists from Lamont versus other institutions sailing on Lamont's research ships. The establishment of UNOLS went a long way towards opening up access to ship time to researchers from nonoceanographic institutions. Data from Woods Hole and Scripps, although not as easily interpreted in terms of trends in MG&G, show that this changeover in institutional affiliation of the chief scientists happened somewhat earlier outside of Lamont.

This change was inevitable and brought a much larger talent pool to the table to compete for funding and ship time. The system became more open and more accountable. Cruises were more carefully planned, and no funds were wasted taking observations unnecessary to test the hypothesis at hand. But much was lost along the way as well. Without omnibus grants in the hands of the leaders of oceanographic institutions, there was no opportunity to put together larger projects that cut across disciplines by a few people with great vision. With less institutional funding, there was no incentive to work with colleagues at one's own institution as opposed to those across the country. The institutions became less cohesive. Since researchers from one oceanographic institution were likely to be scheduled for ship time on another institution's vessel, less attention was paid to maintaining and improving the home institution's assets. "Expeditions" became a string of unrelated legs with completely different science parties and objectives. With more PIs competing for the funding pool, the success rate dropped, such that researchers were writing more proposals to raise the same amount of research funds. The sharp curtailment of ONR support for MG&G that occurred soon after only made matters worse.

To some extent, this changeover in the support pattern in MG&G happened at a fortunate time. The reconnaissance sampling of the geology and geophysics of the oceans had already been completed, and it was time for more focused hypothesis testing in targeted areas, the type of research for which NSF funding is ideally suited. I wonder whether the same was true for the other oceanography disciplines, for which the critical observations were not as routinely measured or as easily archived during the early days of wideopen ocean exploration as they were for MG&G.

The Growth of Special Initiatives

I view the growth of special initiatives in MG&G as an attempt to allow for the earlier type of coordinated planning in spite of the system that predominantly funded a single PI or small group of PIs for one month of ship time to address one problem. Initiatives such as the Ridge Inter-Disciplinary Global Experiments (RIDGE), provided a mechanism to tackle bigger science questions in a systematic way, while still maintaining the openness of the system and the advantages of peer review. This initiative has been immensely successful by any measure, integrating midocean-ridge-related research throughout the oceans and across the disciplines of geology, geophysics, chemistry, and biology. The down side is that RIDGE has been so very successful in terms of discoveries and in capturing the attention of the community that it is in danger of reducing the breadth in interests for the RIDGE generation of students. I recall spending summers at Woods Hole when it was difficult to find a seminar that was not RIDGE-related or a graduate student that was not working on a RIDGE problem. Special initiatives are a superb mechanism for enabling research larger than that supported by a single grant, but they should not be allowed to dominate the field. (RIDGE-related research accounts for all of the targeted funds in the special program and competes successfully for about 35 percent of the core funds.) Nor should they continue for so long that several generations of students learn of nothing else.

Special initiatives have provided a forum for planning larger research programs that has replaced the internal planning that used to occur within the confines of the oceanographic institutions. The big difference is that we all must spend endless hours on airplanes instead of wandering down the hall. Of course, planning was not as extensive in those days as it appears to be today. Denny Hayes recalls having been chief scientist on the *Vema* in 1968 for support of the deep-sea drilling leg to date the basal sediments along the South Atlantic profile to be drilled on Leg 3. The site survey was being accomplished, literally, a few days ahead of the drilling. At one point, Denny jumped from *Vema* into a Zodiac with rolled seismic records under his arm to deliver the data (and I believe some whiskey) to the *Glomar Challenger*. Dick Von Herzen, co-chief scientist on the drill ship, recalls happily taking delivery of the data and whiskey, and reciprocating with some beef—high seas barter in the far South Atlantic. These days, planning is so extensive, time-consuming, and exhaustive that it has led one jaded investigator on soft money to remark, "It is cheaper for NSF to pay us to plan than to pay us to do science."

EPILOGUE

When I was asked to review the history of marine geology and geophysics from the perspective of NSF sponsorship, I firmly believed that I would end up regretting the assignment. It was sure to be a time-consuming task with low prospects for gaining personal or professional satisfaction from the result. However, as I became more involved in putting together my notes for this paper, my view took an about-face. I came to realize that as the director of the only oceanographic institution in the nation that can still set its own ship schedule, determine its own research priorities, and commit itself to high-risk, long lead time, interdisciplinary research, it is essential that I understand what sort of science the NSF and the ONR of the 1950s and 1960s were best suited to accomplish, as contrasted with the type of science that succeeds today and indeed during the entire tenure of my own research career. The Monterey Bay Aquarium Research Institute must go after the problems that go beyond what can be addressed by the individual investigator with a three-year grant and one month of ship time. We should

seek out those vaguely defined areas of ocean science still in search of a fundamental paradigm on which to base testable hypotheses. And we should work to develop those research tools that no one else is so bold to propose for seagoing research.

ACKNOWLEDGMENTS

I am indebted to a large number of colleagues who shared with me their memories and their institutional archives. In particular, material in this report reflects information gleaned from Bob Arko, Jim Cochran, Bill Curry, Deborah Day, R.L. Fischer, Denny Hayes, Jim Hays, Charlie Hollister, Ken Johnson, Garry Karner, Walter Munk, John Mutter, John Orcutt, Mike Reeve, George and Betty Shor, Stu Smith, Fred Spiess, Scott Tilden, Dick Von Herzen, and Jeff Weissel. Thank you all for your time, your generosity, and your insights.

REFERENCES

Bush, V. 1945. *Science-The Endless Frontier. A Report to the President on a Program for Postwar Scientific Research.* U.S. Office of Scientific Research and Development, Government Printing Office, Washington, D.C.

Climate Long-range Investigation, Mapping and Prediction (CLIMAP) Project Members. 1976. The Surface of the Ice Aged Earth. *Science* 191:1131.

Cox, A., R.R. Duell, and G.B. Dalrymple. 1964. Reversals of the Earth's magnetic field. *Science* 144:1537-1543.

Hays, J.D., J. Imbrie, and N.J. Shackleton. 1976. Variations in the Earth's orbit: Pacemaker of the ice ages. *Science* 194:1121-1132.

Heirtzler, J.R., G.O. Dickenson, E.M. Herron, W.C. Pitman III, and X. Le Pichon. 1968. Marine magnetic anomalies, geomagnetic field reversals, and motions of the ocean floor and continents. *J. Geophys. Res.* 73:2119-2136.

Isacks, B.L., J. Oliver, and L. Sykes. 1968. Seismology and the new global tectonics. *J. Geophys. Res.* 73:5855-5899.

Menard, H.W. 1986. *The Ocean of Truth.* Princeton University Press, Princeton, New Jersey.

National Academy of Sciences (NAS). 1929. *Oceanography: Its Scope, Problems, and Economic Importance.* Houghton Mifflin Company, Boston.

Mukerji, C. 1989. *A Fragile Power: Scientists and the State.* Princeton University Press, Princeton, New Jersey.

Shackleton, N.J., and N.D. Opdyke. 1973. Oxygen isotope and palaeomagnetic stratigraphy of equatorial Pacific core V28-238: Oxygen isotope temperatures and ice volumes on a 10^5 year and 10^6 year scale. *Quaternary Research* 3:39-55.

Shor, E.N. 1978. *Scripps Institution of Oceanography: Probing the Oceans 1936 to 1976.* Tofua Press, San Diego, Calif.

Wertenbaker, W. 1974. *The Floor of the Sea: Maurice Ewing and the Search to Understand the Earth.* Little, Brown, and Co., Boston, Massachusetts.

Deep Submergence:
The Beginnings of *ALVIN* as a Tool of Basic Research and Introduction of Featured Speaker Dr. Robert D. Ballard

Sandra Toye

National Science Foundation (ret.)

For oceanographers, names are special—especially when it comes to ships. Research ships carry names drawn from history, or names that reflect the marine environment where they work, or names that suggest noble qualities of character: *Resolution, Endeavor, Challenger, Horizon, Oceanus.*

So how is it that one of the most famous, productive, and glamorous research platforms in our inventory got the kind of name you associate with, say, your uncle?

The year was 1964. The place was Woods Hole, Massachusetts. For the better part of a decade, the Woods Hole Deep Submergence Group had been working with the Navy to build a viable research submarine. At long last, the task was about completed, and people had begun to talk about a commissioning ceremony, which meant that a name must be chosen.

The original Navy project name was *Seapup.* This reflected the stubby look of the sub along with the proper dash of military bravado, and many still favored the name.

Graduate students at the Woods Hole Oceanographic Institution (WHOI) had watched the chubby little boat evolve, and, with the usual perverse humor of grad students everywhere, had a number of pet names, most of which can't be mentioned in polite company. Their favorite was *The Pregnant Guppy.*

Paul Fye was the WHOI director at the time. For those of you who didn't know Dr. Fye, let me say that he was a courtly and rather stern gentleman, not given to flights of fancy. He took the name issue very seriously, consulting widely with colleagues, trustees, and other VIPs. He had settled on the name *Deep Sea Explorer.*

What Dr. Fye didn't know was that the issue was already closed. The engineers and scientists of the Deep Submergence Group, a tight-knit and slightly wacky group, had quietly selected a name and conned the Navy into accepting it. All of the engineering drawings, certification, and commissioning documents for the new sub already bore the name they had chosen. Navy tradition and Woods Hole dignity be damned—the sub was going to be *Alvin.*

Reportedly, Dr. Fye was horrified. To understand why, you must know something about American popular culture in 1964. At the time, the hottest musical group in the country was called Alvin and the Chipmunks. Their specialty was to record songs at 33 rpm and then re-record them at 78 rpm. The results were goofy nasal singsongs that, amazingly, were leading the Hit Parade. Dr. Fye was sure that this strange bunch in the Deep Submergence Group—his own employees, for heavens sake—had gone and named this wonder of technology for a falsetto soprano chipmunk!

Imagine his relief to learn that *Alvin* was at least named for a real human. It was in fact a contraction of the name of one of his staff members, oceanographer Allyn Vine, whose fertile imagination and dogged salesmanship had been instrumental to the project. So *Alvin* it was, and *Alvin* it would remain.

Let us now, like that musical group, Alvin and the Chipmunks, hit the fast-forward button. It was nearly a decade later that DSRV *Alvin*, the National Science Foundation, and the larger oceanographic research community came together.

The year was 1973. The Navy, which had funded virtually all of *Alvin*'s operations since its christening in 1964, was under great pressure from Vietnam-War-era funding restrictions. The Navy informed Woods Hole that the upcoming renewal of the contract would be something called a "no-fund equipment loan." This was a nice way of saying that all of the operational and scientific funding for the Deep Submergence Group was about to end. If Woods Hole could find other support, fine—the Navy would be willing to let it keep the sub. If not, *Alvin* was destined for mothballs or a museum.

Dr. Fye, who was still the director of Woods Hole, appealed the Navy decision, but their problems were overwhelming. The answer was No. He went to the National Oceanic and Atmospheric Administration (NOAA), a relatively young agency with a statutory charge to promote manned submergence—but the entire annual budget for that part of NOAA would scarcely pay for a few weeks of *Alvin* operations—no again. He came to the National Science Foundation (NSF), and the negatives kept coming. The answer again was no. How could this be? How could NSF reject this unique and promising tool?

Truth be told, *Alvin*'s first decade of operations had focused more on engineering and operational matters than on science. Only a few scientists, almost all of them from Woods Hole, had ever used it. Although these few were excited about its potential, the early dives hadn't produced anything worldshaking in the way of research results.

In the oceanographic community at large, most people considered *Alvin* a Woods Hole toy. They derided diving in general as "Gee-Whiz" stuff, okay for popular magazines, but not for serious science. Reflecting these views, NSF advisory committees said that if the Navy wanted to mothball *Alvin*, it would be no great loss to ocean science. NSF should definitely stay out of it.

Dr. Fye was not easily discouraged—he decided to make one more try to keep *Alvin* in business. He had tried diplomacy. He had tried appeal to reason and to the better instincts of the community. This time he would appeal to politics.

Now, we need to understand that in those days, the Massachusetts congressional delegation had more than the usual measure of clout. In the Senate, of course, there was Senator Edward Kennedy. In the House, a fellow named Thomas P. O'Neill just happened to be from Massachusetts. He also just happened to be Speaker of the House.

After a few discreet phone calls, a meeting was set up for Dr. Fye with the "big three" of federal ocean science— the Director of NSF, the Administrator of NOAA, and the Oceanographer of the Navy. At the end of the meeting, it had been agreed that the three agencies would keep *Alvin* going for three years for a last-ditch test of its capacity for research.

This is the point at which the *Alvin* problem landed on my desk. In the summer of 1973, I had been hired by NSF as a program associate in the Office of Oceanographic Facilities and Support. This sounds rather grand, but in reality, a program associate is down near the bottom of the NSF food chain. My job was to do anything that nobody else in the office wanted to do. And nobody else wanted to deal with *Alvin*.

Well, we sometimes forget it, but NSF *is* a government agency, and politics and government do go together, even in the world of science. My bosses gave me unmistakable marching orders: make the *Alvin* agreement work.

I was new and green and not very knowledgeable. But I did know that over the long run, the only thing that will sustain a program at NSF is the support and endorsement of the scientific community. I began to look for allies who would be willing to take on the unpopular task of testing, once and for all, the real fitness of *Alvin* for research.

One ally was UNOLS—the University-National Oceanographic Laboratory System. Itself a young organization trying to establish credibility, UNOLS somewhat reluctantly agreed to take on scientific management of *Alvin* as a national oceanographic facility, thus opening access to scientists throughout the community.

My next ally was, of course, our speaker today, Dr. Robert D. Ballard. Born in Wichita, Kansas, Bob did what all good Midwesterners do—moved to California at the earliest opportunity. There he found one of the guiding passions of his life, the ocean.

After earning his bachelor of science at the University of California at Santa Barbara, Bob, a distinguished ROTC graduate, faced service in Vietnam. After two years in Army intelligence, he made another critical decision—he'd rather be in the Navy! The transfer was made, and a short time later the Navy introduced Lt. (j.g.) Ballard to another of the guiding passions of this career, assigning him to the Deep Submergence Group at Woods Hole, then in the early phases of testing DSRV *Alvin*.

When Bob and I met in 1974, he was a newly minted Ph.D. in marine geology and geophysics from the University of Rhode Island. He had just accepted the first of what would be a career-long series of faculty appointments at Woods Hole. As an aspiring young assistant scientist in the Marine Geology and Geophysics Department, Bob was doing the things that everyone must do at this point in an academic career. He was intent on making his mark in the research community, getting enough grant money to keep his team together, and achieving tenure.

There was nothing in his job sheet that said he had to work with *Alvin*, let alone help NSF and UNOLS develop a scientific constituency for deep submergence. But Bob's earlier career had already convinced him that manned submersibles, properly used in conjunction with remote sensing, were powerful tools of research. During the three years of the initial *Alvin* agreement, he was a tireless missionary to his colleagues.

Over the intervening 25 years, Bob Ballard has been at the center of many of the most exciting discoveries in science, and so has *Alvin*—often together. Bob's richly productive research career includes Project FAMOUS (French-American Mid-Ocean Undersea Study) and the discoveries of the Galapagos Rift communities and the black smokers of the East Pacific Rise.

His list of honors, awards, and publications is as remarkable for its diversity as its extent. His scientific work has been augmented by stunning accomplishments in marine history and archaeology, including discovery of the *Titanic, Bismark,* and *Yorktown*, and the exploration and protection of many other maritime relics.

Upon retiring from Woods Hole with emeritus status last year, Bob became president of the Institute for Exploration and chairman of the JASON Foundation for Education, based at Mystic, Connecticut.

Although many of us have settled down to quieter and less productive lives, Bob has continued, full steam ahead, to explore the oceans. He is still never too busy to share his love of this world with the rest of us. It is a great honor to welcome him to this symposium today.

The History of Woods Hole's Deep Submergence Program

Robert D. Ballard

The Institute for Exploration, Mystic, Connecticut

ABSTRACT

Since its arrival at the Woods Hole Oceanographic Institution in June 1964, the manned submersible *Alvin* has gone from a scientific oddity to an accepted research tool. Over its 35-year history, the deep submergence program at Woods Hole has experienced four distinct phases. Its initial design and introduction into the oceanographic community was driven, like all new paradigms, by a small core of scientists and engineers who saw the unique contribution that a manned presence on the ocean floor could make to marine research. The turning point in the acceptance of manned submersibles came on the heels of the theory of plate tectonics and the first manned exploration of the Mid-Atlantic Ridge during Project FAMOUS (French-American Mid-Ocean Undersea Study) in 1974. What led to the final acceptance of manned submersibles were the discoveries in 1977 and 1979 of hydrothermal vents and high-temperature "black smokers." Since that time, the *Alvin* program at Woods Hole has matured into a highly reliable and productive diving program fully integrated into a series of long-term research programs. Most recently, the manned submersible at Woods Hole has been merged with its newly developed remotely operated vehicle program.

THE EARLY YEARS OF *ALVIN*

The first phase of deep submergence started in the late 1950s when the bathyscaphs *Trieste* and *Archimede* began taking scientists into the abyssal depths of the world's oceans.

In 1956, after several years of bathyscaph operations by the Swiss and French, Jacques Piccard, son of the bathyscaph designer August Piccard, spent 100 days in America, traveling to all the major centers of oceanographic research trying to sell them on the virtues of the bathyscaph. To his pleasant surprise, he found many sympathetic ears eager to enter the world he and only a few others had visited. The visit came to a conclusive end at a National Academy of Sciences meeting in Washington when Piccard and Robert Dietz, an early supporter of the bathyscaph, presented papers on its potential value to deep-sea research.

Convinced of its merits, Willard Bascom spearheaded a resolution that read: "The careful design and repeated testing of the bathyscaph have clearly demonstrated the technical feasibility of operating manned vehicles safely at great depths in the ocean. The scientific implications of this capability are far reaching. We, as individuals interested in the scientific exploration of the deep sea, wish to go on record as favoring the immediate initiation of a national program, aimed at obtaining for the United States undersea vehicles capable of transporting men and their instruments to the great depths of the oceans." This resolution was followed in February 1957 with a contract between Piccard and the Office of Naval Research (ONR) to conduct a series of dives in the Tyrrhenian Sea off Naples so that American scientists could carefully evaluate its potential.

From July to October, *Trieste* made 26 dives carrying acousticians, biologists, geologists, physicists, VIPs, and naval personnel down to depths of 3,200 m. The U.S. Navy and ONR were now convinced that the bathyscaph held great promise and they wanted to support its future.

At the time, the Naval Electronics Laboratory in San Diego, California, was a hub of activity for naval research. There, a close bond existed between the operational navy and the oceanographic community. The largest oceanographic institute, Scripps Institution of Oceanography, was just a short distance away in La Jolla. For that reason, the Navy decided to base the *Trieste* in San Diego, with the goal of using the bathyscaph to dive to the bottom of Challenger Deep in the Marianas Trench in 1960 as a part of Project NEKTON. *Trieste*'s historic dive to the bottom of Challenger Deep in 1960 clearly demonstrated that man could penetrate the oceans to even their deepest depths. But the bathyscaph was large and difficult to operate and maintain in the open sea many miles from its home base.

Global coverage required the capability to carry the diving craft aboard a surface support ship that could transit at high speeds to the dive site and between dives bring the craft back aboard for maintenance and repairs. This dream of a tiny portable submersible was, in fact, already beginning to take shape even before *Trieste*'s 1960 diving campaign in the mind of a young French officer, Jacques Cousteau, who had witnessed the first test dive of the bathyscaph *FNRS-2* off Dakar in 1948. Cousteau's *Souscoup* was the first modern deep submersible to be built. However, its diving capability was limited to 300 m, far too shallow for the oceanographic community.

Just as Cousteau's experience with the French bathyscaph lead to the creation of the *Souscoup*, the Americans diving on the *Trieste* began to think about a similar modern submersible small enough to be carried aboard a mother ship.

No sooner had *Trieste* completed its deep dive in 1960, than the San Diego group including Andy Rechnitzer, Don Walsh, and Larry Schumaker began to dream of its replacement. Listening to these discussions was Harold "Bud" Froehlich, a General Mills engineer who had built *Trieste's* mechanical arm. Soon he was circulating the designs of a small prototype submersible he called the *Seapup* to anyone who was interested.

While this teapot began to boil, another spark was being lit on the East Coast. Charles B. "Swede" Momsen, Jr., the Chief of Undersea Warfare in ONR, the same organization that had sponsored *Trieste,* had received a proposal from J. Louis Reynolds of the Reynolds Metals Company to design and build an all-aluminum submersible called the *Aluminaut*. Momsen was a decorated submarine commander during World War II and was comfortable with its large design. The only problem was that ONR was not in the business of building submarines; it could rent one if Momsen could find scientists interested in using it. Ironically, when Momsen went to the Scripps Institution of Oceanography in San Diego where *Trieste* was now based, he received a cold reception to his idea.

This was not the case when he approached scientists at the Woods Hole Oceanographic Institution (WHOI) on Cape Cod, Massachusetts. In particular, Allyn Vine and WHOI Director Dr. Paul Fye welcomed the idea and offered Woods Hole as *Aluminaut*'s home base. What followed, however, was a long and drawn out series of discussions between ONR, Woods Hole, and J. Louis Reynolds. The sticking point was the ultimate ownership of the *Aluminaut*. Reynolds wanted to maintain title, while ONR wanted the Navy to own it with eventual ownership going to Woods Hole.

As time ticked on, the engineers Woods Hole had hired to operate the *Aluminaut* program began to question its design. They had the same concerns Cousteau had. *Aluminaut* was to be 51 feet long and carry six people. More importantly, like the bathyscaphs, it had to be towed out to sea and could not be brought aboard its mother ship for maintenance and repairs.

Finally after three years and four months of nonstop negotiations, an impasse was reached, and it became clear to Paul Fye that something new had to be tried. The only problem was that others in Washington were trying to pry loose the funds Swede Momsen had been squirreling away for the project.

Knowing other companies were eager to enter the deep-submersible game, Momsen acted quickly and authorized Woods Hole to request bids to build a submersible for the institution. The specifications that went out were not for a submersible like the *Aluminaut* but for a much smaller design, strangely similar to Bud Froehlich's *Seapup*.

Seven companies were sent the request for bid: Froehlich's General Mills, Lockheed, General Dynamic's Electric Boat Division, General Motors, North American Aviation, Philco, and Pratt Whitney Aircraft. Although four of these companies would eventually build their own deep submersibles, only two submitted bids to build what would become *Alvin*; General Mills and North American Aviation. General Mills was the ultimate winner since the Navy felt it was more committed to the project. Ironically, shortly after winning the bid, General Mills sold its division to Litton Industries, which finally built and delivered *Alvin* to Woods Hole in the summer of 1964.

Alvin could initially dive to 1,830 m, far deeper than the *Souscoup*, but clearly not to the abyssal depths of the bathyscaphs. As a result, the early users of *Alvin* were midwater biologists and scientists working on the continental margin.

Frank Manheim, a geologist with the U.S. Geological Survey at Woods Hole, was eager to extend his research on marine snow that had been pioneered by the Japanese. For years, Manheim had filtered seawater obtained from lowered instruments and weighed the dried filters to determine how much marine snow existed per unit volume of the ocean.

When he repeated this procedure using water collected from *Alvin*, he realized that this method of quantifying marine snow was not accurate. On his dive, he had seen a heavy snowfall, but the weight of his filters drawn from water collected by *Alvin* indicated otherwise. In the mud cores he brought up, there was little organic material. This seemed to indicate that the animals were extremely adept at food gathering; they were consuming the fine rain of organic material as soon as it hit bottom.

Studies of the deep scattering layers, also begun by the bathyscaph, continued using *Alvin*. Woods Hole biologists Richard Backus and Jim Craddock used *Alvin* and its highly sensitive CTFM sonar to study the layer. With the lights turned off they homed in on a good-sized blob on the scope. When they turned on the lights, they were surrounded by thousands and thousands of lantern fish. The fish were pointed and moving in all directions—"a fantastic aggregation not a school." The biologists had thought that photophores, which were usually on the underside of the fish, were meant to shine down and blind predators. But the photophores of these lantern fish were aimed in every conceivable direction. Rarely had their nets captured a lantern fish; and yet *Alvin's* net brought up 744 of them on one dive. Small collecting nets mounted on *Alvin* were also successfully used near the bottom by Woods Hole biologist George Grice, who discovered 18 new species of copepods on one dive.

While benthic biologists took advantage of *Alvin's* ability to carry out difficult manipulative tasks in the deep sea, marine geologists took advantage of its high maneuverability. K.O. Emery and a geology group he had formed at Woods Hole began using *Alvin* to explore the submarine canyons off the Northeast coast. Here a great series of canyons cut across an extensive continental shelf. But geologists were not only interested in understanding the origin of submarine canyons and the role they played in transporting sediments across the continental shelf to the deep sea, these canyons also provided them with knowledge about Earth's recent history.

The continental shelf in most parts of the world consists of horizontal layers of sedimentary rock laid down one layer on top of another over millions of years. As submarine canyons form, they cut down into these layers, exposing in their walls the geochronology or geologic history of the region. Using the submersible's ability to maneuver, geologists were able to sample these outcrops and add further detail to the recent stratigraphic history of the continental shelf.

Similar investigations were carried out in the Straits of Florida and the Bahama Banks, as well as submerged terraces such as the Blake Plateau. Instead of studying sedimentary layers of rock deposited by river outflow, geologists were able to investigate layer upon layer of limestone formed in place by coral growth and erosion.

But submarine canyons and steep vertical scarps were not the only geologic features on the continental margins explored in the 1960s using manned submersibles. A popular winter diving area for *Alvin* was the Tongue of the Ocean between New Providence and Andros Islands in the Bahamas. The vertical walls of this 2000-m trough consists of fossiliferous limestone providing geologists with the opportunity to look back into the early carbonate geology of this region.

In addition to using *Alvin* to study the natural history beneath the sea, in one case it was used to investigate human activity on the continental shelf during the Ice Age. Of particular interest was the work dealing with submerged shorelines by K.O. Emery, Robert McMaster, and Richard Edwards. The Ice Age led to the dramatic lowering of sea level on a worldwide basis: 15,000 years ago, off the East

Coast of the United States, it was between 70 and 130 meters below its present position.

As sea level rose with the melting of the continental ice sheets, a series of ancient shorelines was created and later flooded, forming relic features on the present continental shelf. During an *Alvin* dive in 1967, Edwards and Emery encountered a submerged beach and oyster reef formed 8,000 to 10,000 years ago off Chesapeake Bay. On the ridge top inland from the submerged beach they found oyster shells thought to be kitchen middens created by the early humans that must have inhabited this area at the time. Although little has been done since, the continental shelves of the world may prove to contain significant archaeological sites awaiting future discovery.

Just as the scientific community was gaining confidence in *Alvin's* ability to dive routinely and safely, disaster struck. At the end of its 1968 dive season, *Alvin* was in the process of being launched on dive 307 when suddenly its forward cables broke, dropping the submersible into the sea. Quick action saved the crew and passenger, but *Alvin* disappeared beneath the waves, falling 1,585 m to the ocean floor. There it remained until 1969 when a heroic salvage operation returned it to Woods Hole.

As engineers accessed the damage done during her 10 months underwater a startling discovery was made (Jannasch and Wissen, 1970).

Alvin broke surface again in September 1969 after resting almost one year on the ocean floor. In the excitement over her successful recovery, the oceanographers almost overlooked the striking outcome of *Alvin* degradation experiment: the food in the box lunch was practically untouched by decay, although containing the usual amount of bacteria.

The broth, although being the most perishable material, was perfectly palatable. Four of us are living proof of this fact. The apples exhibited a pickled appearance. But the way the salt water had penetrated into the fruit tissue indicated that the membrane functions were hardly affected. Enzymes were still active, and the acidity of the fruit juice was not different from that of a fresh apple. The bread and meat appeared almost fresh except for being soaked with seawater.

In conclusion, the food recovered from *Alvin* after ten months of exposure to deep-sea conditions exhibited a degree of preservation that, in the case of fruit, equaled that of careful storage, and in the case of starches and proteins appeared to surpass by far that of normal refrigeration.

The ocean floor as a giant refrigerator was an image that continued to be reinforced as the deep sea began to yield more and more of its preserved human history. The same year that *Alvin* was lost, she dove on a World War II Hellcat fighter plane that was ditched by its pilot in 1944. Resting in 1,524 m of water, it was in excellent condition.

In marked contrast to these images of a frozen deep ocean setting in which biological processes move at a snail's pace is the work by Dr. Ruth Turner of Harvard's Museum of Comparative Zoology. In 1972, Dr. Turner used *Alvin* to place a series of wooden panels into the bottom at a depth of 1,830 m. After 104 days of exposure, several were recovered (Turner, 1973):

> The wood was so weakened as a result of the activity of wood-boring bivalve mollusks, that it began to fall apart while being picked up by the mechanical arm of *Alvin*. The minute openings of their burrows covered the surface, averaging about 150 per square centimeter. . . .
>
> High population densities, high reproductive rates, early maturity, rapid growth, apparent ease of dispersal, and the ability to utilize a transient habitat make these wood borers classic examples of opportunistic species, the first recorded for the deep sea.

The site at which Dr. Turner carried out her initial wood borers experiments was known as DOS #1 for Deep Ocean Station number one, the first long-term bottom station established in the deep sea. Although others have subsequently been established, scientists continue to return to this site even today. The site was selected in 1971 as the first permanent bottom station because it lay along a line between Woods Hole and Bermuda, where benthic biologists had conducted deep-sea dredging operations for more than six years.

In fact, the first dive *Alvin* made for science took place in 1,785 m of water off Bermuda on July 17, 1966 with biologist Robert Hessler aboard. Hessler wanted to learn more about the benthic world he had been studying for years using deep-sea dredge hauls (Ballard, in press):

> I was awed by the tremendous vertical precepts, and I finally understood why we had so much difficulty ever taking any samples from that area. . . .
>
> That dive really taught me something. From then on whenever I lowered a dredge into the ocean, I could close my eyes and picture what the bottom of the deep sea looked like.

Hessler and benthic biologist Howard Sanders were impressed by the diversity of life in the deep ocean sediments. A student of Sanders, Fred Grassle, began to use *Alvin* to quantify these early observations. Returning to DOS #1, Grassle disturbed small patches of occupied seafloor with No. 2 fuel oil and fertilizer and left trays of sterilized, uninhabited mud to measure its colonization. Others put down small instrumented jars to measure respiration and found that deep-sea animals needed ten to a hundred times less oxygen than their shallow-water counterparts. Microbiologists, like Holger Jannasch, injected organic material into the seafloor as the field of benthic biology moved into a more quantitative phase in its history.

PROJECT FAMOUS—THE FIRST COMPREHENSIVE USE OF MANNED SUBMERSIBLES IN A MAJOR SCIENCE PROGRAM

In 1971, manned submersibles entered a new phase in their application that would eventually dominate their use.

Prior to this time, submersibles were used primarily by biologists and geologists in sedimentary settings ranging from soft mud bottoms to calcium carbonate terrains. But in 1971, *Alvin* began a comprehensive mapping program in the Gulf of Maine. Unlike most continental margin settings, which consist of thick sedimentary wedges, the Gulf of Maine is a seaward continuation of the Appalachian Mountain range. Instead of soft sediments, it consists of crystalline igneous and metamorphic rock dating back hundreds of millions of years.

To carefully map this area required the submersible to implement traditional field mapping techniques used by geologists on land. The creation of such maps requires the collection of geologic information in three dimensions, including not only surface exposure but subsurface structure and composition. First and foremost was the need for detailed bathymetric maps of a region measuring more than 100,000 km² at a contour interval of at least 20 m or better. Fortunately, extensive bathymetric data existed for this area.

Such bathymetric data provided a picture of the regional morphology but not its internal structure and composition. To provide this information required extensive surveying using seismic profiling techniques. The database used in this study included tens of thousands of kilometers of seismic survey lines collected over a period of more than eight years prior to and during the actual diving program. This coverage led to the creation of a three-dimensional picture of not only the regional bedrock geology but also the sedimentary basins contained within it and equally important where the bedrock geology was exposed as outcrops.

Once such outcrops were pinpointed samples had to be collected to determine the bedrock composition. Unfortunately, the entire Gulf of Maine had undergone extensive glaciation during the Ice Age and the retreating glaciers covered the area with glacial deposits of varying thickness, deposits that bear no relationship to the underlying rock formations. As a result, traditional dredging operations could not be carried out since they almost always resulted in the collection of glacial erratics instead of the more difficult to sample bedrock outcrops. Here, careful coordination was required between surface seismic profiling activities needed to pinpoint bedrock outcrops and subsequent dives by *Alvin* to sample them. This mapping program in the Gulf of Maine set the stage for a new phase in the use of *Alvin*.

Although research programs carried out by scientists around the world began in the 1950s and continue to this day using manned submersibles, the 1970s marked a fundamental change in their use. This shift in focus came as technological improvements in deep submergence engineering made it possible for manned submersibles to go much deeper than before. Principal among these improvements was the fabrication of higher-strength steel and titanium pressure spheres. Two vehicle programs led the way, one in the United States and the other in France. America's *Alvin* was modified to carry a titanium pressure hull with an initial operating depth of 3,050 m, while France's CNEXO (Centre

National pour L'Exploitation de Oceans) brought into service the new deep submersible *Cyana* with a similar depth capability.

Associated with this shift in emphasis came an entirely different group of scientists who began pondering the very origin of the ocean floor and the role it played in global geology. In the late 1960s, geophysicists began a revolution in the Earth sciences by advancing a new theory to explain the observed structure of the ocean floor. In so doing, they began to explain the position of the continents rising out of our global sea.

Years earlier, a German meteorologist named Alfred Wegner had advanced what geophysicists at the time regarded as a poorly supported theory. He called it "continental drift" and drew his supporting evidence from the continental land masses. Fitting Africa and South America, he went on to compare their similar interlocking geological features. But unable to explain how continents could actually drift apart, he died in ridicule, and continental drift entered modern geology textbooks not as a unifying theory but as one held up to scorn. Beginning in the 1950s, however, this theory underwent a rebirth as geophysicists began to probe the ocean depths in earnest. Plunging heat-probes into the ocean floor sediments far from land, they were surprised to observe unusually high readings.

Although they were well aware of a central ridge in the middle of the Atlantic Ocean, which the Germans had mapped prior to World War II, it wasn't until after the war that they realized it was seismically active along its entire length. Drs. Bruce Heezen and Maurice Ewing of Columbia University's Lamont Geological Laboratory used the global distribution of earthquakes to propose the existence of a continuous range running through the major ocean basins of the world, which they termed simply the "Mid-Ocean Ridge." Reconnaissance dredging operations followed, which recovered basaltic rock samples along its summit indicating the presence of active volcanism. Various explanations were advanced to explain these observations but the real breakthrough came when Drs. Vine and Matthews published magnetic maps of a segment of the ridge, the Carlsberg Ridge. There for all to see was a systematic series of parallel strips of ocean floor having alternating magnetic polarities. More importantly, they were symmetrical or mirror images of one another, with the line of symmetry being the very center line or axis of the Mid-Ocean Ridge. From these observations came the theory of seafloor spreading later modified to be plate tectonics and Earth scientists "had a new game of chess to play." (P. Hurley, pers. comm.) And play they did.

The formal body of scientists grappling with this revolutionary theory had its roots in the International Council of Scientific Unions. This same body had initiated the International Geophysical Year in 1957 to 1958. By the late 1960s, this interest in plate tectonics was incorporated in the Geodynamics Project, "an international program of research on the dynamics and dynamic history of the earth with emphasis on deep-seated foundations of geological phenomena. This includes investigations related to movements and deformations, past and present, of the lithosphere, and all relevant properties of the earth's interior and especially any evidence for motions at depth." (Ballard, in press).

In late 1971, Dr. K.O. Emery received a letter from Dr. Xavier Le Pichon, a student of Dr. Maurice Ewing and a strong supporter of plate tectonics. Le Pichon had been urged to write to Dr. Emery by Dr. Charles "Chuck" Drake, who like Dr. Le Pichon, was a graduate of Lamont and Chairman of the Geodynamics Project. Dr. Drake had made an earlier dive in the Puerto Rico Trench aboard the French bathyscaph *Archimede* and saw its potential as a geological mapping tool.

Dr. Le Pichon, now in charge of a major new marine laboratory in Brest, France, the Centre Oceanologique de Bretagne (COB), wanted to use the *Archimede* and the new French submersible *Cyana* to investigate the Mid-Ocean Ridge. But he wanted it to be a joint program between France and the United States. He was keenly aware of Woods Hole's submersible *Alvin* and knew it was already proving itself as an emerging geological mapping tool. In his letter, Dr. Le Pichon briefly explained what he had in mind—a detailed mapping effort in the Mid-Atlantic Ridge Rift Valley—and asked Dr. Emery if he thought submersibles were up to the challenge and if Emery was interested in joining the program.

In the final draft of the letter sent to Dr. Le Pichon, Dr. Emery strongly supported such a program but declined to participate. Emery was a continental geologist comfortable with its submerged seaward limits, but he was not a "hard rock" geologist who wanted to venture into the Mid-Ocean Ridge. The next logical person for Le Pichon to turn to was Dr. James Heirtzler, chairman of Woods Hole's Department of Geology and Geophysics. Heirtzler was also a Lamont graduate, and although he was not a field geologist, he was a strong advocate of plate tectonics and a pioneer in the field of marine magnetics.

But before such a major program could receive the funding it required, a considerable amount of support within the Earth sciences community was required. France's new CNEXO to which Dr. Le Pichon's laboratory reported could make decisions without significant outside review or approval. This was not the case in the American system and, in particular, the National Science Foundation (NSF). Clearly, NSF was the obvious source of funding for what would eventually be called Project FAMOUS.

Working under the broad umbrella of the Geodynamics Project, Dr. Heirtzler worked with Drs. Drake, A.G. Fisher, Frank Press, and M. Talwani to organize the Mid-Atlantic Ridge Workshop with the official endorsement of the Ocean Science Committee of the National Academy of Sciences. If the program Le Pichon and Heirtzler had in mind was endorsed by the Academy, it was a strong candidate for NSF funding.

The meeting that took place during the week of January 24, 1972, became known as the Princeton Workshop after the Ivy League school where it was held. NSF's International Decade of Ocean Exploration (IDOE) Office provided the financial support for the meeting, which was an obvious good omen for future funding. More than 40 scientific leaders of the Earth sciences community from nations around the world including the United States, France, the United Kingdom, Canada, and the Netherlands attended.

The final report resulting from this workshop entitled *Understanding the Mid-Atlantic Ridge—A Comprehensive Program*, (NRC, 1972) after numerous formal presentations, much debate, and late night dinners, contained five "high-priority field projects" including the following recommendation:

> Interdisciplinary surface ship surveying and sampling on a small scale over critical areas on the Mid-Atlantic Ridge should be followed up in the most critical subareas by more detailed geological and geophysical investigations, using the capabilities of deep-towed vehicles and submersibles.

The report containing this recommendation also contains the following notice:

> The study reported herein was undertaken under the aegis of the National Research Council with the express approval of its Governing Board. Such approval indicated that the Board considered that the problem is of national significance, that elucidation or solution of the problem required scientific or technical competence, and that the resources of the NRC were particularly suitable to the conduct of the project.

With this official endorsement for a comprehensive American Mid-Atlantic Ridge program, including the use of manned submersibles, Dr. Heirtzler and others could now move forward in formalizing a major joint program with the French, which would become Project FAMOUS.

Although Project FAMOUS would be best known for its first use of a manned submersible in the Mid-Ocean Ridge, it did, in fact, involve every major technological tool then being used by marine geophysicists and would set the example for subsequent Mid-Ocean Ridge investigations. The area selected for this intense investigation of the Mid-Atlantic Ridge was between 36 and 37° N latitude for a long list of reasons. We wanted to be in a region of favorable weather and near a good logistical support base, which in this case was Ponta Delgada in the Azores, but far enough away from its "hotspot activity" to ensure that we were investigating "a typical spreading segment" of the ridge.

The chief scientists of Project FAMOUS were, to no one's surprise, Drs. Jim Heirtzler and Xavier Le Pichon. Prior to the actual joint diving operations, which took place in the summer of 1974, they published the goals of the project in an issue of *Geology* (Heirtzler and Le Pichon, 1974):

> Among questions that we hope to answer are the following:
> What is the detailed age distribution of the surface rocks in

this zone? What is the relative importance of primary constructional features and secondary tectonic ones in shaping the morphology? Is the structure of the boundary zone steady state? And if not, to what extent? What is the distribution of the different types of igneous rocks with respect to the tectonic and volcanic features within the zone? Are large portions of the oceanic crust exposed along the obviously faulted scarps of the Rift Valley? What is the thickness and exact nature of the layer at the origin of the Vine and Matthews magnetic lineations? Are there metamorphic rocks (zeolites, greenschist, or even amphibolites facies) within the zone? What is their distribution with respect to the different tectonic features? Many other questions can be asked concerning the tectonics of the transform-fault area. One of the most important is how localized is the zone of shearing? Is there really a transform fault or a zone of transform faulting? What is the distribution of the ultrabasic rocks that occur within this zone? Is there volcanic activity within the transform fault? And so forth.

To address this long list of questions using traditional field mapping techniques, first and foremost required excellent topographic maps. This was particularly true for the Mid-Ocean Ridge given its complex morphology, but also because its morphology was a direct reflection of the volcanic and tectonic processes taking place within its rifted inner valley.

The first dives into the rift valley in 1973 used the French bathyscaph *Archimede*. In all, seven dives were made by the *Archimede* covering a 5-km^2 area of the central high and the adjacent eastern marginal high. The central high was found primarily to be a constructional volcanic feature not significantly altered by subsequent tectonics. The central zone of extrusive lava flows was found to be bounded to the east, in the area the *Archimede* investigated, by steep vertical scarps up to 100 m in height thought to be volcanic flow fronts.

These preliminary dives clearly revealed that despite the tremendous lateral dimensions of the North American and African crustal plates, the actual zone of injection that includes surface lava flows is extremely narrow and ideally suited to submersible investigation. Had the boundary separating the spreading plates been broad, as reason might have led one to believe, the investigation of such a region by a manned submersible might have been an utter failure.

With these initial promising results, Project FAMOUS moved into its final phase in the summer of 1974 with the coordinated diving programs of the submersibles *Alvin* and *Cyana* and the bathyscaph *Archimede* and continued surface ship studies.

Never before had three deep diving submersibles carried out such a coordinated effort. Having more than one vehicle diving in the area added to the overall safety of the operations as described later, but it also had its drawbacks.

A critical aspect of the FAMOUS dives in the rift valley was a precise knowledge of where the vehicles were at any one time when observations were made, photographs taken, or important rock samples collected. For this reason, each vehicle had its own network of bottom-moored acoustic transponders. At the end of each dive, we were able to produce an edited plot of the submersible's *x-y* track across the rift valley floor. Adding the depth and altitude of *Alvin* along this track, we were able also to produce a bottom depth profile for the dive. Using these two plots and a transcription of the science divers' observations, we were able to produce a series of geological traverses across the rift valley floor. These annotated profiles included a wide range of observations dealing with the various volcanic and tectonic features we observed as well as the sediment cover, which reflected the age of the terrain.

In all, the American submersible *Alvin* conducted 17 dives, while the combined efforts of the French submersible *Cyana* and the bathyscaph *Archimede* completed 27 dives. Each vehicle was assigned to a particular operational area within the inner rift valley and the bounding transform faults. *Alvin*'s work area included a central volcanic high called Mount Pluto and the southern portion of Mount Venus to the north. The *Archimede* overlapped *Alvin*'s coverage of Mount Venus, working north up the rift valley toward transform fault A, which was the primary operational area for the submersible *Cyana*.

When the expedition ended and the final results were published in two volumes of the Geological Society of America Bulletin (1977, 1978), our detailed knowledge of the process of seafloor spreading had taken one giant leap forward illustrated by the following text that appeared in *Science* in 1975 (Ballard et al., 1975):

> Observations confirmed that Mt. Venus and Mt. Pluto are the sites of most recent volcanic activity. The flanks of these hills consist of broad, steep-fronted flow lobes with relatively little sediment cover or attached organisms. The flow fronts consist of tubular lava extrusions elongated downslope, resembling in some respects terrestrial pahoehoe lava.

> . . . in all traverses from the center of the valley outward to the flanks, we were impressed by the rapid increase in sediment cover and bottom life and by the intense tectonic degradation to which the extrusive lava forms were subjected. Generally, within 300 m of the valley center to the west and within 500 m to the east, most of the delicate extrusive forms had been destroyed, the flows were sliced and offset by numerous faults, and the surfaces were reduced to broken, jumbled lava blocks and extensive talus fans at the base of fault scarps.

> In contrast to recent volcanic activity, which appears to be concentrated in a narrow central zone, recent tectonic movement is evident throughout the entire width of the inner rift valley floor. Faults and fissures are numerous, striking 020 degrees parallel to the rift axis.

> Intrusive sills and dikes are exposed only at the base of one 300-m scarp on the west wall. Most fault displacements are

less than 100 m and expose only breccia, truncated lava pillows and tubes.

> In general, faulting appears to be a continuing process, while volcanic activity is episodic.

Simple and logical as these observations may seem, they confirmed the process of seafloor spreading, providing the first systematic documentation of a process that had global significance. Manned submersibles had finally come of age.

On the way back from the FAMOUS research site in the Mid-Atlantic Ridge, the *Alvin* was used to carry out a series of dives along the New England Seamount Chain, revealing ancient volcanic terrain covered by a thick layer of manganese and phosphorite similar to that encountered on the Blake Plateau.

Although Project FAMOUS was capturing the headlines in the early 1970s, scientists continued to use manned submersibles for their more traditional applications on the continental margins. The benthic biology community continued its studies of wood-boring organisms as well as efforts to quantify the biomass within deep-sea sediments and their rate of recolonization. Spurred on by the sandwich recovered from inside the lost *Alvin*, scientists expanded their research to include the decomposition of solid organic materials in the deep sea, with an eye toward the implications of using the ocean as a future dump site.

Geologists also continued using manned submersibles to study the carbon stratigraphy of the Bahama Platform, including its potential for hydrocarbon deposits and the occurrence of "lithotherms," deep-water coral structures that trap bottom transported sediments forming long linear ridges beneath the Gulf Stream in the Straits of Florida.

But clearly, Project FAMOUS had ushered in an entirely new phase of scientific use of manned submersibles, in particular, *Alvin*. Several factors were responsible. The first was the increased diving depth of *Alvin* from 1,800 to 3,050 m. The second was the integration of the manned submersible into a larger context, namely the lengthy preparation of a research site prior to the actual diving program. This preparation included the collection of detailed bathymetric maps and geologic traverses across the proposed study area using deep-towed vehicles such as *Deep Tow* and *Angus*. Most importantly, however, was the emergence of plate tectonics. In the final analysis, manned submersibles were in the right place at the right time.

THE DISCOVERY OF HYDROTHERMAL VENTS AND THE ACCEPTANCE OF *ALVIN* BY THE SCIENTIFIC ESTATE

For most of the scientists participating in Project FAMOUS, it was an unqualified success. But for one group, it was a bitter disappointment. Dr. Dick Holland had led a team from Harvard and Woods Hole that was keenly interested in finding underwater hot springs along the axis of the rift valley but failed to do so. Years before, Dr. Clive Lister of the University of Washington observed a deficit of conductive heat release near the Mid-Ocean Ridge, which he argued supported the existence of hot springs within the axis. Scientists speculated that the Mid-Ocean Ridge owes its vertical relief to the fact that it is swollen with heat energy—that the ridge, unlike mountain ranges on land, is in essence a blister on the surface of the Earth.

Since new oceanic crust is being generated along the axis of the Mid-Ocean Ridge, it is by definition the youngest in age and for this reason should be the hottest. As the process of seafloor spreading continues with the injection of new crustal material along the ridge axis, older oceanic crust is pushed to the side, forming two giant diverging geologic conveyor belts carrying crust away from its site of creation. As it is transported away from the ridge axis, the crust slowly cools, and cooling causes the crust to contract. In essence, scientists were saying that the ridge's vertical profile represents a theoretical cooling curve.

If this hypothesis was correct, it should be possible to correlate the amount of heat coming out of the ocean floor with the distance from the ridge axis at which the measurement is made. The farther away from the ridge, the lower the heat probe reading should be. Well, this is exactly what scientists found, except, as Lister pointed out, along the axis itself. Although heat probe measurements made near the axis were high, they weren't as high as they theoretically should have been. A significant amount of heat was missing. What process was taking place along the axis of the ridge that was removing this otherwise uniformly released heat energy?

The only logical answer was hot springs. We all knew that the ridge must be underlain by magma chambers at a relatively shallow depth of 1 to 2 km. We also knew that these magma chambers contain molten rock at a temperature of 1,200 to 1,400°C. During Project FAMOUS, we discovered that the central volcanic terrain was fractured by numerous fissures and faults, which made it very permeable. Clearly, cold bottom waters within the rift valley at a temperature of 3 to 4°C could easily enter the ocean floor and must penetrate to the hot rock region surrounding the magma chambers below.

Once heated and thermally expanded, these highly enriched geothermal fluids should rise back to the surface of the rift valley floor, exiting as hot springs along its axis. But Holland's team had been unable to detect any temperature anomalies within the FAMOUS study area. Either hot springs didn't exist there at the time of the study or they didn't exist at all.

However, a growing group of marine investigators was warming to Lister's theoretical argument favoring the existence of hot springs along the axis of the Mid-Ocean Ridge. Paralleling Lister's geophysical line of reasoning was one emerging from the field of geochemistry. In 1965, Scripps graduate student Jack Corliss was completing his thesis work based upon the analysis of basaltic rock samples dredged from the Mid-Atlantic Ridge. This analysis clearly suggested that seawater was seeping downward into the newly formed ocean floor, penetrating the hot rock surrounding the magma chamber, leaching out various chemicals to form hydrothermal fluids that then flowed back to the surface of the ocean floor. The driving force of this internal circulation system was the buoyancy of the heated fluids and the tremendous geothermal gradient separating the shallow magma chamber from the cold bottom waters within the rift.

In 1975, the year after Project FAMOUS, two scientists following these two different lines of reasoning joined forces to propose an expedition to the Mid-Ocean Ridge using manned submersibles that would put these theories to test. They were Dr. Richard von Herzen formerly of the Scripps Institution of Oceanography and now at Woods Hole and Dr. Jerry van Andel also formerly of Scripps and now at Oregon State University. Van Andel had been Corliss' thesis adviser at Scripps and was well aware of his line of argument suggesting the existence of hot springs along the axis of the ridge. Von Herzen's specialty was heat flow, and he fully understood Lister's line of reasoning. Being at Woods Hole, von Herzen was keenly aware of *Alvin*'s recent successes during Project FAMOUS. More importantly, van Andel had been one of the principal diving scientists during FAMOUS and knew first hand that *Alvin* was up to the challenge.

The result of this collaboration was a proposal to NSF, which had sponsored the FAMOUS Project, to search for hot springs not in the Atlantic Ocean but in the Pacific along a segment of the Mid-Ocean Ridge called the Galapagos Rift. There were several reasons for picking this site. To begin with, Oregon State University already had a large program in the Pacific called the Nazca Plate Project funded by NSF. Second, the spreading centers in the Pacific were much faster than the spreading center of the Mid-Atlantic Ridge. The faster the spreading rate, the more heat energy was being released along the ridge axis and the greater was the probability of finding hot springs.

During the subsequent cruise in the summer of 1976, a variety of instruments were used to investigate the inner rift valley of the Galapagos Rift, including sediment traps, water chemistry samplers, and the *Deep Tow* system from Scripps, which has a side-scan sonar and bottom camera and lighting unit. To everyone's satisfaction, the expedition succeeded in detecting temperature anomalies within the near bottom waters of the rift, which were marked by a long-term acoustic transponder.

The stage was now set for the final phase of the program, a dive series by *Alvin* to pinpoint the suspected hot springs. This was scheduled to take place during the winter of 1977. Leadership for this effort was transferred from van Andel to Corliss when van Andel accepted an appointment at Stanford University. But concern over Corliss' lack of diving experience led van Andel to ask me to take Corliss on our Cayman Trough expedition, which was about to take place during the winter of 1976.

The Cayman Trough program had two primary purposes. The first was to use the submersible *Alvin*, the bathyscaph *Trieste II*, and the towed camera system *Angus* to investigate a small spreading center situated inside the transform fault system separating the North American Plate from the Caribbean Plate. Known as a "leaky transform fault," this boundary had a slight opening motion that led to the formation of an east-to-west spreading center bordered to the north and south by the steep walls of the Cayman Trough. The result was one of deepest spreading centers in the world, with the central volcanic axis occurring at a depth of 6,100 meters. The walls of the fault scarps within the trough also provided an opportunity for petrologists to obtain samples of the oceanic crust using the submersible *Alvin*.

The second reason for the Cayman Trough program was to maintain the momentum created by the success of the FAMOUS Project, while others like van Andel had an opportunity to define diving programs for funding given the scientific community's new positive attitude about submersibles.

The Cayman Trough investigation provided Corliss with an excellent opportunity to learn first-hand how to conduct a sophisticated research program using manned submersibles. It also provided him with the opportunity to make his first submersible dive.

By the time the Galapagos Hydrothermal Expedition got underway in February 1977, I had been asked to be co-chief scientist of the expedition with Dick von Herzen—not because I was a major scientific leader for this research program but because of my experience at conducting submersible programs. The real scientists behind the program were Jack Corliss and Jack Dymond from Oregon State University and John Edmond from the Massachusetts Institute of Technology. Jerry van Andel was also on the expedition, primarily to help these inexperienced scientists with the actual diving program to be carried out aboard *Alvin*'s support ship *Lulu*.

The expedition's destination was a point 640 kilometers west of the coast of Ecuador, along the rift that separates the fast-spreading Cocos and Nazca plates in the Pacific Ocean. Our plan was to concentrate on the sites where seafloor temperature anomalies recorded by the earlier Scripps-Oregon State-Woods Hole expedition had suggested the existence of hydrothermal vents.

Woods Hole's research vessel *Knorr*, with *Lulu* under tow, began the expedition at Rodman Naval Base in the Panama Canal, but after several days at sea, it was decided to break the tow and let the two ships proceed under separate power to the dive site. This way, the faster *Knorr* could arrive ahead of *Lulu*, install a network of acoustic transponders within the rift valley, and conduct some preliminary reconnaissance runs with the towed camera system *Angus*.

The year before, I had been successful in convincing the

U.S. Navy to conduct a detailed Sound Acoustic Surveillance System (SASS) sonar mapping effort of the Galapagos Rift dive area similar to the survey it had conducted in the FAMOUS area. The FAMOUS expedition and the 1976 program in the Cayman Trough had set a new standard for bathymetric detail that all future submersible diving programs would now seek to emulate.

Once the ship arrived in the area, *Knorr*'s echo-sounder was used to collect a series of profiles perpendicular to the rift axis. Using this information, the buoy left the previous year by Corliss, and satellite navigation, we did our best to tie our present location to the estimated location of the thermal anomalies detected the year before.

Woods Hole's *Angus* camera sled was now lowered into the rift from the research vessel *Knorr*. *Angus* was equipped not only to take thousands of color pictures but also to register temperature changes as minute as one five-hundredth of a degree Celsius. *Angus*'s sensitive thermistor at first recorded no variations in the near-freezing temperatures just meters above the ocean floor. Then, as the first day's run neared its halfway point in the early evening of February 15, recorders on board *Knorr* received an acoustically telemetered signal from *Angus*, revealing a sudden spike in water temperature, lasting less than three minutes. Since the time and temperature data were precisely keyed to the frames of film exposed by *Angus*'s cameras as it canvassed the bottom, we were able to review the pictures taken at the exact moment of the temperature spike. But first, *Angus* had to be hauled back to the surface and the film developed.

All were eager for the first visual evidence of the hypothesized thermal vents—but nothing could have prepared us for what *Angus* had photographed, one and a half miles beneath the surface. The 122-m-long roll of color film revealed a bed of clams—hundreds of clams clustered in a small area on the lava floor of the rift—thriving as if they were in an environment no more hostile than a sunny mudflat on the New England coast. We couldn't help but wonder what these large clams were doing in such numbers at that depth, in that eternal darkness.

The next step in the research plan called for the deployment of *Alvin* to whatever promising sites *Angus* might reveal. It was February 16 when *Lulu* with *Alvin* aboard arrived in the dive area, and we lost no time in getting the submersible into the water at sunrise the next day, February 17.

After a descent lasting an hour and a half, pilot Jack Donnelly brought Jack Corliss and Jerry van Andel to a point less than 275 m from the clam beds and began the drive along the lava floor to the site. Along the way, the bottom appeared as might have been expected: fresh but relatively barren lava flows.

But when *Alvin* reached its goal, the scene the scientists observed through the viewports was remarkably different. Water that *Alvin*'s sensors measured at 12°C, shimmered up from cracks in the lava flows and turned a cloudy blue as manganese and other minerals, which had been carried from deep within the vents, precipitated in the cooler surroundings. Clams, giant specimens, measuring a foot or more in length, along with similarly outsized brown mussels, appeared to be bathed by the simmering water. *Alvin*'s robotic arm, which had been expected to grasp only rock samples from the bottom of the Galapagos Rift, now was pressed into service to grasp samples from this most remarkable community of shellfish. When we planned this cruise, our thoughts had been so far from biology that we had brought no preserving medium along. Some of these samples thus made the trip back to shore immersed in vodka.

Over the coming days, the expedition's researchers took turns scouring the rift for similar signs of life. With guidance from *Angus*, these rovings in *Alvin* bore rich rewards. We eventually identified five sites that teemed, or had recently teemed, with creatures as bizarre as they had been unexpected. We termed our initial find "Clambake I" and also located a site we called "Clambake II," where a change in conditions had killed off the big bivalves and left only a midden of shells. The "Oyster Bed" was our label for a patch of mussels, misidentified as oysters—our flawed attempts at taxonomy went temporarily unchallenged, since there were no biologists along on the expedition—and another site was dubbed the "Dandelion Patch," because it was home to a population of hitherto unknown animals resembling bright yellow dandelions, attached to the bottom not by stalks but by delicate fibers. Finally, there was the "Garden of Eden," lushest and most varied of these strange oases. Here were the dandelions, along with white crabs, limpets, small pink fish, and clusters of vivid red worms that protruded from their own long, stalk-like white tubes. The tube of a specimen later brought to the surface measured more than two meters in length, with the animal itself filling more than half of its elongated tube.

The obvious question in everyone's mind was what enabled these colonies of creatures to flourish at such depths, in an atmosphere totally devoid of sunlight? The answer quickly came, on board *Knorr*, with the analysis of water samples taken by *Alvin* at the vents surrounded by the oases. The first thing noticed, when the sample containers collected by *Alvin* were opened, was a pervading odor of rotten eggs: hydrogen sulfide. This was the clue that enabled us to piece together the chemical and biological processes that made possible the huge clams, tube worms, and other life forms in such high concentrations.

The earlier suggestions of Lister had been shown to be true. The deep fissures in the floor of the rift allowed cold seawater to penetrate Earth's crust, down to the level of hot, newly formed layers of rock surrounding the magma chamber. The temperature of the water rose as it flowed deeper, and its chemical composition changed. The seawater exchanged some of its chemicals with the subsurface rock and leached out others. The sulfate in the water was changed to hydrogen sulfide—hence the telltale smell in the lab. Finally, the heated water rose back to the seafloor through other fissures in the crust and raised the ambient temperature of the vent oases to the surprising levels recorded by *Alvin*.

Living inside the macrofauna of clams and tube worms were hydrogen sulfide-oxidizing bacteria that formed the basis of this unique food chain. In the case of the clams, the available nutrients were abundant enough to lead to gigantism. Clambake II, in the light of this analysis, appeared to have been an oasis chilled and starved into extinction as the recycling of seawater through the vents had ceased for some unknown reason.

We had discovered something new upon Earth. Prior to our investigation of the hydrothermal vents along the axis of the Galapagos Rift, all forms of life had been assumed to be dependent upon photosynthesis, the process by which sunlight is metabolized to sustain the growth of plants and animals. Even the holothurian, living at great depths in a sunless world, depends for its survival upon organic material that drifts down from the sunlit surface. But within the vent field, for the first time, was evidence of a community of animals subsisting on a process of chemosynthesis, beginning with the metabolizing of hydrogen sulfide by microorganisms. They were, after all, creatures that needed no sunlight at all for survival and that owed their existence to the warmth and chemical sustenance of Earth itself.

Two years later in 1979, marine biologist Fred Grassle and I co-led a second expedition to the undersea oases of the Galapagos Rift. Fourteen other biologists accompanied us—this time, there would be no relying on vodka for preserving specimens. We also brought a film crew from the National Geographic Society, which chronicled our discoveries in the television special "Dive to the Edge of Creation." This time, the challenge we faced was quite different from our 1977 task. Then, we were looking for hydrothermal vents and had no idea of the oases. Now, we were trying to locate the same sites we had visited before, in a place where there were no identifying landmarks either above or beneath the surface.

As before, we deployed *Angus* as our eyes and temperature sensor prior to a manned investigation in *Alvin*. Reviewing the thousands of frames exposed by *Angus*'s cameras on the sled's first run along the rift floor, we began to resign ourselves to a long search. Then, with about four frames to go, we found what we were looking for. *Angus* had photographed a clutch of our mysterious dandelions, and we knew we were in the right spot.

Taking our turns in *Alvin*, we explored a string of new vents and their surrounding oases, including the largest discovered on either of the two expeditions—an otherworldly habitat for tube worms 2 to 3 m long. And with our complement of biologists and biochemists, we were able to achieve a far more sophisticated understanding of the processes involved in sustaining the creatures of the oases and to make an attempt at classifying them.

Beyond a doubt, it was the chemosynthesized nutrients that made the oases possible. The warmth of the water itself was not a primary factor; there are animals that survive the near-freezing temperatures even at the deepest reaches of the sea, fed by organic material drifting down from the surface. Here, though, the secret of abundant life was a cornucopia of locally derived nutrients. The concentration of food at the oases far surpasses the amounts available elsewhere on the sea bottom. One of our colleagues estimated that the waters surrounding the vents contain 300 to 500 times the nutrients found at nearby sites lacking the benefit of the mineral-rich flow from the vents.

The driving force behind the unique vent communities is the rapid growth of the chemoautotrophic bacteria that are able to use the dissolved oxygen and carbon dioxide present in oceanic bottom waters to oxidize the reduced inorganic compounds (i.e., H_2S, S, S_2O_3, NH_4, and NO_2) dissolved in the hydrothermal fluids coming out of the vent openings. This chemosynthesis process has been known by microbiologists for many years, but it wasn't until the discovery of the Galapagos Rift hydrothermal vents in 1977 that scientists realized it could form the basis of an entire ecosystem.

Although this process takes place in total darkness, it is still tied to the sunlit surface. For the chemosynthetic process to occur, the bacteria require free oxygen to oxidize the reduced inorganic compounds coming out of the vents. This free oxygen has been generated by green plants as a by-product of the photosynthetic process. An interesting question is: What would happen if the sun suddenly turned off? Clearly, the vent communities would continue to thrive until the free oxygen in seawater was exhausted. But even after that point in time, anaerobic chemosynthesis would persist.

There are many forms of bacteria involved in the chemosynthetic process, which occurs in three basic settings: (1) within the subterranean vent system cutting deep into the volcanic terrain, (2) in large microbial mats covering its surface, and (3) within the internal structure of various symbiotic organisms living around the vent openings. The benthic animals that make up the vent communities have a fascinating strategy for survival. We now know that hydrothermal vents are highly ephemeral or short-lived. They turn on and off in a matter of a few years or tens of years. As a result, vent animals have an "r-type" survival strategy. They are able to settle quickly out of the water when a vent turns on, grow fast, reproduce early, and easily disperse their offspring into the water column to find new vent settings.

The vent communities discovered in 1977 by chemists and geologists and revisited in 1979 by biologists are characterized by large organisms situated in diffuse zonations centered around discrete vent openings where the temperature is the hottest. In the case of the Galapagos vents, the maximum exiting temperature measured was 17°C and the dominant macroorganism living near the vent opening is the giant red tube worm *Riftia pachyptila*. These spectacular organisms form large clusters or hedges standing 2 to 3 m in height. One of the large populations was termed the "Rose Garden." Without eyes, mouth, or digestive tract, the worm's red tip or obturaculum absorbs food and oxygen

from the water by means of hundreds of thousands of tiny tentacles arranged on flaps on the exposed portions of its body. These are the critical compounds needed by the bacteria living inside its body for the chemosynthetic process. Since these ingredients come from both the anaerobic vent fluids (i.e., hydrogen sulfide) and the ambient bottom water (i.e., oxygen and carbon dioxide), the worms position their red tips in the area of mixing just above the vent opening, clustering in thickets to direct the vent fluids up past the tip of the tubes. Sexually differentiated, they most probably broadcast eggs and sperm into the water.

Living directly inside the vent opening itself are a variety of limpets (i.e., *Archaeogastropoda*) that are also observed living on the white base of the tube worms. Living in close proximity to the vent openings in some cases are large beds of mussels (*Mytilidae*) attached to the volcanic substrate, as well as other organisms like the tube worms, by strong byssal threads.

In our investigation of the Galapagos Rift vents communities in 1977, the organisms that I found as impressive as the red tube worms were the giant white clams (*Calyptogena magnifica*) that covered the fresh lava flows. We commonly saw these clams wedged down inside a small fissure cutting across the volcanic terrain, parallel to the rift valley axis. Their anterior end pointed down and their hinge point up, an ideal feeding position with the hydrothermal fluids flowing up past them. The clams of the rift were noteworthy not only for their gargantuan size, but also for the intense blood-red color of their flesh—as with the tube worms, this coloration is due to a high amount of hemoglobin, the pigment of human blood. In numerous cases, you could see that as the clams grew, their enlarging shells conformed to the jagged outline of the fissure opening, wedging them in place.

This vent species was also a critical indicator of past vent activity. Unlike most vent organisms that quickly vanished after a vent turned off, the large white clam shells persisted for many years before finally being dissolved by the ambient bottom water, which is undersaturated by calcium carbonate. In fact, an inactive vent characterized by a cluster of dissolving clam shells was first seen in a deep-tow survey along the Galapagos Rift in 1976 but was not recognized for its importance until after an active vent was found by *Angus* and investigated by *Alvin* in 1977.

Other important organisms living in and around the Galapagos vents are a variety of anemones (*Actinarians*), brachyuran crabs (*Bythograea thermydron*), galatheid crabs (*Munidopsis*), jellyfish called "Dandelions" (i.e., *rhodaliid siphonophores*), and an highly unusual worm (*Enteropneust*) clustered in what resembled piles of "spaghetti." The blind white crabs that frequent the oases and feed upon dead mussels and clams are apparently members of a heretofore unknown crustacean family.

What about the so-called dandelions? Animals despite their plantlike appearance, these turned out to be a new siphonophore, related to the Portuguese man-of-war but spending their lives attached by their threadlike filaments to the rock formations of the bottom. Each of the creature's "petals," dissection showed, has a different purpose. Some capture microorganisms; others digest them; and still others are involved in reproduction. All surround a buoyant pocket of gas, which allows the animal to bob at the end of its tethers.

As our time on the rift went on, we collected new species of leeches, worms, barnacles, and whelks. We even took away some 200 strains of bacteria, which were brought alive to Woods Hole for whatever clues they might offer to the basis of this remarkable food chain. Throughout our observations—whether they involved the humblest microorganisms or the most extravagantly sized and colored worms and bivalves—we were tantalized by the thought that surely such phenomena could not have been confined by evolution only to this obscure stretch of the Galapagos Rift. At how many other places on the bottom of the oceans do such communities thrive, and how many other yet-unknown species draw life from the interplay of seawater with the steaming, mineral-rich depths of Earth's developing crust?

Before the 1979 return trip to the Galapagos Rift took place, plans were already underway for a major expedition to the East Pacific Rise by many of the same French and American scientists who participated in Project FAMOUS. After Project FAMOUS was completed, the French were eager to conduct another large joint program with the United States on the Mid-Ocean Ridge. Since Project FAMOUS was conducted on a slow-spreading segment of the ridge where the plates are moving apart at a rate of 2.5 cm per year, the French wanted to compare what they had learned about the volcanic and tectonic processes of the Mid-Atlantic Ridge with a faster-spreading ridge in the Pacific Ocean.

Spearheaded by Dr. Jean Francheteau, the French chose the East Pacific Rise (EPR), where the plates separate at a range of 6 to 12 cm per year. Based on a series of studies of the rise conducted by U.S. oceanographers, the French selected a segment of the EPR at 21°N latitude, at a spot off the Mexican coast where the Pacific and North American plates diverge. As the program took shape, the French asked a number of U.S. scientists, including myself, if we would be interested in such a joint investigation.

While it had been fitting for Woods Hole to play the lead role in Project FAMOUS since the program was conducted in the Atlantic Ocean and involved the use of its submersible *Alvin*, Woods Hole was not the logical choice for the East Pacific Rise program. The Pacific Ocean was the territory of the Scripps Institution of Oceanography. And since scientists at Scripps had carried out most of the research on the East Pacific Rise on which the French were basing their study, it was decided at a workshop held in La Jolla, California, that Scripps would be the lead U.S. institution for this joint program. Dr. Fred Spiess of Scripps would play the role Jim Heirtzler had played during Project FAMOUS.

As with FAMOUS, the French wanted to carry out the first series of dives of the East Pacific Rise using their submersible *Cyana* and I was invited to participate. This initial dive series was scheduled for February 1978. The French named their phase of the program RITA, for the two transform faults (*Ri*vera and *Ta*mayo) that bounded the spreading segment of the EPR to be investigated. Their plan was similar to the approach they had taken during FAMOUS—to conduct a series of long dive traverses at right angles to the axis of the rise. This meant that they would be diving across time lines, beginning in the center of the rise where young lava is flowing out onto the seafloor and exploring in both directions away from this central zone of injection. They would head east toward the coast of Mexico and the North American plate, and west toward the Pacific Ocean and the Pacific Plate.

The French dive series in 1978 using *Cyana* at 21°N was highly successful, completing 21 dives. Like the FAMOUS study area, the central volcanic axis was found to be relatively narrow, flanked on either side by older tectonically altered terrain characterized by fissures and small-scale fault scarps. Reconnaissance dives were made between the EPR crest and the Brunhes-Matuyama reversal area 21 km to the west (three dives) and in the Tamayo transform (six dives). The extrusion zone is narrow (0.4 to 1 km), like that of slow-spreading centers. The extension zone, bracketed by a nearly continuous bottom traverse, has a half-width of 7-8 km. The Brunhes-Matuyama area was thus tectonically dead.

In contrast, the extension zone of slow-spreading centers is thought to be wider, although there are no field observations. Hydrothermal activity is demonstrated by colored deposits of rocks, tall cones of variegated deposits, and fields of giant clams (dead). Intense hydrothermal activity is probably a general feature of the EPR in contrast with its scarcity in the FAMOUS rift. Large areas of the young seafloor are covered by pahoehoe flows (sheet flows) and by lakes with pillars, expressing the greater fluidity of EPR extrusives compared with Mid-Atlantic Ridge (MAR) pillows and perhaps reflecting readier access to a larger magma pool.

A fascinating find associated with the East Pacific Rise expedition as well as those to the Galapagos Rift in 1977 dealt with the undersea lava flows encountered along these faster-spreading centers. During Project FAMOUS, the dominant extrusive lava form was an endless variety of pillow lavas, which scientists considered to be the classic underwater flow form. But when submersibles began diving in the faster-spreading centers of the Pacific, we encountered an entirely different type of lava feature termed "sheet flows." Unlike pillow lavas, which consist of a network of small lava tubes intertwined like a pile of spaghetti with individual "pillows" budding off from lava tubes, sheet flows form vast lakes or pools of molten lava. Fluctuations in the level of these lakes—caused by drainback into the magma chamber deep beneath the ocean floor—are indicated by "bathtub" rings around the perimeter of the lake. Clearly, the faster spreading rates associated with the East Pacific Rise and Galapagos Rift are commonly characterized by high volumes of sheet flows that flood large areas of the inner rift valley.

Another strange feature found within these lava lakes is lava pillars standing within the lake that resemble "tree molds," which are common in active volcanic areas on land such as those in Hawaii. When lava flows into a forested area on land, the molten rock is quenched when it comes into contact with the moist surface of the tree. Although the tree is consumed by fire, leaving only remnants of charcoal, a hollow cylindrical column of rock is formed; whose interior lining commonly preserves an imprint of the tree's bark. When the eruptive cycle ends and lava flows back into the magma chamber, a forest of tree molds is left standing as mute evidence of the forest that once stood there.

The lava pillars discovered within the lava lakes of the East Pacific Rise and Galapagos Rift are formed in a similar way. Prior to an eruptive cycle along a given spreading segment of the rift axis, the older lava terrain is characterized by a complex and dense network of fissures that are thoroughly permeated with seawater. When the eruptive cycle begins, sheet flows issue from only a few of the fissures within the fractured floor and spread out laterally covering a much larger area. As a result, the remaining water-filled fissures are capped by the flows, trapping large volumes of water beneath them. This seawater becomes heated and seeks to escape upward. Passing through the layer of molten lava contained in the lakes within the rift, this superheated water rapidly quenches the lava through which it passes. Hollow vertical chimneys of solidified rock form within the lava lake. As the level of lake drops, these chimneys remain as pillars of rock commonly supporting a thin canopy or crust of quenched lava running around the perimeter of the once-liquid lava lake. In appearance, it resembles "Yorkshire pudding."

Although the French dive series in 1978 did not result in the discovery of any active hydrothermal vents, it did locate one inactive site characterized by an accumulation of large white clam shells that were badly dissolved. During another dive, the scientist aboard *Cyana* came across some unusual chimneys on the older flanking volcanic terrain, which were sampled. After later analysis onshore, this sample was found to be 100% sphalerite or zinc sulfide containing 10 percent iron, 50 percent zinc, and 1 percent copper, with trace concentrations of lead and silver. The French had discovered an ore deposit on the East Pacific Rise that must have been formed under very high temperature conditions.

Although the highest temperature measured at the vent sites along the Galapagos Rift in 1977 was 23°C , laboratory analyses of the collected water samples suggested that the initial starting temperature of the hydrothermal fluids as they left the reactive zone around the magma chamber was between 350 and 400°C. Clearly, the discovery in 1978 of

high-temperature mineral deposits by the French indicated that high exiting temperatures for hydrothermal vents might actually be possible.

The American phase of the joint U.S.-French investigation of the East Pacific Rise took place in late 1979. At the time, we were just beginning to understand how narrowly confined the central volcanic axis of the Mid-Ocean Ridge truly was, given the significant lateral dimensions of the crustal plates it was creating.

Some of the scientists participating in the expedition, particularly those from Scripps, were convinced that the zone of volcanic extrusion was wide and that there were significant areas of off-axis eruptions taking place several kilometers from the central rift valley. These scientists, headed up by Dr. Fred Spiess, were interested in the seismic velocity, density, porosity, and permeability of the upper oceanic crust. They wanted to know about the fine-scale motions of the seafloor on a time scale of months to years. To this end, they also wanted to use *Alvin* to carry out scientific experiments and instrument deployments, rather than serve as a vehicle for qualitative observation. Previous *Deep Tow* lowerings had located a reasonably flat area to the west of the central axis known as "Tortilla Flats," where they hoped to locate fresh lava flows using their *Deep Tow* system and then visit the site with the submersible *Alvin*.

Deep Tow's primary sensors were a side-scan sonar, temperature probe, and magnetometer: indirect geophysical devices designed to paint a broad regional picture of the seafloor. Although it did have a black-and-white slow scan television camera and a black-and-white still camera, it spent little time in close visual contact with the bottom. Day after day passed as *Deep Tow* surveyed the area, but no active venting was located as the so-called Tortilla Flats proved to be old in age, covered by a thick blanket of sediments.

Another team aboard the *Melville*, including Jean Francheteau and me, was convinced that the zone of volcanic activity was narrowly confined along the central axis and that it was within this narrow zone of recent volcanism that active venting would be found. During our 1977 and 1979 expeditions to the Galapagos Rift, we had discovered that the active hydrothermal vents lay along a straight line apparently associated with the eruptive fissure responsible for the youngest flows within the rift valley. Once a vent was found, it became relatively easy to find additional vents along any particular fissure system by simply driving along the fissure, parallel to the rift axis.

Our tool for this search effort was *Angus*, and we patiently awaited our chance to go into the water. *Angus*, unlike the *Deep Tow* system, was designed by geologists to remain in constant contact with the bottom. It was designed to take a head-on collision with the rugged volcanic terrain and survive, making it possible to enter the narrow axial graben bound on either side by steep fault scarps.

After extensive *Deep Tow* coverage failed to locate any hydrothermal activity, *Angus* was finally permitted to enter the water. Unlike the traverses made by submersibles that are perpendicular to the axis of the rift, *Angus*' traverses were basically parallel to the axis. This was not based on geological reasoning but on operational necessity. Traverses perpendicular to the axis were the most desired. But *Angus* needed to be within a few meters of the ocean floor to obtain the high quality of color images we sought. Since the fault scarps bounding the rift to either side run parallel to the axis, *Angus* tow lines were best run in the same direction to avoid countless collisions with the bottom. It was also along this same strike that we felt active vents would be found. As a result, the first *Angus* trackline resembled a slalom run, as the vehicle was towed from side to side down the strike of what we hoped was the central volcanic axis.

No sooner had *Angus* begun its first run before the temperature sensor on the vehicle indicated it had passed through an active vent area. Repeating its performance in the Galapagos Rift, the sled was recovered so that the color film could be processed in the portable laboratory that had been brought on the expedition for that purpose. But a review of the color film taken across the vent field revealed a scene different from those observed along the axis of the Galapagos Rift. Initially the scene was the same, as frame after frame showed the vehicle passing over a young volcanic terrain characterized by a fresh glassy lava surface.

The first indication of an approaching vent was not a rise in temperature but the appearance of small white *Galathea* crabs dotting the otherwise barren flows. Quickly this gradient of crabs increased, giving way to the larger and more densely packed sessile organisms, in particular, large white clams so typical of the Galapagos Rift vent fields. As the center of the field was approached, "milky" water could be seen along with an increase in the amount of particles in the water. But unlike the Galapagos vents, the center of the vent was not covered with tube worms and large clams. Instead, we saw a large yellowish-brown deposit of sediments largely devoid of life.

The coordinates of this vent site were transmitted by radio from the *Melville* to Jean Francheteau aboard *Alvin*'s support ship *Lulu*. Since *Alvin* and *Angus* shared a common network of bottom transponders, it was easy to vector *Alvin* to any site discovered by *Angus*. Clearly, the role of submersibles was changing with the reconnaissance and regional mapping efforts falling more and more upon towed vehicle systems such as *Angus*. Towed vehicles had been used for several years to conduct regional mapping programs but it wasn't until 1977 to 1979 that joint operations between towed vehicles and manned submersibles became so closely choreographed.

What made this possible was the speed at which the photographic runs conducted by *Angus* were processed. Not only were tens of thousands of frames of color positive film quickly developed in a portable processing van for immediate viewing, but the edited tracks were also quickly plotted. This provided the geologists onboard with the opportunity to immediately generate detailed annotated traverses across the ocean floor. These traverses were then superimposed over the detailed bathymetric database to produce preliminary geologic maps. Each lowering added more and more detail to the evolving map of the area. The goal of this process was to space the *Angus* lines at just the right interval to permit the correlation of observations from one line to the next but not so closely as to produce highly redundant and, therefore, wasteful coverage.

On previous programs, a year or more had passed between the collection of towed vehicle data and follow-up dives by the submersible. Or even worse, the submersible conducted its own reconnaissance traverses working independent of towed vehicles. This was certainly the case with Projects FAMOUS and RITA.

On April 21, with the coordinates of an active hydrothermal field as their dive target, *Alvin* was lowered into the water. Dudley Foster was the pilot on this dive and as he dove over the fresh lava flows he began to see small white crabs on the horizon; he was reminded of similar scenes months before in the Galapagos Rift. But as he entered the vent field, it didn't feel the same. The water was much cloudier than usual. Then suddenly a tall chimney-like spire came into view; belching out its top was a dense black fluid resembling bellowing clouds of smoke. It looked like a steel factory as *Alvin* maneuvered above it for a closer view. But driving in midwater was proving difficult for Dudley; something was pulling him toward what he now called a "black smoker."

The pulling force proved to be the updraft or chimney effect caused by the rising black fluid. The black smoker was pulling water in from the side. And since *Alvin* was neutrally buoyant, it was also being pulled toward the smoker. Driving was made even more difficult as Dudley passed over the smoker and visual contact was lost in a thick cloud of black particles. Suddenly, he bumped against the chimney, which fell over like a giant fallen tree.

Ironically, this made the situation much better as the black fluid was now flowing out of the base of the broken chimney instead of its top. Dudley could now turn on his variable ballast system and take in water, making *Alvin* negatively buoyant as it slowly landed on the bottom. Using his lift props, he now climbed a gentle mount surrounding the fallen chimney. Clearly, these structures were fragile since the entire mound, which was some ten meters in diameter and a few meters tall, consisted of numerous broken chimneys that had fallen before.

Since he was the first human to see such a feature, chimneys appeared to fall over naturally without the help of submersibles. As he approached the fallen chimney with black fluid flowing out of its base, he could see the chimney was hollow, lined by mineral crystals that reflected in the submersible's lights. Now that the submersible was resting firmly on the bottom, Dudley could bring his manipulator into play. Resting in *Alvin*'s science tray was a temperature probe attached to a long plastic tube with a "T" handle that *Alvin*'s mechanical arm could easily grasp.

Lifting it from the tray, Dudley rotated the probe to the right and positioned it just above the cloudy vent opening. The temperature readout inside the pressure hull shot up, and when Dudley inserted it inside the vent, it went off scale. Now Dudley grew nervous. The probe had been used in the Galapagos Rift to measure the exiting temperatures of the vents. Never had it risen above 23°C, comfortably within its 100°C range.

Clearly, this vent was "hot," but how hot? Dudley's fears heightened when he removed the probe and found that its plastic holder had completely melted. His first thought was of his forward viewport, which was only a few feet from the vent opening and made of the same material as the melted probe.

This may have represented a major discovery for scientists but it was also very dangerous for submersibles. Dudley slowly pulled back, dropped his ascent weights, and brought the submersible back to the surface. Once safely back on *Lulu*'s cradle, Dudley saw how lucky he had been that day. Inspecting the fiberglass fairing near the lower viewports, Dudley found that the submersible's skin had melted.

The next day, when Francheteau and I dove in *Alvin*, we were much more cautious when approaching a black smoker, but the thrill was just the same. Francheteau said it best in his wonderful English, "They seem connected to hell itself." This time we were better equipped with a probe that could measure much higher temperatures, in our case, an incredible 350°C or 662°F , hot enough to melt lead, let alone our Plexiglas viewports out of which we were staring in utter amazement. Here in 3,000 meters of water we had visual proof of what geophysicists and geochemists had only theorized. Here also was a crystal clear explanation of what had eluded chemists for centuries, a logical explanation of the ocean's chemistry.

What Jean and I were watching was part of the same process of recycling seawater that fueled the food chain on the oases of the Galapagos Rift. It was superheated water that was funneling out of the mouths of the chimneys—water blackened by its concentrated solution of minerals from deep within the Earth's crust. The construction of the chimneys themselves was a testament to the mineral richness of this subterranean broth; as the fountains of returning seawater cooled, they precipitated material that built the flue pipes ever taller.

During the 1977 dives on the Galapagos Rift, we had already seen the effect of hydrothermal vents (in that instance, not full-scale black smokers) on seafloor animal communities. Now, observing the cycling of seawater through the perforated juncture between two crustal plates, we began to speculate upon the broader relationship between the ocean and the crust that they largely conceal. Some scientists have since speculated that all of the water in the seas may seep down into the hot lower crust and back up through the vents,

over a cycle lasting 10 million to 20 million years. As we saw in the black smokers, the minerals carried back up to the seafloor precipitate and harden into ore deposits—one explanation, perhaps, for the presence of such deposits on dry land that was once covered by the ocean.

Following the discovery of high-temperature hydrothermal vents on the East Pacific Rise at 21°N by the towed camera system *Angus* and the submersible *Alvin*, all hell broke loose. Not only did this discovery prove that the vent communities in the Galapagos Rift were not unique, it also demonstrated that the precipitation of polymetallic minerals within the vent system could result in the exiting of high-temperature fluids directly from the ocean floor and the surface accumulation of important mineral assemblages.

The potential consequences of these discoveries had a profound impact upon many fields of marine research, in particular the fields of biology, chemistry, geology, and geophysics. Just as the theory of plate tectonics had mobilized the Earth sciences in the early 1960s, the discovery of hydrothermal vents in the Galapagos Rift and East Pacific Rise mobilized the field of oceanography. All of a sudden, a large number of marine scientists who had never been in manned submersibles or been interested in the spreading axis of the Mid-Ocean Ridge were submitting proposals to their various funding agencies to investigate deep-sea vents. Some used the importance of these discoveries in basic research to justify their requests, while others argued the commercial potential of the mineral deposits forming around the higher-temperature vents and still others argued their importance to national interests—whatever it took to get them into this new and exciting game.

The initial phase of follow-up studies began in full force in 1980 with an expedition to the Galapagos Rift by the National Oceanic and Atmospheric Administration (NOAA) scientists under the leadership of Alex Malahoff. This expedition resulted in the discovery of major polymetallic sulfide deposits and increased interest in their commercial potential.

In May and June of that same year, Jean Francheteau of CNEXO, France, invited me to participate in an explorer's dream: a three-month-long journey down the East Pacific Rise aboard their premiere research ship the N/O *Charcot*. Taking advantage of the latest American technology in bottom mapping, the French had purchased the first unclassified multi-narrow beam sonar system called a shipboard multi-transducer swath echo sounding system (SEABEAM) and mounted it on the hull of the *Charcot*. For the first time, the scientific community could survey potential dive sites along the Mid-Ocean Ridge quickly and in great detail without having to rely upon the Navy as we had done in the FAMOUS area, Cayman Trough, and Galapagos Rift.

The timing could not have been better. By now, it was clear to Jean and me that there were a variety of factors controlling the distribution of hydrothermal vents along the axis of the Mid-Ocean Ridge. Clearly, they were situated in the youngest volcanic terrain characterized by the central axis.

The magma chambers feeding the most recent flows were also the obvious heat source for the active vents. The faster the spreading rate, the more likely were vents to be found, shifting the focus of our studies from the slow-spreading Mid-Atlantic Ridge to the faster-spreading East Pacific Rise.

But our studies along the axis of the Galapagos Rift in 1977 and 1979 and our studies at the East Pacific Rise at 21°N in 1978 and 1979 revealed a significant along-strike variation. This was even true for the axis of the Mid-Atlantic Ridge in the FAMOUS where venting had not been found. The ridge is divided into a continuous series of spreading segments bound at each end by transform faults that offset the ridge to either side. As the intersection between the axis and transform fault is approached, the depth of the axis begins to increase. Since the topography of the ridge is the result of thermal expansion, the higher the elevation of the axis, we reasoned, the more likely were we to find hydrothermal activity.

With the SEABEAM system now installed on the N/O *Charcot*, Jean and I could run along a major length of the East Pacific Rise testing our model in our search for new sites of hydrothermal venting. From May until July 1980, the *Charcot* slowly zigzagged down the axis of the East Pacific Rise at 22°N to its fastest-spreading segment at 22°S near Easter Island. From these survey lines, we could clearly see individual spreading segments along the strike of axis, each having topographic highs where we felt active venting might be found.

Our first chance to test this model came in April 1981 with a cruise aboard the R/V *Melville* to the East Pacific Rise at 20°S. Using the *Charcot*'s SEABEAM maps to guide us, we conducted a series of *Angus* camera runs down the axis of a fast-spreading segment of the ridge near its topographic high and quickly found active hydrothermal vents.

In January 1982, we had another chance to test this model when Jean brought the submersible *Cyana* aboard the N/O *Le Suroit* to dive at 13°N on the East Pacific Rise, a site surveyed in 1980 by the *Charcot*. Once more, the model proved to be an excellent prediction for finding active hydrothermal vents. We even dove in the *Cyana* where we did not expect to find vents near the axis-transform intersection and didn't.

By now, we were not the only team searching for new vent settings on the Mid-Ocean Ridge. Peter Lonsdale from Scripps, who had played a major role in the discovery of the hydrothermal vents in the Galapagos Rift, was using his considerable skills to search for vent sites in the Gulf of California. The focus of his research was a series of small spreading segments in Guaymas Basin. His efforts proved equally successful in January 1982, when a series of dives by *Alvin* located and investigated a number of active vents. What made these vents unique was their occurrence in an area of thick sediments.

At the northern end of the Gulf of California is the Colorado River delta. For millions of years this river has deposited a tremendous volume of organic-rich sediments into the gulf, including Guaymas Basin. As a result, the active spreading axis underlying the gulf is buried under a thick accumulation of mud. Hydrothermal fluids that flow out of fissures cutting across the young central volcanic terrain must then rise hundreds of meters through this sediment cover before exiting into the basin's bottom waters. During this final vertical journey, these superheated fluids interact with the overlying organic sediments, greatly altering their chemistry. Oil seeps of thermogenic petroleum hydrocarbons were commonly associated with active vent sites and the soft sediment surface was covered by extensive bright yellow and white bacterial mats.

From 1981 on, the investigation of hydrothermal circulation in the ocean's crust intensified and spread throughout the world. A team headed by Peter Rona of NOAA located hydrothermal vents in the Mid-Atlantic Ridge. Both French and American researchers found additional vents along the East Pacific Rise at 10, 11, and 13°N. From 1982, active hydrothermal vents were discovered on the East Pacific Rise at 13°N followed in 1984 by the discovery of similar vents sites on the Juan de Fuca Ridge and Discovery Ridge off the coast of Washington and British Columbia.

As more active hydrothermal sites were discovered on the East Pacific Rise, the search broadened to include other geologic settings. Dives by *Alvin* in the Marianas Back-arc basin successfully located active vents. These discoveries were followed by expeditions to the Mid-Atlantic Ridge, which located active vent sites at 26 and 23°N. More recently, hydrothermal vents have been found to the north on the Mid-Atlantic Ridge at 37°17.5′N and 37°50′N. In these latter instances, the vent sites are near the Azores "hotspot" and associated with large lava lakes.

Once high-temperature vents were discovered along the East Pacific Rise at 21°N in 1979, additional important vent animals were added to the list. Perhaps the most impressive was a worm dubbed the "Pompeii Worm" for its ability to live in close proximity to the black smokers, where the exiting vent temperature can exceed 350°C. These worms (*Alvinella pompejana*) live in tiny tubes that are constantly being covered by fine-grain minerals precipitating out of the vent waters once the hot fluids come into contact with cold ambient seawater.

The dominant organisms associated with the hydrothermal vent communities of the Eastern Pacific (i.e., Galapagos Rift, East Pacific Rise, Guaymas Basin, and Juan de Fuca/Discovery Ridges) include the long, red-tipped vestimentiferan tube worms, large white bivalve clams, and thick accumulations of mussels. Variation in the vent faunal assemblages is thought to be related to differences in vent flow and water chemistry, with higher concentrations of biomass associated with lower-temperature vents (i.e., 5–200°C) compared to the higher temperature vents (i.e., 200–360°C).

The two giant-sized mollusks mentioned earlier are the clam-like *Calyptogena magnifica* and the mussel *Bathy-*

modiolus thermophilus. The clams live on the outer perimeter of the vent site, commonly found in small crevices where low-temperature vent fluids are coming out of the fractured volcanic terrain. They have a long foot that aids them in moving as well as feeding, primarily for the uptake of vent fluids while their gills absorb oxygen and inorganic carbon from the circulating bottom waters.

The mussels, on the other hand, are commonly found in high-temperature settings near the vent opening and apparently ingest bacteria directly through filter feeding. Such direct ingestion of food is thought to be secondary to their primary source of nutrition from symbiotic bacteria living within their bodies. Other mollusks include limpet-like gastropods and whelks.

The most spectacular organisms associated with many hydrothermal vents are the large white, red-tipped vestimentiferan tube worms, *Riftia pachyptila*. Living in a highly precarious setting of varying levels of oxygen and temperature, this organism is truly unique. It lacks a mouth, gut, and digestive system and relies upon the symbiotic bacteria that make up half its body weight to feed it. Since these tube worms live where reduced vent fluids mix with the oxygenated bottom water, they need to withstand prolonged periods of time when anoxic conditions prevail. As a result, their blood includes human-like hemoglobin, which stores oxygen within their body.

Another fascinating worm living under an even harsher vent setting is the Pompeii Worm, *Alvinella pompejana*. These live in a mass of honeycomb-like tubes near high-temperature vents that they freely move in and out of. Their tubes have even been seen attached to sulfide chimney walls of 350°C black smokers, although they must live in the highly mixed waters having a lower temperature.

Scavenging and carnivorous brachyuran crabs are also associated with the vent communities, as well as numerous other organisms including anemones, siphonophores, fish, shrimp, and so forth, too numerous to describe in any detail here.

In 1984, an entirely different geologic setting was found in which similar organisms are living. Cold water seeps on the West Florida Escarpment in the Gulf of Mexico were discovered that support sulfide-oxidizing benthic communities. Groundwater flowing through porous limestone releases sulfide and methane-enriched water that leads to the growth of chemosynthetic bacterial mats and symbiotically supported communities of large mussels and vestimentiferan worms. These communities also include galatheid crabs, gastropods, sea anemones, serpulid worms, and other organisms typical of warm water vent settings.

Further to the west in the Gulf of Mexico, cold water seeps of hydrocarbons including methane were found to support similar benthic communities. More recently, hydrocarbon seeps off the west coast of California, in the North Sea, and the Sea of Okhotsh have been found to support a similar assemblage of organisms. Even the oily bones of a decom-

posing whale off California provide a home for this unique biological ecosystem. The investigation of seamounts was also expanded to include the investigation of craters, calderas, and pyroclastic deposits on seamounts in the Pacific.

Clearly in years to come, chemosynthetic animal communities will be found throughout the world's oceans and lakes wherever the conditions arise to spawn this unique symbiotic relationship. As this paper is being written there are those who are turning their thoughts to the volcanic terrains of Mars or the ice-capped ocean of Europa. Such thoughts include the continuing debate dealing with the very origin of life on our planet.

ALVIN BECOMES A ROUTINE TOOL IN MARINE RESEARCH

Following the excitement of the later 1970s and early 1980s, *Alvin*'s annual diving program pushed north from the East Pacific Rise off Mexico to include regular visits to the Juan de Fuca, an isolated segment of the Mid-Ocean Ridge, connected millions of years ago to the East Pacific Rise. Dives in the Juan de Fuca and Gorda Ridges off the coasts of Oregon and Washington in 1984 resulted in the discovery of hydrothermal vents and high-temperature black smokers.

With increased funding from the National Science Foundation, the engineers supporting the *Alvin* program now were able to "harden" its capability and make major improvements in its propulsion, electrical, and instrumentation systems. Returning to the Mid-Atlantic Ridge after its lengthy overhaul in 1986, *Alvin* was able to investigate newly discovered hydrothermal vents and a unique benthic animal community dominated by shrimp.

In 1987, *Alvin* crossed the Pacific Ocean for the first time in its history, stopping in the Hawaiian Islands. There scientists investigated Loihi Seamount along the volcanic ridge extending southeast from the big island of Hawaii before continuing west to the Mariana Islands. A team of scientists had discovered hydrothermal vents in the back-arc basin west of the Mariana Islands, and *Alvin* was used to document and sample their chemistry and unique biology. Following what would be its only expedition to the western Pacific, *Alvin* and its support ship *Atlantis II* returned to San Diego for maintenance and repairs.

For the next 10 years, *Alvin*'s diving schedule became fairly routine, journeying back and forth along the West, East, and southern coasts of the United States with frequent visits to the Juan de Fuca Ridge and Oregon coast, the California continental borderland, the East Pacific Rise, Guaymas Basin, Galapagos Rift, Mid-Atlantic Ridge, Bermuda, and the East Coast.

The investigation of hydrothermal vents including coldwater seeps continued to dominate *Alvin*'s use, but other programs emerged as well. These included the continued investigation of seamounts and the investigation and instrumentation of the Ocean Drilling Program drill holes. During this same period of time, many improvements were made to *Alvin*'s various sampling and imaging systems and its operating depth was increased to 4,500 m.

After years of development and use, the unmanned remotely operated vehicle program at Woods Hole was finally integrated into the *Alvin* operational schedule with the arrival of its new support ship *Atlantis*. It is now up to the deep submergence user community to determine the long-term viability of manned submersibles such as *Alvin*.

REFERENCES

Ballard, R.D. In press. *Eternal Darkness*. Princeton University Press, Princeton, New Jersey.

Ballard, R.D., W.B. Bryan, J.R. Heirtzler, G. Keller, J.G. Moore, and Tj.H. van Andel. 1975. Manned submersible observations in the FAMOUS area, Mid-Atlantic Ridge. *Science* 190:103-108.

Geological Society of America Bulletin. 1977. Volume 88.

Geological Society of America Bulletin. 1978. Volume 89.

Heirtzler, J.R., and X. Le Pichon. 1974. A plate tectonics study of the genesis of the lithosphere. *Geology* 2:273-274.

National Research Council. 1972. *Understanding the Mid-Atlantic Ridge: A Comprehensive Program*. National Academy of Sciences, Washington, D.C.

Turner, R. 1973. Wood-boring bivalves: Opportunistic species in the deep sea. *Science* 180:1377-1379.

Creating Institutions to Make Scientific Discoveries Possible

A Chronology of the Early Development of Ocean Sciences at NSF

Michael R. Reeve

Division of Ocean Sciences, National Science Foundation

INTRODUCTION

The historical time line below is intended to trace the emergence and development of ocean sciences within the National Science Foundation (NSF). It focuses specifically on the years up to the time that the Ocean Drilling Program was established in the Division of Ocean Sciences. Since then (1984), the division has remained virtually unchanged up to the time of writing. This account touches on other organizational structures and events to provide a context for the emerging story, and provides the context that links the various contributions in this volume. I have used the resources cited in the next paragraph, internal memoranda available to me and now deposited in the National Archives, and personal recollections of colleagues such as those contributed in this volume. (See also ocean science budgets in Appendix E and organizational charts in Appendix F.)

There are several general histories that speak to the events leading up to the establishment of the Foundation and its early years. *Science—The Endless Frontier* by Vannevar Bush was a report to President Roosevelt in 1945 (Bush, 1945). It was reprinted by NSF in 1990 (NSF 90-8) in a volume that also contained appendices and an extensive commentary by Daniel Kevles (California Institute of Technology) concerning the impact of the report. J. Merton England (the NSF historian in 1982), wrote a volume entitled *A Patron for Pure Science—The National Science Foundation's Formative Years, 1945–1957* (NSF 82-24). Finally, George T. Mazuzan (NSF historian from 1987 until his retirement in 1998) wrote *The National Science Foundation: A Brief History* (NSF 88-16).

Also invaluable was NSF Handbook Number 1, titled *Organizational Development of the National Science Foundation*. It covers the period from the Foundation's establishment in 1950 up to 1984. It is an annual compilation of organization charts, together with a summary of "organizational development," which includes organizational changes, significant legislation and executive orders, and National Science Board actions. A copy of this document currently resides in the NSF library.

IN THE BEGINNING

1950—President Truman signed the National Science Foundation Act on May 10, 1950. The Act provided that the Foundation shall consist of a director responsible for administration and a National Science Board to establish substantive policy and approve certain specified actions. Both the director and the board were to be appointed by the President. Beyond this, the act specified structure to the extent of four divisions: (1) Medical Research; (2) Mathematical, Physical, and Engineering Sciences; (3) Biological Sciences; and (4) Scientific Personnel and Education. The Act also specified the establishment of divisional committees to make recommendations to, and advise and consult with, the board and the director on matters relating to programs of their own divisions. The President appointed 24 board members and convened the first meeting at the White House on December 2, 1950.

1951—At its second meeting on January 3, 1951, the board established the four prescribed divisions. On April 6, the President appointed Alan T. Waterman as the first and only director to serve two consecutive full terms, each of six years. Waterman was formerly Chief Scientist at the Office of Naval Research (ONR). Assistant directors were appointed to three of the divisions. The appointee for the Biological Sciences Division also acted for the Medical Research Division, until the two divisions were combined a few months later.

The first four programs were established in each of the Divisions of Mathematical, Physical and Engineering Sciences (MPES) and Biological and Medical Sciences (BMS). The board also appointed divisional committees.

87

1952—NSF made its first awards in this year. Among some 100 awards, two could be identified as ocean related. These included a one-year grant of $4,700 to Dr. Robert Ginsberg at the University of Miami for studies on the "Geological Role of Certain Blue-Green Algae." Dr. Ginsberg is still an active faculty member at the Rosenstiel School of Marine and Atmospheric Sciences of the University of Miami and still submits proposals to NSF. He also graced the assembly with his presence at this symposium.

1953—The Earth Sciences Program was established within the MPES. This was to be the ancestral home of all the nonbiological ocean sciences (chemistry, physics, geology and geophysics). Support for biological oceanography can be traced to multiple origins, spreading across all the BMS programs, although most predominantly from the Developmental, Environmental and Systematic Biology Program. The Foundation made about 5 awards related to ocean sciences out of a total of about 175.

The original NSF Act of 1950 contained a limitation of $15 million that could be appropriated annually to the NSF. This budgetary limit was removed by amendment of the original act on August 8, 1953. This was a very important change as reflected by the fact that the fiscal year 1999 budget stands at $3.7 billion, and the budget for the Ocean Sciences Division alone is now $215 million.

THE BEGINNINGS OF BIG OCEANOGRAPHY

The National Academy of Sciences asked NSF to seek funds for and administer the U.S. component of the ICSU (International Council of Scientific Unions) International Geophysical Year (IGY) program.

1955—NSF took up the IGY challenge and was appropriated $2 million for fiscal year 1955. The Office for the International Geophysical Year was established within the Office of the Director, in response to the provision of funds by Congress for this first major interdisciplinary program that NSF was entrusted to administer.

1957—The NSF appropriation grew to $37 million in fiscal year 1956, much of which was not expended until fiscal year 1957. In that fiscal year, some $15 million was expended for IGY, compared to a total of about $20 million for all other research projects and facilities support. In oceanography-related fields, NSF awarded about $1.25 million to several programs, which included the Deep Current Program, Atlantic and Pacific Oceans ($786,000), Pacific and Atlantic Ocean Island Observatories ($223,000), CO_2 Analysis and Radiochemistry of Sea Water ($174,000), and Arctic Oceanography and Sea Ice ($51,000). By comparison, the Earth Science Program, which was responsible for funding nearly all of nonbiological oceanography, expended about $165,000, mostly on ocean-related geology and geophysics,

as estimated by reading grant titles from the fiscal year 1957 annual report. A further $331,000 in IGY funds in oceanography would be awarded in fiscal year 1958.

Beyond the Foundation, the National Academy of Sciences Committee on Oceanography (NASCO) was established to formulate recommendations concerning long-range national policy for the development of oceanography, to encourage basic research in the marine sciences, and to provide advice to government agencies on various oceanographic problems. The evolution of this important committee over the subsequent 42 years can be followed up to the present-day Ocean Studies Board, which organized and hosted the symposium that forms the basis of the present volume.

On May 1, 1957, NSF reported back to Congress, as requested, regarding the desirability of constructing and equipping a geophysical institute in the Territory of Hawaii. The report was positive but carried the provision that Congress should appropriate the full cost rather than it being a part of the Foundation's regular budget.

1958—On August 4, 1958, the Office of IGY was redesignated as the Office of Special International Programs and established the U.S. Antarctic Research Program. This office eventually evolved into the Office of Polar Programs, a separate NSF entity that would also begin to fund oceanography, often in joint ventures with the Ocean Sciences Division (formed in 1975) up to the present time.

1959—Beyond NSF, but within the federal government, the Interagency Committee on Oceanography (ICO) was set up by the newly formed Federal Council for Science and Technology, as the first attempt to recognize this fledgling scientific discipline, aspects of which were on the agendas of several agencies at the time. The ICO was charged to develop a National Oceanographic Program, which included reviewing activities and plans of individual agencies, coordinating budget planning, and considerations of special problems important in advancing oceanography. The initial goals of the ICO were to introduce, as fast as possible, more ships, facilities, and manpower. This goal was, in the words of Harve Carlson, NSF division director of Biological and Medical Sciences and ICO chairman in 1965, "impressively met." The National Academy of Sciences published the NASCO report *Oceanography 1960–1970* (NAS, 1959).

THE ERA OF RAPID EVOLUTION

Within the Division of Biological and Medical Sciences, a Biological Facilities Program was established in 1959. This program was to be very influential in the subsequent development of biological oceanography, marine biology, and the beginning of the academic fleet. This history is addressed by Mary Johrde in her contribution to this volume.

1961—November 8, 1961, may be identified as the first beginnings of the future integration of NSF programs relating to the oceans. A memo to files by Harve Carlson, director of BMS, reported on a meeting "to bring about better communication between interested divisions and offices within the Foundation in reference to oceanography." The attendees identified a list of discussion items for future meetings, which included drawing up a 10-year "program for oceanography," providing regular input to NSF representatives on the ICO (i.e., the associate director for research with Carlson as alternate), and issues relating to ICO and NASCO.

A second meeting was held on December 29, 1961, and a twice-monthly meeting schedule was set up through April 1963. New issues not mentioned at the previous meeting included coordination of the planned International Indian Ocean Expedition, the question of who makes international commitments involving universities (ICO, NSF, or State Department), anticipation of congressional problems (the Magnuson Act and other oceanography-related bills), and ships and ship titles. A bill had been passed directing the establishment of a position of assistant director for oceanography in the White House Office of Science and Technology, but it was vetoed by President Kennedy.

1962—In March 1962, a contractor was selected for Phase 2 of Project Mohole (Deep Crustal Studies of the Earth), and the position of NSF managing coordinator for Project Mohole was selected, reporting to the Associate Director for Administration. On May 4 the Foundation's Mohole Committee was established. Initially funded at $1.65 million, the project was expected to "require between 3 and 5 years to complete." The detailed story of this and subsequent programs of deep seafloor drilling is told by Edward Winterer in this volume.

On March 27 the NSF produced an internal report entitled "10-Year Projection of National Science Foundation Plans to Support Basic Research in Oceanography." The projection of plans was made "without regard to possible budgetary restrictions," but was "meant to convey some notion of the magnitude of effort required. . . ." There were four sections with budgets rising from 1962 to 1971 in Physical Oceanography, Biological Oceanography, Antarctic Program, and the International Activities. Table 1 includes re-

search, facilities, and "all other aspects of oceanography." Physical oceanography was defined to include "all physical, chemical and geological phenomena."

The budget numbers for fiscal year 1962 are realistic since they are not out-year projections or based on wishful thinking, but they are approximate, since grants were included under "biological oceanography" at the judgment of program managers of several different programs.

The year 1962 saw the initiation of the second large-scale ocean-related program following on from IGY. It was the International Indian Ocean Expedition (IIOE). Conceived in 1958 within ICSU, it was based on the premise (NSF, 1962b) that the Indian Ocean was the least understood ocean, biologically and physically, but there were indications that it might have a biological productivity higher than either the Atlantic or the Pacific Oceans. This was contrasted to the fact that "many inhabitants of the surrounding region suffer from severe dietary protein deficiency." Also, the seasonal reversal of monsoon winds made it a "huge natural laboratory for observing the effects of wind stress on oceanic currents." The NSF budgets for fiscal years 1962 and 1963 for IIOE were $2.1 million and $4.4 million, respectively. By comparison with the numbers in Table 1, it is clear that these funds represented a very significant infusion of new support for oceanography.

On April 13, 1962, NSF Director Alan Waterman signed directive O/D-102, which officially established the NSF Coordinating Group on Oceanography (CGO). "In addition to its general responsibilities," it was specifically tasked with coordination of oceanographic facilities; conversion, construction, and operation of ships; and the International Indian Ocean Expedition.

Within two months, Randal Robertson, the Associate Director for Research and chairman of CGO, established an Ad Hoc Panel on Grants and Contracts for Ship Construction, Conversion and Operations, whose initial assignment was to assemble existing agreements and background information, and recommend a set of procedures to be adopted.

The Division of Mathematical, Physical and Engineering Sciences redesignated its program areas as "sections" on October 29, 1962. The Earth Sciences Program Office became the Earth Sciences Section, and four programs were established: Oceanography, Geophysics, Geology, and Geochemistry.

Beyond NSF, NASCO now decided to prepare a report giving the best estimates of the possible actual worth to this country from the planned National Oceanographic Program, particularly an expanded research effort.

1963—In two meetings of the CGO (January 25 and March 27, 1963), committee members wrestled with definitions of "oceanography" and "oceanographic manpower." It was noted that only in MPES was there a single program, and hence a "line item," for the support of all oceanography. Various programs of BMS supported marine-related biol-

TABLE 1   NSF Ten-Year Budget Projections (million dollars)

Year	Physical Oceanography	Biological Oceanography	Antarctic Programs	International Activities
1962	6.5	10.0	1.54	1.5
1971	41.9	30.2	2.60	3.2

NOTE: See Appendix E for actual budget figures.
SOURCE: NSF (1962a).

ogy, and here the problem of definition was acute. Also, support came from the Division of Scientific Personnel and from the Offices of International Activities and the Antarctic Program. The NSF budget for oceanography was estimated by the committee to be $11.86 million for research, $7.3 million for shore facilities, $3.5 million for ships, $4.0 million for the International Indian Ocean Expedition and $0.14 million for a data center, totaling $26.80 million.

To determine manpower, the NSF asked the International Oceanographic Foundation (Miami) and a working group guided by Joel Hedgepeth, Walton Smith, Donald Pritchard, and Fritz Koczy for assistance. They agreed that the International Register, which contained some 5,000 oceanographers, was an unrealistically high estimate.

Athelstan Spilhaus, chairman of NASCO, proposed to the annual meeting of the American Fisheries Society the creation of Sea Grant Colleges analogous to Land Grant Colleges.

1965—Regular meetings of the CGO had lapsed. On June 9, 1965, Harve Carlson made a request to the Associate Director for Research that "we again set up regular meetings of the Coordinating Group on Oceanography." He reasoned that as the NSF member of the ICO, he was being asked for increasing amounts of data and for policy decisions regarding the role of NSF in interagency ocean issues (e.g., should Mohole and the "long core vessel" be included in the ICO budget?). Carlson noted that "with our buildup in oceanography, a rather major responsibility has grown over the last three to five years." The associate director responded affirmatively, naming Carlson his vice-chair.

On November 19 the Division of Environmental Sciences was established. It was formed to contain the Office of Antarctic Programs, Atmospheric Sciences Section, and Earth Sciences Section, both from MPES.

1966—President Johnson signed a bill into law that created what was to become known as the Stratton Commission, after its chairman Julius Stratton. The purpose of the temporary council and commission was to study the national oceanography program and propose revisions.

Meanwhile, in July 1966, the President's Science Advisory Committee (PSAC), chaired by Gordon MacDonald, issued "what may well be the single most influential design for reorganizing the oceanography program (*Science*, July 22, 1966) *Effective Use of the Sea* (PSAC, 1966). It called for the establishment of a new oceanography agency, but did not seek to put either NSF or ONR research into it. It downplayed an Academy report *Economic Benefits of Oceanographic Research* (NAS, 1964) that had apparently grossly exaggerated the economic benefits of a national oceanographic program.

The NSF Coordinating Group on Oceanography met on August 8, 1966, and Randall Robertson (its chair) reported that the director had asked him to establish a task force to analyze the PSAC report. The major topics of discussion were the concepts of a new oceanography agency (referred to here as a "wet NASA" [National Aeronautics and Space Administration]) and the regional fleet. The former was deemed "not appropriate at the present time" but the latter was looked on with favor.

By August 24 the ill-fated Project Mohole was halted as necessitated by congressional denial of funds, and the office was officially closed on December 31, 1966.

The National Sea Grant College Program Act was signed into law on October 15, and the Office of Sea Grant Programs was established as an organizational component of NSF reporting to the Associate Director for Research.

1967—On June 8, 1967, NSF issued an important notice announcing it was ready to receive Sea Grant proposals. NSF announced the first Sea Grant awards—nine grants totaling nearly $2 million—on February 21, 1968.

THE EMERGENCE OF THE DIVISION OF OCEAN SCIENCES

In March 1967, the Oceanography Program was raised to the section level within the Division of Environmental Sciences with the Physical Oceanography Program, Submarine Geology and Geophysics Program, and Oceanographic Facilities Program managed by Mary Johrde (see her contribution in this volume). Thus can be identified a recognizable cluster of programs (although without biology) out of which the modern Division of Ocean Sciences evolved.

1969—A further step was taken toward the evolution of the Division of Ocean Sciences when the Biological Oceanography Program was added as a new unit within the Environmental and Systematic Biology Section (Biological and Medical Sciences Division), which also included the Environmental Biology and Systematic Biology Programs.

The NSF Director, as a member of the Marine Council, responded in March to the Vice President's request for input on the Stratton Commission report. The director's message contained many cautions on the establishment of the National Oceanic and Atmospheric Administration (NOAA), advising that it required further study. He was unenthusiastic about the transfer of Sea Grant to such a new agency.

In April, Edward Todd (NSF Deputy Associate Director for Research) reported on a briefing by ONR for NSF staff on the "oceanographic ships problem." Feenan Jennings (this volume) represented ONR. They agreed to a further meeting to discuss the division of ship operation support between NSF and ONR, the impact on operational costs of the addition of large AGOR-class oceanographic vessels to the fleet by ONR, unilateral planning by each agency for fleet replacements or additions, and the recommendations of the Marine Science Commission with respect to "University-National Laboratories" and regional fleets.

PLATE 1   A hydrothermal vent community with the giant tubeworm (*Riftia pachyptila*), mussels (*Bathymodiolus thermophilus*), and crabs (*Bythograea thermydron*) at the Rose Garden hydrothermal vent on the Galapagos Rift at a depth of 2,500 m. Photo by Fred Grassle.

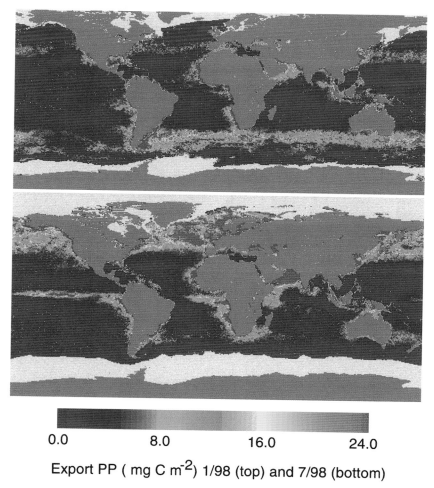

PLATE 2   Global estimates of export (or new) primary production for January 1998 (top) and July 1998 (bottom). Total primary production was estimated using SeaWiFS estimate of surface chlorophyll, sea surface temperature, surface irradiance and a temperature-dependent model of P^B_{opt} (Behrenfeld and Falkowski, 1997). Total primary production was converted to export primary production by the Eppley and Peterson (1979) relationship. Figure by Paul Falkowski.

0.0   8.0   16.0   24.0

Export PP (mg C m^{-2}) 1/98 (top) and 7/98 (bottom)

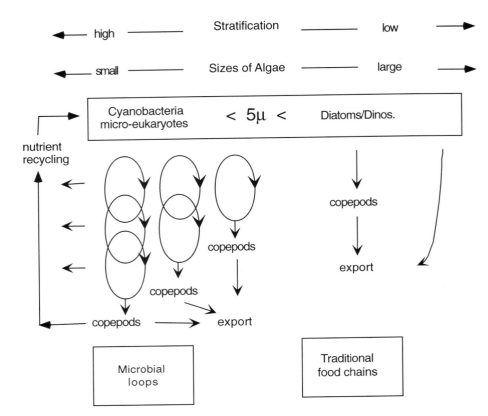

PLATE 3a   A simple representation of microbial loops and the changes in food web structure with decreasing rates of nutrient input from right to left. This figure, based on Azam et al. (1983) and Cushing (1989), is reprinted from Figure 1 in Steele (1998), with permission from Proceedings of the Royal Society of London.

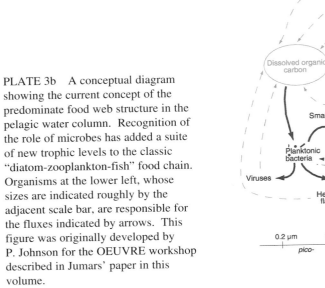

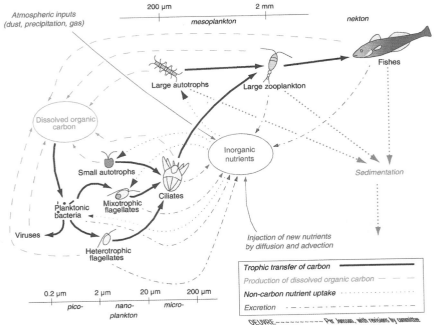

PLATE 3b   A conceptual diagram showing the current concept of the predominate food web structure in the pelagic water column. Recognition of the role of microbes has added a suite of new trophic levels to the classic "diatom-zooplankton-fish" food chain. Organisms at the lower left, whose sizes are indicated roughly by the adjacent scale bar, are responsible for the fluxes indicated by arrows. This figure was originally developed by P. Johnson for the OEUVRE workshop described in Jumars' paper in this volume.

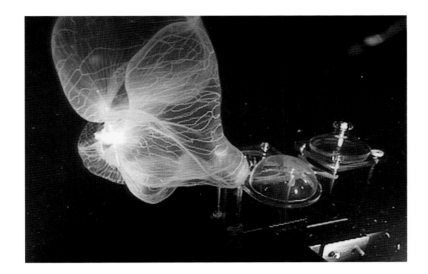

PLATE 4   One of the large gelatinous organisms, *Deepstaria enigmatica*, that have been recently found to be very abundant in mesopelagic waters of the world ocean. This medusa was photographed in Monterey Bay by Kevin Raskoff © MBARI, 1998.

PLATE 5   The Red Sea parted, allowing Moses and the Israelites to escape the pursuing soldiers of the Pharaoh (By permission of Pictures Now! Powered by Wood River Media, Inc. 1998 Wood River Media. June 1998. http://www.lycos.com/picturethis/religion/judaism/history/bible_stories/crossing_the_red_sea/310521.html).

PLATE 6 (drawn by Mike Dormer)(top) I confess to a certain nostalgia for the days when the ONR princes dispensed their largess to a few ocean courtiers. (bottom) 50 years later, our courtier has a problem.

On October 27, 1969, William McElroy, who had become the third NSF director in July, effected a major reorganization of NSF based on provisions of the National Science Foundation Act of 1968. Four assistant directorships were established for Research, Education, Institutional Programs, and National and International Programs. The Office of Assistant Director for Research had five Divisions reporting to it (Biological and Medical Sciences; Engineering; Social Sciences; Environmental Sciences; Mathematical and Physical Sciences) and also included the Office of Interdisciplinary Research.

Oceanography (except biological oceanography) remained in the Environmental Sciences Division, but there were several marine-related elements in the Office of the Assistant Director for National and International Programs. These included the Sea Grant Program, Antarctic Programs, Computing Activities, Science Information Service, International Programs, and National Centers and Facilities Operations. The last was responsible for the Ocean Sediment Coring Program, National Center for Atmospheric Research (NCAR), and the astronomical observatories. The Office of Polar Programs (OPP) and Office of the International Decade of Ocean Exploration (IDOE) were added on December 19, 1969, and March 21, 1970, respectively, following a letter from the Vice-President, which confirmed NSF as lead agency for IDOE and for the extension of Arctic research. The IDOE was the third major influx of funds that supported ocean sciences at NSF (after IGY and IIOE). The history of IDOE is covered by Feenan Jennings in this volume.

1970—The Biological Oceanography Program did not remain long in the Division of Biological and Medical Sciences. It was transferred into the Oceanography Section in the Division of Environmental Sciences on July 24, 1970. The brief tenure of Sea Grant at NSF ended when the Office of Sea Grant Programs was transferred to NOAA on October 3.

On October 1, responsibility for oceanographic ship operation support was transferred from the Research Directorate to the National and International Programs Directorate to provide initial program development in support of the National Oceanographic Laboratory System concept.

1971—The position of project officer for the National Oceanographic Laboratory System (NOLS) was established, and the Office for Oceanographic Facilities and Support was established to implement management support for the NOLS concept on March 30, 1971. The history of events relating to ship operations may be found in the contributions of Johrde, Toye, and Byrne in this volume.

IDOE was organized into four programs. These were Environmental Quality, Environmental Forecasting, Seabed Assessment Program, and Living Resources. The Ocean Sediment Coring Program was located within National Centers and Facilities Operations.

1974—On July 1 (the first day of fiscal year 1975), the Marine Chemistry Program was established in the Oceanography Section. This completed the four subdisciplinary science structure that has remained stable through fiscal year 1999.

1975—On July 10, 1975, NSF underwent a major reorganization into seven directorates, including the Directorate for Astronomical, Earth and Ocean Sciences (AAEO). Many sections became divisions, including those in this directorate. The Division of Ocean Sciences (OCE) was formed. The Offices of IDOE and Oceanographic Facilities and Support moved into OCE from the Office of National Centers and Facilities Operations (NCFO), and the Ocean Sediment Coring Program moved from NCFO to the Division of Earth Sciences.

Atmospheric Sciences was added to the directorate on September 30 to create AAEO, together with OPP. The Division of Ocean Sciences consisted of the Oceanography Section, the Office for Oceanographic Facilities and Support, and the Office for the IDOE.

1976—On April 19, 1976, the Marine Science Affairs Program was established within IDOE, and the Office of Polar Programs was redesignated as a division.

1978—The Office of the IDOE was redesignated the IDOE Section on March 8, 1978, with five programs (Environmental Forecasting, Environmental Quality, Living Resources, Marine Science Affairs, and Seabed Assessment).

1979—The Office of Oceanographic Facilities and Support was restructured into (1) the Office of the Head, Oceanographic Facilities and Support; (2) the Acquisition and Maintenance Program; and (3) the Operations Program.

1980—The 10-year mandated period for IDOE officially ended. Unlike the IGY and IIOE, however, the funds, which had been incorporated into the base of the Division of Ocean Sciences, remained there.

The Ocean Sediment Coring Program in the Earth Sciences Division was disestablished on October 1. The program's functions were redefined and reassigned to the Division of Ocean Drilling Programs, which was simultaneously established in AAEO and was comprised of (1) Office of the Division Director, (2) Science Section, (3) Engineering and Operations Section, and (4) Field Operations.

1981—On January 30, 1981, the Acquisition and Maintenance Program of Office of Oceanographic Facilities and Support (OFS) was divided into two programs—Ocean Technology, and Fleet Maintenance and Upgrading.

A MATURE SCIENCE

OCE was restructured on July 26, 1981, following the end of IDOE to accommodate its programs and funding. The research support functions of the IDOE and Oceanography Sections were merged into one section (Ocean Sciences Research Section [OSRS]) with eight programs:

1. Biological Oceanography
2. Oceanic Biology
3. Chemical Oceanography
4. Marine Chemistry
5. Seafloor Processes
6. Submarine Geology and Geophysics
7. Physical Oceanography
8. Ocean Dynamics

From the outset, there were in reality only four programs, in the four component subdisciplines, and four separate budgets, but for the next two years these eight programs appeared on official listings, creating considerable confusion within the community.

The Office for Oceanographic Facilities and Support was reorganized into the Oceanographic Facilities Support Section with the Oceanographic Technology Program and the Operations Program.

On August 3, 1981, the Office of Scientific Ocean Drilling was established within the Office of the NSF Director and the Ocean Drilling Program was moved into it from AAEO.

1982—The Oceanographic Facilities Program was established in OFS on April 12, 1982. The Office of Scientific Ocean Drilling was transferred, intact, from the Office of the Director to AAEO on November 14.

1983—Funds were added to the Oceanographic Technology Program to support a technology development component.

1984—The Ocean Drilling Program was established within OFS, and the section was renamed Oceanographic Centers and Facilities Section (OCFS). A little later, the nominal eight programs of OSRS were formally integrated by discipline, resulting in the Biological, Chemical, and Physical Oceanography Programs and Marine Geology and Geophysics Program.

Since 1984, the structure of the division has remained unchanged up to the time of writing, with the minor exception that the technology development component in OCFS

was transferred into OSRS as the Ocean Technology and Interdisciplinary Coordination Program. The transfer was made because over the previous decade it had become clear that technology development was mostly in service of research and frequently grants were jointly funded with one of the research programs. Thus, since 1993, there have been five programs in OSRS.

There will be much history eventually written about the development of the U.S. Global Change Research Program from 1984 to the present, and the gradual but very significant increase in budgets within OCE over this period. Only one element of this program, however, has yet been brought to completion (the Tropical Ocean and Global Atmosphere program; see Lambert in this volume), and several are still in their early stages. Some thoughts on the influence of this fourth wave of major ocean programs (after the International Geophysical Year, the International Indian Ocean Expedition, and the International Decade of Ocean Exploration) can be found in the recently published volume *Global Ocean Science* (NRC, 1999).

REFERENCES

Bush, V. 1945. *Science—The Endless Frontier. A Report to the President on a Program for Postwar Scientific Research*. U.S. Office of Scientific Research and Development, Government Printing Office, Washington, D.C.

England, J.M. 1982. *A Patron for Pure Science—The National Science Foundation's Formative Years, 1945–1957*. NSF Report 82-24. National Science Foundation, Washington, D.C.

Mazuzan, G.T. 1988. *The National Science Foundation: A Brief History*. NSF Report 88-16. National Science Foundation, Washington, D.C.

National Academy of Sciences (NAS). 1959. *Oceanography 1960-1970*. National Research Council, Washington, D.C.

National Academy of Sciences (NAS). 1964. *Economic Benefits from Oceanographic Research, a Special Report*. National Research Council, Washington, D.C.

National Science Foundation (NSF). 1962a. *10-Year Projection of National Science Foundation Plans to Support Basic Research in Oceanography*. Unpublished document.

National Science Foundation (NSF). 1962b. *Twelfth Annual Report*. National Science Foundation, Washington, D.C.

National Science Foundation (NSF). 1984. *Organizational Development of the National Science Foundation*. NSF Handbook Number 1. Unpublished document.

National Research Council (NRC). 1999. *Global Ocean Science: Toward an Integrated Approach*. National Academy Press, Washington, D.C.

President's Science Advisory Committee (PSAC). 1966. *Effective Use of the Sea*. Panel on Oceanography. President's Science Committee, Government Printing Office, Washington, D.C.

Ocean Sciences at the National Science Foundation: Early Evolution

MARY JOHRDE

National Science Foundation (ret.)

My comments focus on the first two decades of the National Science Foundation (NSF) and on the growth of facilities support, since ships and related shore facilities were essential to the development of ocean science—also because that's what I did at NSF.

I arrived at NSF about 40 years ago in early November 1958, as program assistant in a newly established Specialized Facilities Program in the Biological and Medical Sciences Division (BMS). BMS was one of two original divisions; the other was Mathematics, Physics and Engineering (MPE), which made up the research portion of the organization. NSF began in 1950, early in the post-World War II era, in a very modest fashion. At the outset it had a mandated budget ceiling of only $15 million. This ceiling was removed in August 1953 (fiscal year 1954), but still the annual budgets remained small.

Where was oceanography in this fledgling agency? Research grants for various phases of ocean-related sciences were handled by relevant disciplinary programs. In MPE the effort was somewhat focused within the Earth Sciences Program, which supported research in geology, geophysics, and geochemistry, key areas of interest to ocean science. In BMS, support was more diffuse and the definitions of what constituted ocean science were more difficult to pin down.

According to Dick Lambert, who has done an exhaustive job on the nonbiological NSF ocean science research support for 1950-1980, in fiscal year 1952 there were 3 Earth Sciences (ES) grants awarded of which one was for oceanography; by fiscal year 1954 there were 27 ES grants of which five were oceanography—not exactly a big splash for a beginning.

Though early support levels were minimal, by the mid-1950s NSF participated in such interagency activities as securing support for a National Academy of Sciences Committee on Oceanography (NASCO). The first major interagency role was assignment of responsibility in 1955 for the International Geophysical Year (IGY). Funding for IGY was handled as an entirely separate appropriation, and a separate office was established in NSF attached to the director's office. The magnitude of this assignment is reflected by the fact that IGY budgets equaled total NSF budgets for each of the fiscal years 1955-1958, and IGY's overall total of $55 million equaled NSF funding for the agency's first six years. The IGY field program began officially in July 1957.

The growing field of oceanography received a boost from IGY, but NSF-supported programs awaited the impact of two other events: the offspring of IGY, namely the International Indian Ocean Expedition (IIOE) and the national reaction to *Sputnik*, the Soviet satellite that beat us into space on October 4, 1957. This led not only to the formation of the National Aeronautics and Space Administration (NASA) and the space program but also to immediate talk about "inner space" and the importance of the oceans. By May 1959 the Federal Council for Science and Technology was in operation and had set up an Interagency Committee on Oceanography (ICO) to develop an annual National Oceanographic Program. We were into an era of proliferating structures for the coordination of scientific activities at the federal level. The President's Science Advisory Committee had a Panel on Oceanography whose first major report recommended federal reorganization and the formation of a "wet NASA." Agency budgets expanded, staff related to oceanography increased, and internal organizational structures grew more complex. Internal coordination became the rage with its own alphabet of committees, panels, groups, and so forth. NSF reflected it all.

This was the world into which the BMS Specialized Facilities Program emerged with its rather freewheeling approach to support for construction of buildings, boats and ships, equipment, boat basins, operational support for ships, summer research training at field stations, museum collection maintenance, and much more. A handful of such grants were made by BMS as early as fiscal year 1955, paving the way for initiating the program in fiscal year 1959. It was the kind of program that could only have existed early in the history of the agency and within the portion (BMS) that felt the need to play "catch up" with the larger division (MPE), where several major facilities projects had already been launched. It was also based on the conviction that biologists were going to need to do "their own thing" with respect to marine science facilities, including ships. BMS had declined, in the late 1950s, to participate with MPE in planning for a major new oceanographic ship—the *Atlantis II* for Woods Hole Oceanographic Institution (WHOI). Catching the groundswell of the post-*Sputnik* push in oceanography and further motivated by the prospect of IIOE, BMS Specialized Facilities Program was soon "specializing" in ship construction and conversion.

IIOE was the answer to the "what-next" question following IGY. It was planned and managed by NSF programs directly involved in oceanographic support. The Earth Sciences Program had added John Lyman as associate program director for oceanography in the late 1950s and Dick Bader as an assistant program director by mid-1961. In addition to the new facilities program, BMS brought in marine biologist Dixy Lee Ray as a special consultant to the division director of BMS, John Wilson. Reorganizations, expansion, and program additions were occurring throughout the Foundation. The BMS Facilities Program began under the direction of Louis Levin, then deputy division director, who subsequently moved to head the newly established Office of Institutional Programs in May 1961. Harve Carlson, returning from the Office of Naval Research (ONR) in London moved into Levin's position only to move up in November 1961 to become assistant director of BMS when John Wilson became deputy director to Alan Waterman. Jack Spencer was recruited to head the facilities program. In May 1961 an independent Office of Antarctic Programs (OAP) was separated from the director's office and became involved in oceanography.

Probably the first grant for IIOE was awarded in fiscal year 1961 to NASCO to perform certain aspects of external coordination. Internally, the BMS independence about ships for biologists required that active coordination between MPE and BMS be developed at the outset of planning for IIOE. In September 1961, a first meeting regarding ships for IIOE was attended by John Wilson, Carlson, Lyman, and Ray. By then Lyman and Ray had already associated long enough to have developed their infamous coordinating style from which many quotable quotes emerged. It was fortuitous that Dick Bader arrived when he did and picked up the role of facilities coordinator. In October 1961, he began action to establish a facilities panel for ES Oceanography. It became a memorable road show in 1962, as it undertook a series of site visits to virtually all academic institutions then engaged in ocean-related programs. The purpose was to assess the quality of staff, research, and training activities, and above all the extent and nature of equipment and facilities available for these programs.

In short order, a specialized BMS Facilities Program had three ships underway: construction of the *Eastward* at

Duke University Marine Lab to be a research and training ship for biologists from any institution; conversion of the motor sailer *Te Vega* at Stanford's Hopkins Marine Station, for research and training and use as an adjunct vessel for the IIOE; and consideration of the former presidential yacht *Williamsburg* for the major biological ship for Indian Ocean cruises. The objective had been "no new ships solely for IIOE," but attempts to find a suitable leased ship for the biologists had failed. The *Williamsburg* stirred controversy from all sides. I first met Dick Bader at a meeting about the *Williamsburg* chaired by Dixy. He and I rose from opposite sides of the table and questioned the same well-advertised problems about this old vessel. Henceforth, we worked together on planning for the ship's conversion, which led to my participation as a BMS representative on Bader's facilities panel and to the transfer of funds back and forth between the BMS and ES programs for joint support of ships and other items. BMS built one more significant ship, the *Alpha Helix*, which was used for expeditionary field biology and medical research sometimes unrelated to the marine environment, but for most of the 1960s ships provided to academic institutions were conversions and jointly funded by BMS and MPE.

The *Atlantis II*, of course, was really a ship from the late 1950s; WHOI had the initial grant in hand in December 1959, but IIOE probably smoothed the way for significant cost overruns and problems along the way. This was the first ship specifically designed for research at sea within the category now known as the academic fleet. Upon completion and commissioning in 1963 the *Atlantis II* headed directly for the Indian Ocean. And ultimately *Anton Brunn* (aka *Williamsburg*) and *Te Vega* followed.

By 1965, Earth and Atmospheric Science Sections were split off from MPE and in combination with Office of Antarctic Programs formed a new Division of Environmental Sciences (DES). But informal coordination between DES and BMS was still the basis for a semblance of unity for ocean science in NSF. In 1966 following Dick Bader's departure, I moved into an associate program director position in Earth Sciences. Taking a page from Dick's book, my first attempt was to establish a single NSF Advisory Panel for Ship Operations, to create formal procedures for block funding for ships and guidelines for evaluating and managing this very essential support for oceanography. The panel and annual review process were in place by the time the Oceanography Program in DES had evolved into a section (1967) and BMS had finally expanded its organization with the establishment of a Biological Oceanography Program in a newly created Environmental and Systematic Biology Section (1968). Ed Chin was the first program director and Jean DeBell was associate program director; I was now program director for an Oceanographic Facilities Program, DES.

A closing footnote on BMS Specialized Facilities: in the eight years that I was associated with the program, awards totaling $13.9 million were made for oceanographic items and $22.9 million for all other categories. For six of the eight years the program had separate funds for operational support of facilities and these too were divided between oceanographic ($9.1 million) and all other ($5.3 million). It was a program of considerable significance for those years, and to its everlasting credit it provided the first block grant (fiscal year 1959) for ship operations support, as a format for future ship support. In spite of major difficulties in adapting the old ship for biological cruises, the *Anton Brunn* served well for two years in the Indian Ocean and during a third year off the west coast of South America. The *Eastward* gave biologists an opportunity to learn oceangoing techniques and to work effectively with other disciplines. With its interinstitutional programs, *Eastward* could be said to have led the way toward the University-National Oceanographic Laboratory System (UNOLS). And finally, *Alpha Helix* managed to have a second career as a general purpose ship for the University of Alaska.

In 1968 further consolidations were needed to create a better structure for ocean sciences inside NSF. It is of interest to note, however, that ocean-related activities were scattered far more widely and variously among the agencies around town. Congressional cries for a more coordinated effort and a single or predominant agency culminated in passage of the Marine Resources and Engineering Development Act of 1966, which established yet another council and a Commission on Marine Sciences, Engineering and Resources, the Stratton Commission.

The January 1969 report from this study group recommended the formation of the National Oceanic and Atmospheric Administration (NOAA) and, among many other items, that a small group of (academic) institutions be designated by the federal government as University-National Laboratories (UNLs) and be equipped to undertake major marine science with broad continuing support. Also, support for these UNLs should not preclude support for other existing institutions. Understandably this recommendation set off wild hopes, fears, and expectations among the academic oceanographic institutions (see paper by John Byrne and Bob Dinsmore, this volume). Would there truly be sub-

stantial sustaining support for a few? And what would happen to all the others?

The debate about how to respond to the report was underway in October 1969, when William McElroy, who became the third NSF Director in July, set in motion a major reorganization of NSF based on provisions of the National Science Foundation Act of 1968. Four assistant directorships were established: Research, Education, Institutional Programs, and National and International Programs. The latter became the home of Antarctic Programs, International Decade of Ocean Exploration, and Oceanographic Facilities and Support (OFS) among others. OFS became the site for developing the UNOLS concept, which derived from the commission report's UNLs. UNOLS stands for University-National Oceanographic Laboratory System, as you know, and it attempts to achieve some of the goals set forth by the Stratton Commission.

Early in the UNOLS debate while Scripps still clung firmly to the hope of achieving national laboratory status, Paul Fye at WHOI read the handwriting on the wall and decided to take a lead position with respect to whatever this UNOLS thing was going to become. He lent Art Maxwell of his staff to the planning process and recruited Capt. Robertson Dinsmore, then retiring from the U.S. Coast Guard, as a member of WHOI's facilities operations. Art Maxwell had been a participant on Dick Bader's facilities panel; Bob Dinsmore had spent his last Coast Guard tour in Washington on the oceanographic scene. Thus, when UNOLS became a reality, Paul Fye was in position to host the executive offices and offer Bob Dinsmore as first executive secretary. He wisely declined federal support for Bob's salary, saying the executive secretary of UNOLS should belong to the community. It was Bob who put UNOLS together as an operating organization. His knowledge of ships, oceanography, and above all the Washington scene, was invaluable to the process.

I leave the rest of the NSF organization story to Sandra Toye and the UNOLS story to John Byrne and Bob Dinsmore.

Ocean Sciences at the National Science Foundation: An Administrative History

SANDRA TOYE

National Science Foundation (ret.)

ABSTRACT

This paper traces the National Science Foundation's (NSF) organizational decisions from 1950 to the present and the place of the ocean sciences within that administrative structure. With its interdisciplinary character and its reliance on small and large-scale research, facilities, and instrumentation, oceanography has often been a test case for NSF management. Today, an oceanographer approaching the National Science Foundation for support will have no trouble finding a point of contact: one telephone call or a visit to one suite of offices brings the scientist to the Division of Ocean Sciences. Here the scientist can explore NSF programs and policies; research projects of any size; ship operations; instrumentation; international or interdisciplinary programs.[1]

CURRENT NSF ORGANIZATION FOR THE OCEAN SCIENCES

The administrative structure of the Division of Ocean Sciences is comprehensive and straightforward. The Ocean Sciences Research Section supports research projects large and small in physical, biological, and chemical oceanography; marine geology and geophysics; and oceanographic technology. The Oceanographic Centers and Facilities Section manages NSF support for research ship operations and construction, specialized facilities operations, the Ocean Drilling Program (ODP), and instrumentation and technical support.

Moreover, the Ocean Sciences Division is part of a larger organization, the Directorate for Geosciences, which encompasses closely related fields—the Earth and Atmospheric Sciences. The intellectual ties among these fields are mirrored in an administrative structure that ensures coordination and integration in such inherently interdisciplinary activities as climate research. This strengthens the National Science Foundation's ability to represent the geosciences properly in interagency and international forums, and in policy and budget negotiations with the Administration and Congress.

But it hasn't always been this way. For most of the 50-year period covered by this symposium, NSF management of the ocean sciences has been fragmented. Major aspects of the field have been lodged in different parts of the Foundation with sharply different management styles, and often with differing scientific views and objectives. The story of how the ocean sciences came together within NSF reflects many larger trends and issues in Foundation management philosophy. It also tracks the evolution and maturation of the ocean sciences themselves.

THE 1950S, NSF'S FIRST DECADE OF INDIVIDUAL INVESTIGATOR SUPPORT: OCEAN SCIENTISTS SEEK A NICHE

NSF's Initial Organization for Research Support Proves to Be Lasting

On May 10, 1950, President Harry Truman signed into law the act that created the National Science Foundation. For the first two years, the focus of the Foundation was getting itself organized, which it did with remarkable care and foresight. The decisions made in that period still govern the fundamental operating style of NSF fifty years later.

The support of individual investigators was identified as the fundamental research mission of the agency, with sup-

---

[1]The single exception is polar oceanography, which is managed by the NSF Office of Polar Programs.

port to be provided through permissive grant mechanisms rather than contracts. Advisory committees and peer review of proposals ensured strong input from the external community. NSF's first research grants, 28 awards ranging from $780 to $50,000, were made in February 1952, completing the expenditure of NSF's $3.5 million budget. Among these was an award for oceanographic research.

The administrative decisions taken in NSF's early years were also destined to endure. Education programs and science information activities were to be centralized, with a single office or division managing the assigned programs to all institutions and across all fields of science and engineering.

Research project support was to be handled differently. Responsibility was distributed between two research divisions—one for Mathematical, Physical and Engineering Sciences (MPE) and one for Biological and Medical Sciences (BMS). [2] Within each division, programs would be established to handle proposals from a given discipline or subdiscipline.

From the outset, there was a significant difference in the two sets of disciplines that shaped the program-level definitions of the two divisions. MPE served intellectually related but operationally separate communities of practitioners— physicists, chemists, Earth scientists, mathematicians, astronomers and engineers—with distinct research agendas and arrays of instruments and equipment. Moreover, these scientists generally held positions in university departments corresponding to the MPE program boundaries.

In contrast, BMS covered essentially one very large discipline. By necessity, BMS programs were defined by the thrust of the proposed research activity—whether it addressed regulatory or molecular or developmental aspects of the organism or system being studied. Researchers who shared appointments in the same department and used the same research equipment might draw support from different BMS programs.

The initial organization of NSF research support existed virtually unchanged into the 1960s. The underlying philosophy of managing research by academic discipline was even longer lived: it remains the organizing principle of NSF today. Despite the obvious strength and endurance of the disciplinary concept, it did, and still does, pose difficulties for the assessment and management of research that does not fit within the prescribed program boundaries.

Oceanography: Below the NSF Disciplinary Horizon

For oceanography, an inherently interdisciplinary field, NSF's early organizational choices created problems that

would not be fully rectified for 25 years. Each proposal for ocean research would compete in the larger field in which it had its intellectual roots. This meant that oceanography proposals were sometimes handled by program managers and reviewed by intellectual peers who might have little or no exposure to the unique demands and opportunities of ocean research. The problem was particularly acute for field programs with expensive requirements for research vessel time and other specialized facilities and instruments.

One subset of the ocean sciences did find a receptive home in NSF. The MPE Division's Earth Sciences Program handled proposals in geology, geophysics, and geochemistry. These fields were at the threshold of the intellectual revolution of plate tectonics. Marine practitioners of the geological sciences were deeply involved in this revolution, and research conducted at sea was at the heart of the ferment. Thus, from the outset, oceanographers were influential players as grantees as well as advisors and reviewers.

The organizational misfit between the ocean sciences and NSF's administrative structure during the 1950s did not become a policy issue for several years. Coming out of World War II with strong ties to the Navy and the Atomic Energy Commission, oceanography found its principal needs well supported by those agencies. NSF, as a newcomer with heavy obligations to other fields, was initially not a significant player in oceanography. By the end of the decade, that situation would begin to change.

The First Watershed: The International Geophysical Year (IGY 1957-1958)

The IGY is often cited as NSF's most enduring venture into "big science," resulting in the permanent addition of international cooperative programs and the U.S. Antarctic Program (USAP) to the Foundation's research portfolio. But the IGY was equally important for the changes it brought about in NSF's outlook toward support of individual investigators, particularly in the environmental sciences.

In 1955, largely at the urging of the National Academy of Sciences, NSF was selected as the lead agency for planning and managing U.S. participation in the IGY. Given the multi-disciplinary nature of the project, it was clear that the IGY would not fit in either of the existing research divisions. A special coordinating Office for the IGY was set up in the Office of the NSF Director, a pattern that the agency would follow repeatedly as new programs were assigned to it over the next decade.

The IGY itself was unquestionably "big science"; it would ultimately involve 30,000 scientists and technicians from 66 countries in a comprehensive study of Planet Earth. The budget for U.S. participation in the 18 months of field operations totaled $43.5 million. Despite the size and complexity of the IGY, it fit well with NSF's interests and priorities. It was essentially science-driven, despite its political and diplomatic aspects. For all its size and coordination, it was not

---

[2]NSF's charter prescribed a separate Division of Medical Science, but this entity was folded into the Division of Biological Sciences almost immediately.

a single large project so much as an aggregation of complementary projects, offering opportunities for involvement to scientists in many fields.

The IGY also had administrative features that made it easy for NSF to accommodate. First, it was time-limited: theoretically, at least, it created no long-term commitments for NSF. Even more important, the IGY budget was funded entirely by "new money"—appropriations over and above those for ongoing NSF programs.

THE 1960S: NSF OCEAN RESEARCH IN THE WAKE OF THE IGY

The impact of the IGY experience extended into NSF's traditional research support structure for "small science." Despite the finite limits of the IGY itself, its field programs produced new ideas and generated data that resulted in research proposals long afterward. NSF's role in the IGY made it the natural recipient of proposals of this sort. Furthermore, IGY scientists had enjoyed both the intellectual enrichment and the logistical and financial feasibility offered by coordinated programs. They continued to propose cooperative field programs and other forms of collaborative research that NSF's disciplinary program structure was not equipped to handle.

The IGY greatly increased the visibility and reputation of environmental research. It became apparent that these fields often had research objectives and requirements that were fundamentally different from those of the larger disciplines in which they were intellectually based. As we have seen, MPE's basic structure made it possible for an emerging discipline to argue for a program of its own. By 1959, the Atmospheric Sciences enjoyed separate program status in the MPE division.

Outside NSF, in the aftermath of the IGY, oceanography was widely recognized as a legitimate academic discipline with its own set of research imperatives. The National Academy of Sciences Committee on Oceanography (NASCO) and the Interagency Committee on Oceanography (ICO), part of the new Federal Council on Science and Technology, were actively engaged in policy recommendations to expand ocean research and education.

Within NSF, however, oceanography was still not a recognized discipline, and ocean research was still dispersed across the agency. Throughout the early 1960s, NSF created a succession of internal coordinating groups to respond to the growing external requirements of NASCO and ICO as well as to deal with the unique operational and logistic needs of the growing cadre of NSF-supported oceanographers. In 1963, NSF expenditures for oceanography totaled $26.8 million, a sum that exceeded the budgets of many established NSF programs. The agency's inability to deal with ocean research and policy in a coherent way made for difficulties in dealing with the Academy, other federal agencies, and the science community itself.

1960s Reorganizations Bring a Degree of Unity to NSF Ocean Sciences Program

The early 1960s were a period of expansiveness and optimism about government programs in general and science and technology in particular. It was the era of the space program and intense competition with the Soviet Union for scientific dominance. NSF was the recipient of responsibility for many of the new programs. By 1962, the Office of the NSF Director was crowded with a plethora of special offices that had been created as ad hoc responses to new program responsibilities.

The time had come to fold these programs into NSF's line organizations, which had been largely unchanged from its establishment in 1950. As part of the agency-wide consolidation, the formerly independent BMS and MPE Divisions, along with several other research support programs, were brought together under a new organization, headed by an Associate Director for Research (AD/R).

MPE, now a Division of AD/R, took advantage of the new situation to restructure its portfolio. Because of its disciplinary substructure, MPE was able to react to the emerging identity of the environmental sciences as fields in their own right. The former Earth Sciences Program was elevated to the status of a section, putting it on an organizational par with Mathematics, Physics, Chemistry, Engineering and Atmospheric Sciences. The new Earth Science Section established four component programs: Geology, Geophysics, Geochemistry, and *Oceanography*. For the first time, the field could point to a "home" in NSF, but it was not comprehensive, covering only submarine geology and geophysics (SG&G) and physical oceanography.

Over the next five years, the evolution and elevation of the environmental sciences accelerated. In 1965, AD/R created a Division of Environmental Sciences. The new division subsumed the Atmospheric Sciences Section and also assumed responsibility for the Office of Polar Programs, which had been shifted among several organizational settings in its short lifespan. In 1967, the Division of Environmental Sciences created a fourth section—Oceanography. The section included physical oceanography, SG&G, and in a significant departure from previous structures, an oceanographic facilities program.

BMS did not use the 1962 reorganization as an opportunity to rethink its structure. At that point, BMS had nine major program areas, and oceanography proposals were handled in several of them. In 1965, the BMS programs were reorganized into a more hierarchical structure, with sections having responsibility for several related programs. This eased coordination problems somewhat, but biological oceanography proposals still straddled too many organizational lines. Finally, in 1968, a Biological Oceanography Program was established in the Environmental and Systematic Biology Section.

Oceanographic Facilities Support: A Special Problem

NSF's charter does not mention equipment or facilities. From the outset, however, NSF policymakers decided that instrumentation and facilities were an inherent part of NSF's mission to support research. In the first round of research grants in 1952, some of the budget was earmarked to help institutions acquire instrumentation for shared use.

In oceanography, shared-use facilities, particularly research vessels, are an inextricable aspect of the enterprise. This was among the characteristics that made ocean science research proposals difficult for NSF program managers to handle. In an era when $15,000 was considered a generous grant budget, reviewers and program managers were hard-pressed to deal objectively with ship costs that might double or triple the budget of a project grant. Fieldwork would often be whittled back in budget negotiations to a point that undermined the research objectives. Sometimes program managers refused to pay for ship time at all, leaving researchers to get aboard a research vessel as best they could. Even in programs or sections that dealt primarily with oceanography proposals, ship costs were an unwelcome demand on research budgets, and funding for them was uncertain and uncoordinated.

Dealing with cooperative field programs and shared-use instrumentation was a particular problem in the life sciences. The BMS Program structure, as we have seen in prior sections, tended to cut across subdisciplines or academic departments. It was an effective way to compare the merits of competing research ideas, but it did not provide a good setting for looking at cooperative projects or shared-use research equipment. In 1958, responding to criticisms that largely originated with oceanographic institutions, BMS created a small fund for Special Programs and Instrumentation to deal with such proposals. By 1960, Facilities and Special Programs graduated to full program status in BMS. Interestingly, NSF's first grant for research ship operations came from this program.

NSF's diffuse program management was also a problem for the institutions that operated research ships. Sending a ship to distant waters is a complex and expensive operation, requiring months and sometimes years of preparation. It is only worthwhile if there is a body of research large enough to share the costs and justify the commitment. When the proposals for a single cruise or expedition were under review in many different NSF programs, with independent management styles and funding schedules, it was difficult to gather the critical mass of approved projects in a time frame that matched the planning period required for ship commitments.

In the early years, this problem was minimized because the Office of Naval Research (ONR) was the major funder of oceanography. Administratively, ONR used block-funded contracts that covered all of the research, instrumentation, and ship costs for its projects at a given institution. This provided a sufficient framework for ship operators to set plans for cruises into distant water. With these commitments in place, scientists could approach NSF for grants that might add to or complement the cruise objectives.

Throughout the 1960s, as NSF support for ocean research became a more significant fraction of the total funding for the field, the facilities support issue became more pressing. Dealing with the ship support problem was part of the mission of all of the internal NSF ocean science coordinating bodies mentioned in the preceding section. When an Oceanography Section was created in the Division of Environmental Sciences in 1967, an Oceanographic Facilities Program was part of its portfolio.

Research Ship Construction and Conversion in the 1960s

In 1962, AD/R established an Ad Hoc Panel on Grants and Contracts for Ship Construction, Conversion, and Operations to advise NSF on a set of procedures for handling these areas. Given the lack of focused NSF programs in ocean science and the chronic problems of paying for ship time for researchers, it is surprising to find that proposals for ship construction found support at NSF in this era.

In fact, NSF funded the construction of three oceanographic research ships and the conversion of several others during the early 1960s. BMS Facilities and Special Programs funded the construction of R/V *Eastward* (Duke University) in 1962 and R/V *Alpha Helix* (Scripps Institution of Oceanography) in 1965; MPE's Earth Sciences Program supported the design and construction of Woods Hole Oceanographic Institution's *Atlantis II* (1963). The justification for the funding of these ships underscores the fragmentation of NSF's treatment of the ocean sciences during that period. Each was proposed as a specialized facility, outfitted for the needs of the supporting discipline.

All three ships eventually became general-purpose oceanographic vessels, but the two BMS-funded ships kept their ties to special biological programs for more than a decade. Because ship construction was funded by standard research grants, NSF exercised little management direction of the design and construction projects, and the completed ships became the property of the institutions that built them. In later decades, this policy would be criticized, and NSF would alter its procedures to be more proactive in managing construction projects, retaining title, and assigning ships to operators through special contracts.

Project Mohole 1957-1967

From its start in 1957 until the project office closed its doors in 1967, Project Mohole was among NSF's most controversial undertakings. Mohole had its roots in the review of regular NSF disciplinary science projects. At an Earth Science Advisory Committee meeting in 1957, the concept of drilling through Earth's crust into the Mohorovicic discontinuity was aired as a serious proposal. Over the next few years, the Earth Sciences Program supported grants for feasibility studies and field tests. By 1960, the project required funding and management oversight beyond what the program could provide.

In 1960, NSF was no longer a newcomer to large scale facility-based operation, having run the IGY. In addition, the agency had been funding construction and operation of astronomy centers and physics facilities for several years. These projects had been marked by their fair share of management problems. Despite these experiences, when Mohole reached developmental stage, NSF still had no policies or procedures in place for management of, or even decision-making about, projects that required large-scale capital investment and long-term operational commitments. Every decision was an ad hoc matter, usually requiring personal involvement of the NSF Director. For each such project, NSF would seek "new money": an additional appropriation outside its ongoing budget. Fortunately, in the expansive 1950s and 1960s, funds were generally made available.

The early Mohole studies and initial field tests had been carried out by AMSOC, a special committee of the National Academy of Sciences. But their charters precluded both the Academy and the Foundation from direct operation of projects. The NSF Director had recently been involved in disputes between the academic consortium managing one of the astronomy construction projects and the commercial subcontractors actually fabricating the equipment. Perhaps because of this experience, NSF decided to contract directly with a commercial firm for the technically demanding developmental phases of Mohole. After an extremely contentious competition of nearly two years' duration, in 1962 NSF entered into a contract with Brown and Root, an engineering firm with no experience in scientific management.

By this time, strains of every kind had begun to afflict the project. Scientific disagreements had emerged about the extent and phasing of the developmental work; the contract competition drew political fire; personality disputes had emerged; and finally, NSF decisions about management of the project were criticized. The underlying management concern, from NSF's viewpoint, was to maintain accountability and control over the very large contract budget and the challenging engineering problems. From the point of view of Mohole's proponents in the community, the issue was to maintain clear and competent scientific oversight.

NSF established the position of Managing Coordinator for Project Mohole, and appointed an engineer with the requisite technical experience to the job. The Mohole Project Office was attached to the Office of the NSF Director, but because of the huge budget and administrative implications of the contract, the coordinator actually reported to NSF's Associate Director for Administration. Scientific guidance was to come from the Academy, with a NSF program officer from Earth Sciences acting as Science Coordinator in-house.

Policy guidance came to the Managing Coordinator from a committee comprised of the Associate Director for Administration, the Science Coordinator, and the NSF Director's executive assistant. Quite apart from the scientific, technical, and political problems confronting the project, it would be hard to conceive a more unworkable managerial scheme.

For the next three years, Project Mohole pursued a mercurial course; sometimes appearing to be well underway, only to fall prey to cost overruns, technical barriers, and scientific disagreements. In 1966, the Congress denied NSF's request for further funding of the project. One year later, the project office closed its doors and Mohole entered the history books.

The Origins of the Deep Sea Drilling Project

Oceanographers were among Mohole's leaders from the project's conception to its demise; but perhaps because of the lack of a clear disciplinary identity for oceanography in NSF at the time, the field escaped much of the blame for its failure. Indeed, in NSF's institutional lore, oceanographers are credited with having "rescued something of great value" from the traumatic Mohole experience—the Deep Sea Drilling Project (DSDP).

Like everything about Mohole, there were arguments about the origins of DSDP. Some saw it as a preliminary test program for Mohole technology; others considered it a worthy project in its own right. For oceanographers whose research interests lay in the oceanic sediments and underlying crust, the idea of a separate ocean sediment coring program gained momentum.[3]

In 1963, in the midst of the Mohole controversy, NSF's second director took office. He was disturbed by the disunity of the academic leadership of the project and the lack of scientific capability of Brown and Root. In one of his first meetings with Mohole's proponents, the director urged the development of a scientific consortium that would eventually take over management of the program. He also expressed enthusiasm about the sediment coring concept, and indicated that NSF might consider it as "a companion program" to Mohole. In Congressional testimony in the fall of 1963, NSF went considerably further, stating that the agency was prepared to support an entirely separate sediment coring program if funding were made available for it.

Proponents of the sediment coring program had made short-lived attempts to organize a management consortium in the previous years. Spurred by NSF support, in 1964 four ocean research institutions created the Joint Oceanographic Institutions for Deep Earth Sampling (JOIDES) and proposed it as the scientific management entity for the new pro-

---

[3]This is the origin of the bureaucratic title given to the DSDP in NSF budget and organization documents—the Ocean Sediment Coring Program (OSCP).

gram. In a departure from other existing academic consortia such as Associated Universities, Inc., which managed research centers as corporate entities, JOIDES did not incorporate, indicating that one of its members would serve as the operational contractor.

In 1966, Congress provided $5.4 million in "new money" to start the ocean sediment program. NSF accepted a proposal from a JOIDES member, the Scripps Institution of Oceanography, to operate the program; scientific guidance would be provided by JOIDES. Two years later, with *Glomar Challenger* as its platform, DSDP began one of the most productive scientific ventures in NSF history (see Winterer paper in this volume).

Sea Grant: 1966-1970

In 1966, Congress passed the National Sea Grant College Program Act. Sea Grant was modeled on the Land Grant concept that had left an indelible mark on higher education a century before. NSF was assigned responsibility for the new program.

Sea Grant included components that cut across every line organization in NSF—education, basic and applied research, institutional support, and public outreach. In earlier times, NSF would have created a special management office reporting to the NSF Director. Given the nature of Sea Grant, that would probably have been a good choice in this case. However, just having undergone a series of reorganizations designed to assign such functions to line operating units, NSF decided to place Sea Grant under the Associate Director for Research.

The Office of Sea Grant invited proposals in 1967 and made its first awards, totaling $2 million, the following year. The program had its critics. Among the most vocal were other marine research institutions that felt that Sea Grant awardees were not always held to the same standards that were exacted in "standard" research support programs. Because of the administrative decision to place the program in the Research Directorate, such comparisons were probably inevitable.

By the closing years of the decade, the new Nixon Administration was weighing the report of the Stratton Commission, a group appointed by the previous Administration to examine ocean policy issues. One of its recommendations was the establishment of the National Oceanic and Atmospheric Administration (NOAA). Sea Grant was among the programs proposed for assignment to the new agency. In the interagency Marine Council, NSF expressed misgivings about creation of the new agency and reassignment of Sea Grant, but was overruled on both points.

THE 1970S: ANOTHER WATERSHED

It is hard to imagine a sharper contrast than between the optimism and expansiveness of the early 1960s and the pes-

simism, disenchantment with government programs, and social unrest of the end of the decade. The primary cause was the Vietnam War. For the nation, the science community, the Foundation, and the ocean sciences, it was a time of profound change.

As the budgetary and social pressures of the Vietnam War increased, Navy support for academic oceanography began to decline. The Mansfield Amendment, attached to a Defense Department procurement bill that took effect in 1970, made it unlawful for the Department of Defense to fund projects in basic science unless they were clearly related to a military function or operation. The chilling effect of the prohibition was felt at once throughout the research community. In the ocean sciences, in the space of a few years, ONR dropped from dominance to a minority position in the support of academic research.

Proposals for creation of NOAA, under discussion for several years, would come to fruition in 1970. It had been argued that one of the possible roles for the new agency would be directing centralized operations of regional research fleets. Although this concept was not embodied in the NOAA legislation, it was still popular in some circles, and would recur in one form or another over the next two decades. It helped to spur NSF to take seriously the recommendations for creation of a National Oceanographic Laboratory System (NOLS) to coordinate academic ship operations.

NSF's Management Style Comes Under Attack

By the end of the 1960s, NSF's management capabilities had come into question in many quarters. Mohole was a public embarrassment, and several NSF education programs had become philosophically and politically controversial. Although they had drawn less public attention, some of NSF's ventures into construction of astronomy facilities had encountered management problems that were well known in the Administration and Congress.

The new Republican Administration was intent on curbing the growth of some of the programs established in the prior decade, and budgets were pressed by the costs of the ongoing Vietnam War. At the same time, the Administration wanted to be sure that civilian agencies picked up some of the research support being dropped by the Department of Defense, particularly research with economic and social relevance. Some Presidential advisors felt that NSF was too passive and not sufficiently concerned with managerial and budgetary realities to be trusted with new programs. A new NSF Director was appointed and given instructions to "clean house."

The Reorganization of 1969-70: Major Changes for the Ocean Sciences

The new NSF Director was given a significant administrative tool in the form of a law that provided him, for the first time, five Presidentially-appointed subordinates—a deputy director and four assistant directors (ADS). He undertook a complete reorganization, bringing in new people to fill the new posts. Assistant directors for Institutional Programs, Education, and Research were named, all with instructions to streamline their respective organizations.

The new Assistant Director for Research (AD/R) decided to unify, at last, the biological and physical sides of ocean sciences. The Biological Oceanography Program was transferred to the Ocean Science Research Section (OSRS) of the Division of Environmental Sciences, joining the existing programs in Physical Oceanography and SG&G. A few years later, Marine Chemistry would be established as a separate program, rounding out the OSRS offerings.

The last of the new AD positions was used to create the Directorate for National and International Programs (AD/NI). A former Chief of Naval Research was appointed to the position and charged with nothing less than revolutionizing NSF's approach to coordinated research and large-scale facilities and centers. The astronomy observatories, National Center for Atmospheric Research (NCAR), and the DSDP were pulled out of their respective research areas and brought together in an Office for National Centers and Facilities (NCF). The Office for Antarctic Programs was transferred from AD/R, as was Sea Grant. The latter, however, would be transferred to NOAA just a few months later.

Responsibilities for oceanographic facilities were also transferred from AD/R, and AD/NI was given the additional duty of Special Project Officer for NOLS—the National Oceanographic Laboratory System—the emerging fleet coordinating entity. A short time later, these functions would be brought together as the Office of Oceanographic Facilities and Support (OFS).

In the late 1960s, the specialized agencies of the United Nations had generated recommendations for a coordinated international research effort in the world's oceans. The idea found strong support among U.S. science advisors. In 1969, the White House announced a special Presidential initiative, the International Decade for Ocean Exploration (IDOE), and assigned responsibility to NSF, along with $15 million in "new money," NSF's favorite currency. The IDOE assignment went to AD/NI.

In many ways, the 1970 reorganization was a major step forward for the ocean sciences. AD/R's unification of all of the sub-disciplines in the OSRS was an essential and overdue recognition of the comprehensiveness of the field, and enabled NSF to interact more rationally with the community. AD/NI's emphasis on management and accountability brought significant improvements to the Foundation's oversight practices for centers and facilities. For OFS and IDOE, AD/NI proved to be an excellent incubator for the special management attention needed to gear up new programs and create the necessary interagency and international linkages.

On the other hand, the separation of facilities and "big science" from the research project support aspects of their respective disciplines was a controversial move. At a practical level, it created bureaucratic coordination problems for NSF staff and the affected research communities. Perhaps of more concern in the long run, the reorganization underscored the long-standing tensions between "big" and "small" science by making them direct competitors for NSF resources.

For the ocean sciences, one of the fields most affected by the split between the project research "base" and the larger programs, the reorganization came at a particularly important time. Between the new funding brought into the field by the IDOE and the ongoing reduction in ONR support, NSF had become the lead agency in the support of academic oceanographic research. To some extent, the Foundation's ability to act effectively in that role was weakened by the divided administrative structure for the field.

The International Decade for Ocean Exploration

This paper will offer only brief comments on some of the organizational aspects of IDOE; Feenan Jennings discusses IDOE's scientific legacy later in this volume. As indicated in the prior section, IDOE was created as a Presidential initiative.

The role IDOE set for itself was the sponsorship of a small number of large-scale long-term research projects, drawing on the expertise of specialists from all disciplines, to address scientifically challenging and socially relevant problems in the oceans. Despite its ambitions to support truly inter-disciplinary work, each of IDOE's four major program areas had strong ties to one of the component fields of Ocean Sciences: Seabed Assessment (SG&G); Environmental Forecasting (Physical Oceanography); Environmental Quality (Marine Chemistry); and Living Resources (Biological Oceanography). IDOE made extensive use of planning workshops to achieve the coordination and integration required to meet the program's objectives.

The workshops were also helpful in developing proposals that would consistently meet NSF quality standards. IDOE was committed to the NSF tradition of peer review, using both ad hoc mail and panel reviews. This multi-level review generally ensured excellence in the research core of IDOE projects. However, in the large-scale programs, there were components such as data archival, site surveys, and instrument development that were essential to the overall scientific objectives, but not particularly exciting in their own right. Reviewers more accustomed to looking at stand-alone proposals were sometimes unduly critical of proposals of this type, assigning tepid ratings that made funding hard to justify.

As originally conceived, NSF would pass along as much as half of the IDOE budget to other federal agencies. But the proposals from mission agencies generally fared poorly under peer review. Moreover, mission agencies found it hard to subscribe to the broader goals of IDOE, tending to limit their proposals to components compatible with their ongoing programs. As a result, academic researchers, with only modest engagement of other agencies, eventually carried out most IDOE programs.

Facilities Programs:
OFS and UNOLS Evolve Together

The establishment of the Office for Oceanographic Facilities and Support (OFS) and the creation of the University-National Oceanographic Laboratory System (UNOLS) coincided with NSF's abrupt assumption of the dominant role in ocean science funding.

Today the "academic fleet" is widely acknowledged as a capable and efficient research establishment. In 1970, the label of "academic fleet" would have been a misnomer; each institution tried to maintain its own operation, competition for funding was intense, and there was little incentive for cooperation in scheduling. In the rapid growth of the preceding decades, it had been almost too easy for institutions to obtain ships. Military surplus ships were converted for research use, as were yachts and tuna clippers. While some of the conversions served well, others were poorly maintained and outfitted, or simply not properly configured for research. Basically, there were more ships in operation than the system could afford.

The task facing OFS and UNOLS was thus not simply to compensate for the decline in Navy funding, but to change the community's way of doing business at sea. Having a ship had become part of an institution's identity as a center for marine research. If institutions were to be persuaded to give up inefficient ship operations, they would have to be convinced that they could still be serious players in ocean research. The key to that assurance would be to ensure that any scientist with a legitimate need for ship time could have access to the supported fleet. That required a change in attitude on the part of operating institutions as well as the scientific community.

The major instrument of that change was UNOLS. As with JOIDES, UNOLS was not an incorporated body, but rather an association of oceanographic institutions—ship operators as well as ship users. One of the participating institutions would serve as home base—the Woods Hole Oceanographic Institution offered to be the initial host. In the ensuing years, UNOLS established credibility with ship operators, the larger research community, and NSF and other federal agencies.

OFS brought NSF attention to bear on the long-neglected area of facilities support. Ship operation funds for all NSF-supported research, whether originating in AD/R or AD/NI, were budgeted and administered by OFS. With the changing roles of federal agencies, interagency coordination took on new importance. OFS chaired the interagency negotiations that kept DSRV *Alvin* in operation and secured access to surplus Navy fuel supplies during the oil crisis in the mid-

1970s. New programs were established for shipboard instrumentation, technician support, and oceanographic technology. The design and construction of several classes of mid-sized and coastal research ships was supported.

IDOE also made important contributions to the new approach to facilities use. Its large-scale coordinated programs became the linchpin of the schedules of the larger ships, enabling institutions to plan distant expeditions. IDOE's team approach introduced many senior scientists to the experience of working on ships other than those of their home institutions. Individual investigators were also indirect beneficiaries, because the schedule lead times provided an opportunity for them to seek support for additional projects in the areas visited by IDOE cruises.

The changes did not occur without debate. Ship operating institutions had to surrender much of their independence in scheduling. Decisions to reduce the size of the fleet were invariably controversial: institutional identity might be at stake, and ships generate emotional ties not often associated with inanimate research equipment. The net outcome of the changes, however, was a more capable and cost-effective fleet. Even more important, the new approach to scheduling did a better job of matching the needs of researchers to the facilities most capable of supporting the projects.

The Reorganization of 1975:
The Ocean Sciences Are Reunified, But at a Price

In 1975, the NSF organizational pendulum swung again. Part of the reason was the continued "big vs. small" science tension, exacerbated by the complaints of the affected disciplines about the bureaucratic divisions between their "base" research programs and the AD/NI portfolio. Another internal pressure was the view that AD/R, encompassing basic research in all fields, had become unmanageable.

The ostensible purpose of the 1975 change was to restore the grouping of like disciplines as the organizing principle for all NSF research activities. Three new research directorates were established: Mathematical and Physical Sciences (MPS), Biological, Behavioral, and Social Sciences (BBS), and the awkwardly-titled Directorate for Astronomical, Earth and Ocean Sciences (AEOS). The AEOS portfolio also included the Office of Polar Programs and the Division of Atmospheric Sciences. The presence of the latter was soon acknowledged by expanding the title to an even more unmanageable formulation, directorate for Astronomical, Atmospheric, Earth and Ocean Sciences (AAEO), promptly dubbed "A-Squared E-O" by NSF staffers.

The naming problem of the new directorate came about because of the Foundation's continued ambivalence about management of large-scale facilities. Logically, the disciplinary concept should have sent astronomy to MPS, where its intellectual roots in physics and mathematics lay, leaving the environmental sciences to form a separate, coherent directorate. But the positive experience of the prior five years in terms of management improvements in the large-scale operations made NSF wary of that move. Instead, the basic research programs in astronomy were brought into AAEO to join with the operation of the observatories. That decision was vigorously protested by the astronomy community, but it would not be changed for more than a decade.

Another anomaly marked the internal organization of AAEO. The DSDP had been among the large-scale programs assigned to AD/NI in 1970. Although it had intellectual content spanning both the Earth and ocean sciences, the DSDP had been staffed and managed by oceanographers from its inception. Nonetheless, AAEO decided to assign it to the Division of Earth Sciences. The rationale was largely bureaucratic: it made for more symmetrical divisions in terms of budget and program structure. Without the DSDP, Earth Sciences would have been the only part of AAEO that did not have both research and facilities elements. The oceanographic community protested the decision, but to no avail.

For the ocean sciences, the reorganization of 1975 produced mixed results. IDOE and OFS were moved intact from AD/NI to AAEO; OSRS was moved intact from AD/R. The resultant Division of Ocean Sciences thus brought together for the first time all of the research support elements, large and small, and all of the facilities programs, with the exception of the DSDP.

This newfound organizational unity might have been the occasion for a surge of energy in NSF's leadership in the field. Unfortunately, the newly-created position of division director would remain vacant for more than two years. In the interim, the IDOE and OFS section heads alternated as acting division director, but neither had any mandate to complete internal organizational changes or to exert NSF leadership externally on behalf of the division.

This period of organizational limbo was particularly unfortunate on the research side, where tensions between proponents of "big" and "small" continued to grow. With IDOE at its mid-point, leadership toward a comprehensive ocean sciences research portfolio, including both individual and coordinated projects, might have invigorated the remaining years of the IDOE. Instead, resolution was postponed for several years.

THE 1980S: A TIME OF CONSOLIDATION—OCEAN SCIENCES INCORPORATE LARGE-SCALE RESEARCH AND OCEAN DRILLING

The End of the IDOE

1980 marked the official end of the IDOE. Contemporary views of its legacy were mixed. Some IDOE programs were acknowledged as tremendously successful, achieving research objectives that could not have been reached without the cooperative planning and management that characterized the program. Others were considered to have fallen short,

not only of IDOE's objectives, but of what might have been accomplished by more traditional individual investigator projects.

As had been the case with the IGY two decades earlier, IDOE had given rise to many ideas for additional research, both large and small in scope. It had encouraged interdisciplinary research, not only within the ocean sciences, but also with other environmental fields. Moreover, the increased funding levels associated with the IDOE would largely be retained by the ocean sciences.

The research support functions were merged into an enlarged OSRS, where the former IDOE sections were restructured and renamed, resulting in an awkward transitional set of eight programs. The distinctions between them were more historic than substantive: "Oceanic Biology" and "Biological Oceanography" coexisted, for example, as did "Chemical Oceanography" and "Marine Chemistry." In 1981, NSF began dismantling the internal management structure for the IDOE. IDOE facility needs had always been handled by OFS, so that element did not require organizational change.

The Ocean Drilling Crisis

By 1980, DSDP had been in operation for 15 years. It had become a truly international program, with foreign participation in every aspect of science planning and operations as well as financial support. Discussion about successor programs had been going on in the community for some time. Three schools of thought had emerged: (1) continue the DSDP with a new ship or rehabilitated *Glomar Challenger*; (2) begin an Ocean Margin Drilling Program (OMDP), concentrating on deep penetration of a small number of drill sites, using a new advanced platform with riser capability; or (3) begin a new generation Advanced Ocean Drilling Program (AODP) with a new platform and revised management structure. A fourth option was to end ocean drilling altogether. That last view had few adherents in the ocean science community, but was seriously considered by NSF management, Congress, and the Administration.

DSDP was granted a two-year extension while discussions of options grew more heated. When discussions of new drilling options began in the 1970s, senior NSF officials indicated that while international JOIDES scientific direction was welcome, the agency would prefer to deal with an incorporated entity as the primary contractor for any new program. The ten U.S. members of JOIDES created JOI, Inc. Even though the ten continued to be members of JOIDES and participated in the ongoing aspects of the DSDP, the creation of JOI caused unease among the international partners.

Initially, JOI undertook a few small service contracts and special studies related to the DSDP and the future options. Soon, however, by virtue of an agreement among NSF, a consortium of U.S. oil companies, and JOI, Inc., JOI became the prime contractor for developmental work for the margin drilling option. Conceptually, OMDP was proposed as an international program; as a practical matter, it soon became apparent that the proprietary interests of the participating oil companies might preclude foreign participation. International unease about the shape of future drilling gave rise to diplomatic complaints and threats to resign from JOIDES and DSDP sponsorship.

Another complicating factor entered the scene when pressure developed to convert the *Glomar Explorer*, the enormous spy ship that had been in mothballs since its reputed intelligence missions some years earlier, to serve as the platform for any future drilling program. In the space of three years, the scientific and political debate escalated to become one of the most contentious in the history of NSF. Ultimately, the OMDP experiment was abandoned for both scientific and technical reasons, and *Glomar Explorer* was rejected as too costly to convert and operate. A final completion date was set for the DSDP, and NSF committed to a new, expanded international drilling program. JOIDES would continue as the scientific monitor, but JOI, Inc., would become the operational contractor. JOI selected Texas A&M University as its primary subcontractor for the Ocean Drilling Program, and the conversion of a large commercial drillship, eventually renamed *JOIDES Resolution*, was soon underway.

NSF tried a series of organizational changes to deal with the tumultuous arguments over the fate of ocean drilling. In 1980, AAEO pulled DSDP out of the Earth Sciences Division and established an Office of Ocean Drilling Programs. The new office was also assigned responsibility for developing the emerging options. Less than a year later, the office was removed from AAEO control, relocated in the Office of the NSF Director, and renamed the Office of Scientific Ocean Drilling (OSOD).

Late in 1982, OSOD was transferred back to AD/AAEO, and a new program director, the third in two years, was named. Six months later, OSOD was assigned to the Division of Ocean Sciences, with instructions to work toward eventual integration of the drilling activity. At the end of 1984, with the ODP just months from its initial cruise, OSOD was disestablished and ODP was folded into the Oceanographic Facilities Section of the Division of Ocean Sciences. The new entity was named the Oceanographic Centers and Facilities Section (OCFS).

With that change, the Division of Ocean Sciences essentially took the form that it maintains today. The only significant oceanographic support managed elsewhere in NSF is for the Antarctic, under the purview of the Office of Polar Programs.

1986-87: The Directorate for Geosciences Emerges

The last significant organizational change affecting NSF Ocean Sciences occurred at the Directorate level. In 1986, a new NSF Director became concerned by continued complaints from the astronomy community about their "misassignment" to AAEO. The AD/AAEO, also newly appointed, concluded that there was merit to the argument of the astronomy community. Moreover, he believed that the environmental sciences had never fulfilled the potential of their organizational co-location, in part because the different interests of astronomy diluted the unified management focus that would be needed. Research thrusts such as global climate and the availability of new satellite and computer technologies called for greater integration across the environmental sciences.

In 1986, the Astronomy Division was reassigned to MPS. Concurrently, the Directorate for Geosciences (GEO) was established, with the focus on "whole earth" research as its unifying principle. With that change, today's management structure for the ocean sciences was essentially complete.

CONCLUSIONS

The Foundation's early and enduring decision to organize research support by discipline was, for many years, a source of difficulty for oceanography. When the Foundation was established in 1950, oceanography was a young and evolving field. Profoundly interdisciplinary, it would not find a unified home in the NSF research support portfolio until 1970, at which time the biological and physical subdisciplines were brought together in an Ocean Science Research Section.

The evolution in the ocean sciences was closely linked to the growth of other environmental sciences. Following NSF's successful management of U.S. participation in the International Geophysical Year (1957-1958), the environmental sciences asserted themselves as separate fields with important research objectives and practices of their own. Environmental sciences received increasing recognition and stature in NSF throughout the 1960s and 1970s, but it was not until much later that they fully came into their own with the establishment of the Geosciences Directorate in 1986.

NSF's early institutional certainty about managing individual project research stands in marked contrast to its ambivalence about the support of large-scale research and facilities. Oceanographers were among the first to challenge NSF reticence in this area, and over the years, ocean science has often been a test case generating new NSF policies and arrangements for the management of "big science" and facilities.

Project Mohole dominated NSF management councils from its inception in 1957 until its demise in 1967. That failure gave rise, however, to NSF's largest and longest-lived experiment in the support of big science, the Deep Sea Drilling Project. Administratively, these programs were set apart from the rest of the ocean sciences until the mid-1980s, at which time the Ocean Drilling Program was finally brought under the purview of the Division of Ocean Sciences.

Following on the experience of the IGY, scientists pushed for large-scale coordinated field projects, often in the world's oceans. In the 1960s, such projects were generally handled on an ad hoc basis. In 1970, with support from the United Nations and the White House, the International Decade of Ocean Exploration provided a long-term administrative home in NSF for large-scale coordinated research. Much of the administrative history of the ocean sciences in the 1980s and 1990s dealt with the integration of the IDOE into the regular ocean science research structure and the development of a balanced approach to both large- and small-scale project research.

Support for research ships and other large-scale instrumentation and equipment became part of NSF's portfolio very early on. Indeed, the need to coordinate ship operations support was the driving force in NSF's early attempts, in the 1950s and 1960s, to deal coherently with the diffuse struc-

ture of research project support in oceanography. By the 1970s, NSF was the lead agency for federal support of the University-National Oceanographic Laboratory System. Oceanographic facilities support and research project support were separately administered in NSF until the Division of Ocean Sciences was established in 1975.

Today's unified Division of Ocean Sciences, encompassing all fields of project research, large and small, facilities programs, and the Ocean Drilling Program, is thus the product of a long and sometimes difficult administrative evolution. Similarly, the co-location of the environmental sciences in the Geosciences Directorate was a long time in coming. These arrangements have now been in place for more than a decade. NSF's administrative history is characterized by change, growth, and experimentation: what might lie ahead for the ocean sciences in the decade to come?

Two Years of Turbulence Leading to a Quarter Century of Cooperation: The Birth of UNOLS

JOHN V. BYRNE AND ROBERTSON P. DINSMORE

Oregon State University (ret.) and USCG/Woods Hole Oceanographic Institution (ret.)

ABSTRACT

It started in 1969 with *Our Nation and the Sea*. This seminal report of the Stratton Commission called for the establishment of University-National Laboratories (UNLs). In order to maintain the United States as the world leader in ocean research, leading academic ocean research laboratories would be recognized as UNLs. They would receive adequate facilities and the assurance of adequate funding.

Building on this recommendation, the National Science Foundation (NSF) in 1970 proposed a National Oceanographic Laboratory System (NOLS). NOLS would be an association of institutions, grouped regionally; ship scheduling, assessment, and planning would be coordinated and NSF would provide management functions matching those of the academic sector. "To the extent possible," members would be assured of multiyear funding for ship operations.

To the oceanographic institutions that were looking for long-term stable funding, the NOLS proposal threatened to vest management authority, if not the actual operation of the ships, in the hands of NSF. These institutions resisted.

Following a year of intense debate, a compromise proposal was drafted by laboratory representatives and NSF staff. This proposal for a *University*-National Oceanography Laboratory System (UNOLS) was adopted unanimously by 18 institutions that operated the 35-ship academic fleet. UNOLS would be cooperatively managed by the 18 institutions; it would coordinate schedules and create seagoing opportunities for any competent and funded ocean scientist. It would assess the adequacy of facilities and make recommendations to the funding agencies for new construction, modifications, and replacements. To the extent possible, multiple-year funding would be ensured for ship operations by NSF and other federal funding agencies.

Since its adoption in 1971, UNOLS has served the U.S. academic ocean community. It has grown to 57 members, of which 19 members operate 29 vessels. It has been central to maintaining and assisting in the operation of the most effective ocean research fleet in the world. It has served as a model of scientific cooperation.

UNOLS AND THE FLEET

The voyage of the HMS *Challenger* in the 1870s was the beginning of modern oceanography—at least some would say it was. If so, then World War II was the beginning of "postmodern" oceanography. Ocean research as we know it at the end of the twentieth century really had its start during World War II and immediately following with the creation of the Office of Naval Research (ONR) in 1946. During the last four years of the 1940s ONR gave oceanog-

raphy in the United States a vigorous thrust toward the development that would take place during the 1950s.

The 1950s—New Beginnings

Efforts by ONR immediately following World War II set the tone and the style of oceanography (see Knauss, this volume). Marked by the creation of the National Science Foundation (NSF) in 1950, the early 1950s were a period of organization and beginnings. Oceanography programs were started or stimulated at a number of universities. Programs at the Woods Hole Oceanographic Institution, the Scripps Institution of Oceanography, the University of Washington, and Columbia's Lamont Geological Laboratory were reinvigorated. In 1957, at the request of ONR, the National Academy of Sciences (NAS) created the NAS Committee on Oceanography (NASCO)[1] to study the needs of oceanography and the opportunities before it.

1957 saw the beginning of the International Geophysical Year, the creation of the President's Science Advisory Committee (PSAC), and with the launching of *Sputnik* the wake-up call for all science within the United States.

The 1960s—A Decade of Promotion

It was 1959 when the course was set for the 1960s. In early January, the Navy released its TENOC report, *Ten Years in Oceanography* (Lill et al., 1959). Scarcely six weeks later the National Academy of Sciences released the report of its Committee on Oceanography, *Oceanography 1960–1970* (NAS, 1959). These two reports raised the hopes and fired the aspirations of ocean laboratory directors throughout the nation.

NASCO called for the federal government to double its support of basic research over a 10-year period. (In 1958, about $23 million was spent for all research in the ocean; about $9 million was considered to be in support of basic research, of which $8 million were federal funds.) Further, the committee recommended a program of ocean-wide surveys, particularly in waters more than 100 miles from U.S. shores; it suggested that private foundations, universities, industry, and state governments take an active part in the expansion of oceanography programs. But its fourth recommendation and the specific recommendations associated with it received the greatest attention. The fourth recommendation called for an increase in financial support of basic ocean research by specified federal agencies; it recommended that the Navy, the Maritime Administration, and NSF finance new research ship construction. It also included the specifics of a plan for fleet expansion:

[1] National Academy of Sciences Committee on Oceanography: Harrison Brown, professor of geochemistry, California Institute of Technology; Chairman; Maurice Ewing, Lamont Geological Observatory, Columbia University, Palisades, New York; Columbus O'D. Iselin, Woods Hole Oceanographic Institution, Woods Hole, Massachusetts; Fritz Koczy, Marine Laboratory of the University of Miami, Miami, Florida; Sumner Pike, Lubec, Maine, formerly commissioner, U.S. Atomic Energy Commission; Colin Pittendrigh, Department of Biology, Princeton University, Princeton, New Jersey; Roger Revelle, Scripps Institution of Oceanography, La Jolla, California; Gordon Riley, Bingham Oceanographic Laboratory, Yale University, New Haven, Connecticut; Milner B. Schaefer, Inter-American Tropical Tuna Commission, La Jolla, California; Athelstan Spilhaus, Institute of Technology, University of Minnesota, Minneapolis; Richard Vetter, (Executive Secretary) on leave from the Geophysics Branch of the Office of Naval Research, Washington, D.C.

A shipbuilding program should be started aimed at replacing, modernizing and enlarging the number of oceangoing ships now being used for research, surveying and development. Specifically in the period 1960-1970 the research, development and survey fleet should be increased from its present size of about 45 ships to 85 ships. Taking into account the replacement of ships which must be retired during the next decade, this means that 70 ships should be constructed at a total estimated cost of $213 million.

Vessel size, construction schedules, and costs were laid out. The seeds of much of what was to come during the 1960s can be found in this NASCO report.

With the TENOC and NASCO reports as ammunition, the selling job for oceanography took off. NSF and ONR increased their budgets for oceanographic research. Federal attention to the oceans was stimulated. A Subcommittee on Oceanography was added to the Federal Council for Science and Technology. In 1966, the National Sea Grant College Program Act was passed. Internationally, new interest in the resources of the sea was aroused by a proposal to the United Nations by the Ambassador from Malta, Arvid Pardo. Pardo proposed that the UN internationalize the deep seabed and that the resources of the seabed (largely manganese nodules) be a part of the "common heritage of all mankind." The resulting United Nations Law of the Sea Convention would continue for years. Also in 1966, the passage of the Marine Resources and Engineering Development Act created the National Council on Marine Resources and Engineering Development ("the council"), with Vice-President Hubert H. Humphrey as its chair. The Vice-President was much more than a figurehead. He was a knowledgeable and active chairman. Ed Wenk, the Executive Secretary, was even more active. Together they stimulated a high level of attention to the oceans at the congressional and federal agency levels. National attention was invigorated by the work of the council and the active role of the vice-president. Congressional attention had risen significantly (Wenk, 1972). The same act that created the council also called for a 15-member Advisory Commission on Marine Science, Engineering and Resources. It would be chaired by Julius Stratton and henceforth would be known as the Stratton Commission. The commission was:

1. to examine the Nation's stake in the development, utilization, and preservation of our marine environment;
2. to review all current and contemplated marine activities and to assess their adequacy to achieve the national goals set forth in the Act;
3. to formulate a comprehensive, long-term, national program for marine affairs designed to meet present and future national needs in the most effective possible manner;
4. to recommend a plan of Government organization best adapted to the support of the program and to indicate the expected costs.

By 1968, annual oceanographic research and education budgets in a number of laboratories were in the multimillion-dollar range, and the academic oceangoing research fleet consisted of 35 vessels greater than 65 feet in length.

The rapid expansion of oceanography during the 1960s had been stimulated by the NASCO and TENOC reports of 1959. During the 1970s and beyond, oceanography would be shaped by the Stratton Commission report, *Our Nation and the Sea*, released in January 1969 (CMSER, 1969a).

Our Nation and the Sea addressed our national capability in the sea, management of the coastal zone, marine resources, the global environment, technical and operating services, and organizing a national ocean effort. The report included more than 120 recommendations: it called for an independent civilian agency to administer federal civil marine and atmospheric programs to be known as the National Oceanic and Atmospheric Agency (NOAA); the appointment of a National Advisory Committee on Oceans and Atmospheres (NACOA); an International Decade of Exploration (IDOE); a Coastal Zone Management Program, including coastal zone laboratories; and in order to maintain U.S. leadership in ocean research, the creation of a number of University-National Laboratories (UNLs).

UNL TO NOLS TO UNOLS

This section describes the transition from University-National Laboratories to a National Oceanographic Laboratory System to a University-National Oceanographic Laboratory System.

The Stratton Commission declared:

1. U.S. leadership in marine science depended mainly on the work of a small number of major oceanographic institutions, such as the Scripps Institution of Oceanography, Woods Hole Oceanographic Institution, and Columbia University's Lamont Geological Observatory.
2. Creation of big science capability in a few efficient centers is more economical than pursuing the major scientific tasks on a scattered project-by-project and facility-by-facility basis.
3. The laboratories must be assured of an adequate level of institutional support for broad program purposes.
4. The laboratories should be located to cover different parts of the ocean efficiently and to be readily available to other scientists and institutions.
5. The direct management of these laboratories should be assigned to universities with a strong interest and demonstrated competence in marine affairs.

The commission went on to suggest that the laboratories would include but not be restricted to the leading laboratories mentioned earlier and that they certainly would be needed on the Atlantic, Pacific, and Gulf Coasts, the Great Lakes, in the Arctic, and in the mid-Pacific.

The commission recommended "that University-National Laboratories (UNL's) be established at appropriate locations, equipped with the facilities necessary to undertake global and regional programs in ocean science, and assured of adequate institutional funding for continuity and maintenance of both programs and facilities."

In the reports supporting the Stratton Commission's final recommendations (CMSER, 1969 a,b), the Panel on Basic Science and Research, chaired by Robert M. White and John A. Knauss (Volume 1), stated that the laboratories selected to be UNLs must make some formal provision for outside investigators. Further, a partnership between marine science and technology should be fostered and engineering competence should be closely aligned with the laboratory or established within the laboratory.

The recommendation and supporting rationale to establish UNLs were to stimulate subsequent discussions, proposals, and lively debate among the directors of existing laboratories and the personnel of federal funding agencies, most notably the NSF. Adrenalin surged in every ocean laboratory director. Each director saw great opportunities for his own laboratory.

A National Oceanographic Laboratory System—Starting the Debate

When the Stratton Commission report was published a distinct operational pattern of oceanographic research had already been established within the academic community. The system had evolved so that each institution doing ocean research did so from its own research vessel or vessels. "If you were going to be an oceanographic research institution, you needed a research vessel" (Knauss, this volume).

In addition to the vessels at Scripps, Woods Hole, and Lamont, research vessels, new or converted, had been provided by ONR and NSF to the Universities of Rhode Island, Miami, Texas A&M, Oregon State, Washington, and Hawaii. In all, the academic oceanographic research fleet included 35 vessels more than 65 feet in length, 15 of them greater than 150 feet in length, and 9 of these longer than 200 feet. For the most part, their operation was funded through block grants by NSF, and early on by ONR. Although the research conducted from these vessels was done primarily by researchers of the operating institution, visiting scientists from other laboratories were often accommodated. Scheduling and operational management of the vessels were in the hands of the oceanographic research institution.

The future of any university laboratory would be ensured if that laboratory were selected to be one of the Stratton Commission's University-National Laboratories. Every laboratory director recognized the opportunity and positioned his or her laboratory to take advantage of it. Their sense of anticipation was high—and optimistic.

Following the release of *Our Nation and the Sea*, federal agencies began gearing up to carry out the Stratton Commission recommendations. At the Department of Interior, organization was initiated to incorporate the new National Oceanic and Atmospheric Administration (NOAA) within the department. (Later circumstances resulted in NOAA becoming part of the Department of Commerce.) NSF saw a role for itself with regard to the University-National Laboratory recommendation, and convened a meeting of representatives of the major oceanographic laboratories in Washington, D.C. on May 13, 1970. At the meeting, NSF proposed a National Oceanographic Laboratory System (NOLS), and on May 25 it followed with a memorandum from William D. McElroy, Director of NSF, with details. Labeled a "discussion paper," the NSF proposal called for mechanisms to enhance the coordination and operation of oceanographic research vessels on a regional basis, with the vessels available to all users on an equal basis. (It had been rumored that during prior years, vessels from several laboratories had sailed to the Mediterranean unbeknownst to each other. The lack of coordination had resulted in the loss of an opportunity to conduct synoptic research and, furthermore, led to expenses that might have been reduced had the operational plans been coordinated.)

The major features of the proposal called for "block funding" of "key capabilities" of NOLS labs; multiple-year funding or some "alternative form" of long-term commitment, the sharing or coordinated use of specialized facilities, and "coordination in the planning and the conduct of research to effect national specialization within a balanced program." Several alternative management structures were proposed, but a grouping in seven regions was favored by NSF. There was an implication in the wording of the discussion paper that Woods Hole and Scripps would be the operators for "worldwide cruises of well-defined and well-reviewed programs of national interest."

The directors of the medium-sized but aspiring laboratories (e.g., Rhode Island, Oregon State, Washington) were concerned. The directors of Woods Hole and Scripps were pleased at the thought of stable long-term support, but were dismayed by the possibility of centralized federal control of their ship operations. One can only imagine the thoughts of those at Lamont—mentioned by Stratton, but omitted by NSF. McElroy called for "frank and informal" responses to the NSF proposal. He would receive them.

To the laboratory directors, who were focused on the possibility of stable financial support, it looked like an attempt by NSF to take over the control—if not the actual operation—of the oceanographic fleet. The suggestion that the fleet be regionalized was an added threat, particularly to the larger laboratories, which sent their vessels to all oceans of the world.

Woods Hole took the lead in reacting. In a graciously worded letter (July 16, 1970), the Director of Woods Hole, Paul Fye "Paul," wrote to William D. McElroy, Director of NSF "Dear Bill". Fye responded to the NSF proposal for a regionalized NOLS. The letter included an addendum as to how Woods Hole would operate within NOLS, should NOLS in the proposed form be adopted.

Woods Hole "enthusiastically supports *the efforts* you are undertaking *to explore new ways* to meet the *growing problems* of oceanographic research" (emphasis by the authors). Fye pointed out that the key factors in the NOLS concept of cooperative planning and cooperative use of facilities were already happening, particularly at Woods Hole. Mention was made of the Joint Oceanographic Institutions for Deep Earth Sampling (JOIDES), the International Indian Ocean Expedition (IIOE), the Global Atmospheric Research Program (GARP), the Geochemical Ocean Sections Study (GEOSECS), and so forth. "These illustrations point out that a great deal of cooperative planning and operation already exist. . . . Perhaps more significant to the NOLS plan is the extensive use of Woods Hole ships by non-Woods Hole staff members . . . we contend that we have an excellent record in making facilities available to oceanographers from outside our Institution."

Fye then went on to indicate that any strategy should ensure that good science would have top priority, that ships would be used efficiently, and that cruises would be planned to achieve optimum scientific results; duplication should be avoided and seagoing opportunities should be provided to competent scientists from laboratories not operating ships. Then, "In order to achieve these objectives, we feel the following conditions are essential. First, *adequate funding must be available on a long-term and flexible basis*. Next, *organization and coordination must be such that they involve minimum bureaucratic procedures*. Further, *the scheduling of ships must be in the hands of the operating institution which has the responsibility to ensure these are managed properly*." (emphasis Fye's). Essentially the labs wanted more and stable funding, few constraints, and autonomy to continue scheduling and operating their vessels.

With regard to regional coordination, Fye recommended that Scripps, Woods Hole, and Lamont form an interinstitutional committee to review scientific programs and ship scheduling for worldwide ship operations, and that similar groups be established for operations for the East Coast, West Coast, Gulf of Mexico, and Great Lakes regions. In the remainder of his four-page letter, he made a strong plea for stable core support for the oceanographic laboratories and that the laboratories be free to manage themselves. He would also take a shot at the funding agencies.

> ". . . each oceanographic laboratory has problems facing it today which the implementation of the continuing core support aspect of NOLS could alleviate. There is a general crisis in ship funding; currently no one has assurances as to whether or not they will be able to operate research ships, submersibles, or aircraft in the next calendar year. Since this involves expenditure of many millions of dollars (3.5 million for Woods Hole next year) any mistake in estimating the available funding could prove to be extremely damaging to the financial status of the operating laboratory. In addi-

tion, research proposals submitted to NSF and other agencies are not approved in sufficient time to permit the development of an optimum or realistic ship schedule. These difficulties now exist within all oceanographic institutions and could be significantly improved if NOLS provided stable funding. It is, of course, most important that our present difficulties in funding and scheduling not be compounded by the imposition of unworkable outside constraints.

Paul Fye's letter framed the debate.

The directors of Woods Hole, Lamont, and Scripps would meet in August at Woods Hole to consolidate their resistance to the NSF plan for NOLS and would request an audience with McElroy. In the meantime, virtually every oceanography laboratory director responded to McElroy with concern and alternatives to the NSF plan (e.g., not seven regions, but two; an unrestricted directors' fund). The Navy weighed in, too. Assistant Secretary of the Navy Robert A. Frosch, in a letter to McElroy (August 10, 1970), urged that any implementation of the NOLS concept recognize the Navy's research needs and permit flexibility of the oceanographic laboratories to respond to these needs.

But the lab directors felt they needed more clout. They turned to the National Academy of Sciences Committee on Oceanography to engage its support. Essentially from its membership, NASCO created a Facilities Utilization Panel to address NSF's NOLS proposal. Not too surprisingly the panel consisted of laboratory directors, former directors, and other leaders.[2]

The panel indicated that the NOLS plan as formulated by NSF "has substantial merit and that its adoption in the *modified* form proposed below will result in a significant advance in the U.S. Oceanographic Programs. . . ." The panel laid out guiding principles, the foremost of which was "to improve the level and stability of federal support for academic oceanography."

Other principles included leaving control of ships programs in the hands of working scientists, building on the ship-operating experience of existing laboratories, enhancing the sharing of facilities among qualified investigators, maintaining mutually agreed upon cooperative arrangements rather than establishing centralized control, maintaining freedom for scientists from any laboratory to work in any geographic area, and involving other federal agencies. The panel also proposed that a review committee reporting to NSF be established to assess the effectiveness of all ship and laboratory programs. In addition, an implementation plan was described that called for cooperative planning based not on the geographic location of laboratories but on common interest in areas of operation or major oceanographic problems. There were no surprises, but this report did come from the Academy.

Although the debate was on, and was intense, innovative ideas were surfacing. The laboratory directors argued for stable core funding (at an increased level) but not fettered by federal control or bureaucracy. On the other hand, NSF was concerned about fragmentation, the random distribution of facilities and scientists, the rising costs of ship operations, the decline of ONR support, the need for greater accountability, and the pressure to accommodate scientists from non-ship-operating institutions. NSF believed these factors called for greater centralization of planning, scheduling, and assessment. NSF had the input from the lab directors McElroy had requested. It was time to act.

Mary Johrde had been given responsibility for NOLS within NSF. Her short version of a NOLS Planning Document was issued in January 1971 with the opening caveat from Benjamin Franklin, "We must all hang together, or assuredly we shall all hang separately." It reflected the tone of concern within NSF. It was vintage Johrde. Her perspective was from Washington; she cared; she was determined. Eight or so lab directors on one side; Mary Johrde on the other—the odds were almost even.

The proposed NOLS plan of January 1971 succinctly defined the NSF position. It described the factors leading to the NSF position and stated firmly NSF's intentions with respect to how the academic oceanographic fleet would be managed and what NSF's role would be.

NOLS (January 1971)

In the "NOLS Planning Document: Short Version," Mary Johrde reviewed the development of federal support of oceanographic vessels (32 vessels operated by 18 institutions), and the factors NSF believed called for a change in management and operation of the facilities (ships, submersibles, aircraft, data acquisition systems, docks, shops, etc.).

The objective, it said, was "to preserve to the maximum extent the independence and integrity of existing oceanographic institutions and concurrently to create a mechanism for cooperative utilization of oceanographic facilities."

Then, "This objective will be achieved by an association of institutions in a national system in which utilization and acquisition of oceanographic facilities will be justified in terms of the facilities requirements of those qualified scientists who can make a contribution to the national oceanographic effort. *Individual institutions will continue to operate facilities, but scheduling, assessment and planning with respect to their utilization and acquisition will be handled cooperatively by the System*" (emphasis by the authors).

Further, only those institutions electing to participate would receive support for acquisition and operation of ships,

---

[2]The foxes were in the hen house. Their report was released in December 1970. The NASCO Facilities Utilization Panel included Richard B. Bader (Miami), chair; Wayne V. Burt (Oregon State University); Peter Dehlinger (OSU and ONR); Paul M. Fye (Woods Hole); Jeffrey Frautshchy (Scripps); John Lyman (formerly U.S. Navy Hydrographic Office and NSF); Robert A. Ragotzkie (Wisconsin); and George P. Woollard (Hawaii).

and for this agreement to participate, NSF "will, *in so far as possible*, express its intent two years in advance to commit NOLS support for operation of ships and other shared facilities" and will urge ONR to do likewise. No promises would be made with regard to facilities added to the mix after the establishment of NOLS.

A description of the NOLS organization then followed. There would be two regions: an Eastern Region and a Western Region; the Gulf of Mexico and the Great Lakes might be subdivisions of the Eastern Region; Hawaii, Alaska, and the Pacific Territories, part of the Western Region. There would be operating committees for each region to schedule the facilities within the region and to assess needs for additional or replacement facilities. In addition, there would be a Central Committee for Planning and Assessment for the entire system.

The NOLS office within NSF would be designed to provide management functions matching those of the academic sector. NSF and ONR would establish joint panels to consider ship operation requirements and ship construction and conversion. NSF would select two "host institutions" to take the lead for organizing meetings, and so forth in the Eastern and the Western Regions; NSF would provide funding for meetings and would approve the nominations by the institutions for the Regional Organizing Committees. The document then went on the describe how NOLS would actually function.

The reactions of the laboratory directors ranged from concern to outrage. In their eyes the NOLS plan was a proposal to take over a significant portion of their management responsibility and authority, the portion that determined where, when, and how they would conduct research in any part of the ocean. Their resistance stiffened.

Again, Paul Fye wrote to Bill McElroy (March 22, 1971): "Dear Bill; I know some of your staff have been puzzled at the strong opposition found within the oceanographic laboratories over the last form of the NOLS plan . . . Our concern with the NSF (January, 1971) statement of the NOLS operational plan is as much with the philosophy on which it is based as with the operational mechanics themselves."

He then went on to review the intent of the Stratton Commission in recommending UNLs, discussed the importance of the relationship "between the creative scientist and the tools of his research," and wrote of the concern about and resistance to the plan on the part of senior scientists at Woods Hole.

> Why is this so when admittedly its [the NOLS plan] purpose is good, its goals are desirable and overall it isn't a bad plan? The fundamental error is that it removes the operational control of research tools further from the creative scientist. Is this necessary to achieve these goals and this purpose? We think not.

We are pleased that NSF has consulted the oceanographers who use the research ships about this plan. We recognize the sincere attempt by members of your staff to understand our objections. I understand that a continuing committee co-sponsored by the Academy and the Foundation will explore ways for further improvement of the NOLS plan.

In his response (April 7, 1971), McElroy stressed the need for "participatory management of oceanographic facilities by the academic community *in conjunction* with the Foundation" (emphasis ours). The groundwork for collaboration was reinforced. McElroy then referred to a joint meeting in April of representatives of oceanographic laboratories and NSF staff to consider changes and possible improvements to the NOLS planning document.

To the credit of Mary Johrde and her NSF colleagues, some type of compromise seemed appropriate. The group of laboratory and NSF representatives met on April 23 and 24, 1971, and drafted a compromise plan. It met again in July and August to refine the compromise proposal and to prepare it for presentation to the academic community.

UNOLS

On August 4, 1971, "A Proposal to Establish a University-National Oceanographic Laboratory System" (UNOLS) was completed and then distributed to the academic oceanographic laboratories. The proposal acknowledged the development of a strong U.S. oceanographic program and of the importance of the academic oceanographic laboratories in this development, but it also recognized factors that could have an effect on the long-term viability of U.S. leadership in oceanography.

> The academic community is also acutely aware that the continued health of the programs depends heavily on its assuming greater responsibility to assist the funding agencies in an appropriate manner in monitoring the utilization of these resources to insure: that there is a proper balance between research and facility support, that available facilities are used efficiently, that scientists from both ship-operating and non-ship-operating laboratories have access to the sea, that needs for new facilities or the phasing out of old or excess ones are assessed and priorities established accordingly, that long-term support becomes an integral part of planning, and that consideration be given to the encouragement of new operating elements only to the extent that a demonstrable need for such exists and sufficient continuing support is available.

> In order to provide a mechanism whereby the academic community can assist the Federal agencies in meeting the responsibilities noted above and at the same time continue the high standards of research that have been exhibited in the past as well as to provide a flexibility of operation allowing for a coordinated approach to some of the future challenges—it is proposed that the academic laboratories organize a system in which they can work cooperatively together and with the funding agencies for the effective use, assess-

ment and planning of oceanographic facilities. The organization will be known as the University-National Oceanographic Laboratory System (UNOLS).

Membership in the system would be open to academic institutions operating federally funded facilities. Facility use would be open to scientists from any institution, primarily for the conduct of federally funded programs. The purpose of the organization would be to provide a formal mechanism for community-wide coordination and review of the use of available facilities, equal opportunity for access to these facilities, community-wide assessments of the current match of facilities to the needs of federally funded oceanographic programs, and appropriate recommendations of priorities for replacing, modifying, improving, increasing, or decreasing the numbers and mix of facilities for the community of users.

There would be a UNOLS committee to monitor the activities of the system, provide advice and assistance to members, and submit reports to the funding agencies. It would consist of seven members, three of whom would be from nonmember institutions. A UNOLS office, with an executive secretary, would be established at a member institution to handle staff duties. Support for the office would be prorated among the funding agencies.

Once a year, UNOLS members would meet to coordinate ship schedules. There would be three separate meetings: one to schedule ship operations in the open ocean (500 miles or more offshore); one for the coastal waters (less than 500 miles from shore) for the East Coast; and one for the West Coast coastal waters. Detailed logistics for preparation and coordination of schedules were suggested.

The UNOLS committee would consist of members elected to three-year terms and would devote its early attention to the effective use of existing oceanographic facilities. It would evaluate the need for replacement and additional facilities and would recommend to the funding agencies on behalf of the oceanographic community consideration of specialized facilities or new concepts in facilities.

Many of the attributes of earlier versions of the NOLS plan were included. There were significant differences, however. There would be one national program, not divided geographically, but considering separately only open ocean, West Coast, and East Coast operations. The overwhelming difference from all the NOLS proposals was that this system would be managed almost exclusively by the institutions themselves. It would be a *University*-National Oceanographic Laboratory System—a cooperative venture.

The drafters of the UNOLS proposal, Dick Barber (Duke University), John Byrne (Oregon State University), Art Maxwell (Woods Hole), Bob Ragotzkie (University of Wisconsin, Madison), and Jay Savage (University of Southern California), presented and discussed the proposal at a meeting of representatives of the oceanographic laboratories at the Lamont Geological Laboratory of Columbia University on September 22, 1971.

UNOLS—The First Year and Beyond

The laboratories that agreed to participate did so with misgivings. UNOLS represented a new way of doing business. Those that participated agreed to give up an element of autonomy for the good of the community. However, because there was the threat of losing funding if they failed to participate, there was a strong incentive to do so. Even so, UNOLS was considered to be an experiment. The proposal was not accepted until a "renewal or dissolution" clause was added to the charter. In order for UNOLS to continue, it would require a renewal every three years by vote of its members. The proposal was unanimously accepted. UNOLS was born.

There were 18 initial members:

Duke University
Florida State University
Johns Hopkins University
Lamont-Doherty Geological Laboratory of Columbia University
Nova University
Oregon State University
Scripps Institution of Oceanography
Skidaway Institute of Oceanography
Stanford University
Texas A&M University
University of Alaska
University of Hawaii
University of Miami
University of Michigan
University of Rhode Island
University of Southern California
University of Washington
Woods Hole Oceanographic Institution

These 18 institutions operated 33 vessels more than 65-feet long (Table 1).

At the first meeting, Art Maxwell of Woods Hole was elected chairman and Jay Savage of the University of Southern California, vice-chairman. The UNOLS committee was also elected at the time and included John Byrne, Oregon State University, chair; John Craven, University of Hawaii; Charles Drake, Dartmouth; David Menzel, Skidaway; Bob Ragotzkie, Wisconsin; Hank Stommel, Massachusetts Institute of Technology; and Warren Wooster, Scripps Institution of Oceanography. Soon after the inception of UNOLS, with Woods Hole as the host institution and Art Maxwell as chair, Captain Robertson P. Dinsmore (Coast Guard, Retired), was selected to serve as the executive secretary. Maxwell, Dinsmore, and Woods Hole would lead UNOLS during the early years of its existence.

The charter of UNOLS was adopted at the first regular UNOLS meeting held at Texas A&M at College Station in May 1972. At the outset, the main function of UNOLS was

TABLE 1   The UNOLS Fleet

Operating Institution	1972 Name	Length (feet)	1996 Name	Length (feet)
University of Alaska	*Acona*	85	*Alpha Helix*	133
Scripps Inst. of Oceanography	*Melville*	245	*Melville*	279
	Agassiz	150	*New Horizon*	170
	Oconostota	100	*Robert G. Sproul*	125
	Scripps	95	*Roger Revelle*	274
	T. Washington	209		
	Alpha Helix	133		
University of Hawaii	*Kana Keoki*	156	*Moana Wave*	210
	Teritu	90		
Oregon State University	*Yaquina*	180	*Wecoma*	185
	Cayuse	80		
University of Southern California	*Velero IV*	110		
University of Washington	*T.G. Thompson*	209	*T.G. Thompson*	274
	Hoh	65	*C.A. Barnes*	66
	Onar	65		
Stanford University	*Proteus*	100		
Lamont-Doherty	*Conrad*	209	*Maurice Ewing*	239
	Vema	202		
Duke University	*Eastward*	118	*Cape Hatteras*	135
Florida State University	*Tursiops*	65		
Skidaway Institute	*Kit Jones*	64	*Blue Fin* (for Georgia System)	72
Johns Hopkins	*R. Warfield*	106		
	Maury	65		
University of Miami	*Gillis*	209	*Columbus Iselin*	170
	Calanus	64	*Calanus*	68
	Iselin	170		
Nova	*Gulf Stream*	55		
University of Rhode Island	*Trident*	180	*Endeavor*	184
Texas A&M University	*Alaminos*	180	*Gyre*	182
Woods Hole Oceanographic Inst.	*Knorr*	245	*Knorr*	279
	Atlantis II	210	*Atlantis II*	210
	Gosnold	99	*Oceanus*	177
	Chain	213	*Atlantis*	274
University of Michigan	*Inland Seas*	114	*Laurentian*	80
	Mysis	50		
Harbor Branch Oceanographic Inst.			*Seward Johnson*	204
			Edwin Link	168
			Sea Diver	113
Moss Landing Marine Laboratory			*Point Sur*	135
University of Delaware			*Cape Henlopen*	120
Bermuda Biological Station			*Weatherbird II*	115
Louisiana U. Marine Consortium			*Pelican*	105
University of Texas			*Longhorn*	105

SOURCE: UNOLS (1972) and Anonymous (1996).

to coordinate ships' schedules and to focus on the replacement of federally funded vessels. Early on, the Research Vessels Operators Council (RVOC), which had existed for some time, was incorporated into UNOLS to serve as an expert advisory group directly involved with the operation of vessels. During the first year, UNOLS' efforts began to focus on the development of coastal ships, uniform standards of operation, foreign clearances, uniformity of technical services, national facilities, and of course, the fleet re-

placement. Attention was also directed to specialized facilities. These included the expeditionary vessel *Alpha Helix*, the deep submersible, *Alvin*, Scripps aircraft, and other unique facilities that would be available to the entire oceanographic community.

During the first years of its operation, UNOLS membership changed. Stanford, Florida State, and Nova dropped out, while Texas, Delaware, and Moss Landing became members. Associate memberships (non-ship operators) were created in order to involve more of the research community.

UNOLS Today

The past 27 years has seen a broadening, strengthening, and maturing of UNOLS. As a concept, UNOLS helped define a new cooperative way of conducting oceanographic research. Together with NSF's International Decade of Ocean Exploration program, a new era of U.S. oceanographic research was initiated—one that provided opportunities for all competent ocean scientists who were willing to engage in cooperative research. Today, UNOLS consists of 57 academic institutions that operate significant marine science programs: 19 of these institutions operate the fleet of 29 research vessels—the strongest, most capable fleet of oceanographic research vessels in the world.

Several of the institutions that dropped out are again members, but not as vessel operators. Over the years, several additional institutions have joined as vessel-operating laboratories. These include the Harbor Branch Oceanographic Institution, the Bermuda Biological Station, and the Louisiana Universities Marine Consortium.

Since 1972 the fleet has changed. Seven of the original thirty-five vessels are still in service, these have been joined by twenty-two new vessels. The size distribution of the fleet is shown in Table 2.

During its more than quarter century of existence, the UNOLS charter has been repeatedly adopted every three years. It has been amended or revised 11 times. Today UNOLS still operates according to the original concept so laboriously formulated in 1970 and 1971; it is larger, more sophisticated, and stronger than ever. As pointed out in the 25-year history of UNOLS, available on the UNOLS Web site (www.gso.uri.edu/unols/25annpap.html), "UNOLS will continue to be a major presence in U.S. oceanography for the next twenty-five years. Today it stands as a model of inter-agency and federal/academic coordination. It has developed a flexible, cost-effective management structure. It emphasizes an entrepreneurial atmosphere to keep the fleet at the forefront of technology while maintaining the cost-effective structure. The close coordination with academic institutions results in substantial cost savings. It encourages the collegial atmosphere that leads to close cooperation between the operators. As a result of these factors, the UNOLS fleet is an integral part of our nation's science program."

The U.S. oceanographic research program is the foremost in the world. UNOLS has been a major contributor to this position of leadership. Moreover, it serves as a model of how scientists and scientific institutions can cooperate to reach the highest levels of scientific achievement.

ACKNOWLEDGMENTS

The authors are indebted to scientific colleagues who have reflected on the days of debate and development of UNOLS. They include Mary Johrde, Art Maxwell, and David Ross. The support of the archivist of the Woods Hole Oceanographic Institution is gratefully acknowledged. Finally, this report would not have been possible without the competent and dedicated efforts of our assistant at Oregon State University, Carol Mason. To all we extend our deep appreciation.

REFERENCES

Publications

Commission on Marine Science, Engineering and Resources (CMSER). 1969a. Pp. 21-22 in *Our Nation and the Sea: A Plan for National Action*. U.S. Government Printing Office, Washington, D.C.

Commission on Marine Science, Engineering and Resources (CMSER). 1969b. Pp. 42-65 in *Panel Report: Science and Environment*. Volume I., U.S. Government Printing Office, Washington, D.C.

Lill, G.G., A.E. Maxwell, and F.D. Jennings. 1959. *The Next Ten Years of Oceanography*. Internal Memo, Office of Naval Research.

National Academy of Sciences (NAS). 1959. *Oceanography 1960 to 1970*. Volume 1: Introduction and Summary of Recommendations. National Academy Press, Washington D.C.

National Academy of Sciences (NAS). 1970. *A National Oceanography Laboratory System*. National Academy Press, Washington, D.C.

Treadwell, T.K., D.S. Gorsline, and R. West. 1988. *History of the U.S. Academic Oceanographic Research Fleet and the Sources of Research Ships*. UNOLS Fleet Committee. Texas A&M University, College Station, Texas. 55 pp.

University-National Oceanographic Laboratory System (UNOLS). 1972. *First Annual Report of UNOLS Advisory Council to Federal Funding Agencies*. Woods Hole Oceanographic Institution, Woods Hole, Massachusetts. 43 pp. + Appendices.

Wenk, E., Jr. 1972. *The Politics of the Ocean*. University of Washington Press, Seattle, Washington. 590 pp.

Unpublished Reports

Anonymous. 1996. The University-National Oceanographic Laboratory System: Celebrating 25 Years as the Nation's Premier Oceanographic Research Fleet. UNOLS Web Site: (www.gso.uri.edu/unols/25annpap.html)

Anonymous. 1998. UNOLS Charter (as of July 15, 1989). UNOLS Web Site: (www.gso.uri.edu/unols/25annpap.html)

Barber, R., J.V. Byrne, A.E. Maxwell, R. Ragotzkie, and S. Savage. Au-

TABLE 2   Size Distribution of UNOLS Fleet

Length	Number of Vessels	
	1972	1996
Over 200 feet	9	9
150-200 feet	6	7
100-150 feet	7	9
65-100 feet	13	4
TOTAL	35	29

SOURCE: UNOLS (1972) and Anonymous (1996).

gust 4, 1971. A Proposal to Establish a University-National Oceano-graphic Laboratory System. 7 pp.

Dinsmore, R.P. 1996. History of UNOLS. Woods Hole Oceanographic Institution, Woods Hole, Massachusetts. 4 pp.

Johrde, M.K. January 1971. NOLS Planning Document: Short Version. National Science Foundation, Washington, D.C. 13 pp.

Correspondence

May 25, 1970. William D. McElroy memorandum to Distribution List. Subject: A National Oceanographic Laboratory System; a discussion paper.

July 16, 1970. Paul M. Fye to William D. McElroy.

August 10, 1070. Robert A. Frosch to William D. McElroy.

March 22, 1971. Paul M. Fye to William D. McElroy.

April 7, 1971. William D. McElroy to Paul M. Fye.

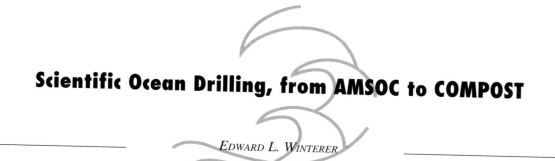

Scientific Ocean Drilling, from AMSOC to COMPOST

Edward L. Winterer

Scripps Institution of Oceanography, University of California, San Diego

ABSTRACT

For more than 30 years, following the abandonment of the bungled Mohole project, designed to drill a hole through the crust-mantle boundary, the National Science Foundation (NSF) has energetically supported and shepherded along a spectacularly successful scientific ocean drilling program that has cored oceanic sediments and crust at more than a thousand places over most of the global ocean. The program has tested major hypotheses such as seafloor spreading, provided the material basis for a increasingly fine-grained geologic time scale, delivered otherwise unattainable data on compositions and processes from levels deep beneath the seafloor, including the oceanic crust, and made possible the elaboration of a detailed global paleoceanographic history, extending back about 180 million years. Early mistakes and fumbles about responsibilities for oversight, funding, management, science operations, and scientific advice were corrected. Short-lived ventures into complicated, very high-tech schemes were abandoned with no harm to the main, continuing scientific thrust of the program. NSF found important funding and participation from other nations and has been responsive to requests from U.S. scientists for funds to carry out site surveys, postcruise studies of cores, and downhole experiments. The crossroad ahead, when present funding expires in 2003, is hazardous. It is a major question whether the very costly and specialized riser-drilling program being planned for a new Japanese vessel, with still-fuzzy definition of the science objectives, can be funded alongside the more flexible style of nonriser drilling that has attracted scientists from such a large range of disciplines.

Proposals and programs for coring into the ocean floor from floating platforms began in the United States in 1957 with a modest planning grant of $15,000 from the National Science Foundation (NSF) to the National Academy of Sciences. Since then, a half dozen ocean drilling programs—some huge, some tiny; some successes, some failures—have been funded, for a total NSF and international expenditure of a billion dollars. My own fervent conviction is that we have received extraordinary value for money. It is my purpose here, in my own idiosyncratic way, to take stock of the successes and failures of these programs, in terms of both their scientific and technical accomplishments and their management structures.

THE MOHOLE PROJECT

Some 41 years ago, Walter Munk, Harry Hess, and a few others, reacting to a long and wearying panel session reviewing good, but normal, science proposals in Earth science for the National Science Foundation, asked themselves if there weren't some truly major question running across subdisciplines that could be posed and answered, even if it took stretching technology and even if it might cost quite a lot of money. Their candidate question was: What is the physical nature of the Mohorovicic seismic discontinuity—the Moho—that marks a change in physical properties that defines the boundary between the Earth's crust and underlying mantle? To learn the answer, they thought it technically possible to drill a hole through the crust and to sample rocks across and at the boundary. Because the Moho beneath the oceans is only about 5 km beneath the deep seafloor, a drill ship that could drill through 5 km of rock in water depths of 5 km would be required. Not easy, not something that industry was actually doing, but something that was technically probably within reach.

The self-constituted, small, and very informal American

Miscellaneous Society (AMSOC), to which Munk, Hess, and Roger Revelle belonged, took the idea under its aegis at a meeting at Munk's home—always characterized as a wine breakfast—and submitted a proposal to NSF in mid-1957 to explore the feasibility of drilling (and coring) a hole to the Moho. To give AMSOC a cover of respectability and fiscal responsibility, the National Academy of Sciences (NAS), at the suggestion of NSF, gave it an administrative home. NSF granted $15,000 (half the amount requested) and the work began, with Willard Bascom, an experienced marine engineer, as executive director. It was he who coined the name Mohole for the project. AMSOC was to provide both scientific advice and management.

One geophysicist, Maurice Ewing, Director of Lamont Geological Observatory, who became an AMSOC member by happening to be in the hallway of the Cosmos Club when AMSOC gathered nearby for a meeting, urged from the start that the single-site Mohole attempt should be preceded by coring at many places through the oceanic sediment cover, thought from seismic data, largely collected by ships of his institution, to be not more than about 1 km thick. He argued that not only would such an intermediate step provide experience in drilling in great water depths, but it would answer fundamental questions about the age of the ocean basins (permanent or young?) and the nature of the rocks below the sediments (harder sediments? volcanic rocks?). AMSOC, reflecting diverse views in the community, was divided on this, but decided temporarily to put aside Ewing's option and to keep the focus on the ultimate Moho target. The debate over this choice intensified over the life of Mohole, and the progressive ascendancy of the gradualists weakened the support in the scientific community for the one-hole approach.

Industry and Congress quickly rallied in support of the concept of very deep drilling, spurred by the public boast of the USSR to start drilling its own hole through the Earth's crust and thus to demonstrate its technological superiority over the United States once again, as had just been done with *Sputnik* (van Keuren, 1995). Riding on this wave, a preliminary notice of a proposal to NSF went forward from AMSOC to NSF in 1958 for $2.75 million, to be available in 1960.

The AMSOC proposal gave three possible types of drill sites: on a continent, on an oceanic atoll, and on the deep seafloor. The seafloor option prevailed, and for this a dynamically positioned, floating rig was deemed most feasible. The hardware part of the Mohole project got underway with some testing to see if a drill vessel could hold position in deep water during drilling, using a dynamic positioning system. AMSOC chartered an industry vessel, *CUSS-1*, which, after some preliminary tests in soft sediments of a Neogene turbidite basin in waters about 1,000 m deep west of San Diego, then drilled a hole 183 m deep in 3,570 m of water off the Mexican island of Guadalupe. The dynamic positioning scheme and the coring of both pelagic sediments and basaltic basement there were successful, opening the way to the more ambitious stage, a hole all the way to the Moho. The cost for this Guadalupe phase of Mohole was about $1.5 million. Enthusiasm was high and work to identify the best drill site proceeded. After extensive studies of existing geophysical records and some new survey work, a panel of geophysicists chose a site on the deep seafloor about 300 km north of the island of Oahu. All that was needed now was the actual heavy-duty drill ship.

NSF next opened the bidding for construction and operation of the Mohole drill vessel. Several consortia of experienced oil companies and shipbuilders submitted bids, but the nod went not to the lowest bidder, but rather to a company with no experience in drilling, the Texas-based major engineering and construction firm of Brown and Root. Partly because of the low evaluation score assigned to its presentation in the first round of bidding and its ascent to the top in several re-reviews, cries of unfair political influence by Brown and Root resounded. A Houston Congressman, Albert Thomas, chaired the committee that controlled NSF's budget and another Texan, L.B. Johnson, was Vice-President. A particular feature of the contract was that Bascom's AMSOC group (now organized as a private company) was to be incorporated into the Brown and Root operation, to keep the contractor oriented toward the scientific goals. Bascom soon jumped ship, declaring that the contractor was not paying much attention to his group's advice. Although AMSOC-NAS was still supposed to be providing scientific advice, AMSOC members were scientists fully engaged in their own projects and, absent Bascom's group, could not or would not assume Moho management reponsibilities. The result was that NSF itself, rather than some academic entity, was managing the project.

The whole dreary tale of the bidding and rebidding process and of the subsequent delays, cost overruns (from original estimates of $14 million to later estimates of about $160 million) and final failure of the project has been recounted in detail, for example in Solow's 1963 article in *Fortune* magazine. After the expenditure of about $57 million, Congress (Representative Thomas, chairman of the committee controlling the NSF money having just died) denied further funding and NSF had to abandon Project Mohole, with no ship built or any ocean crustal hole drilled. One hole, about 300 m deep, was drilled on land into serpentinite (altered mantle?) near the coast of Puerto Rico, as a test of drilling tools. NSF learned the hazards of attempting management by NSF rather than by contractors with roots in the academic community concerned directly with the scientific goals of the project.

In hindsight, given what we know now from three decades of drilling experience in crustal rocks, it is highly unlikely that drilling at the candidate Moho site near Oahu would have penetrated more than a small fraction of the thickness of the oceanic crust. By 1965, the Moho, as a near-term scientific objective, gradually faded from the agenda of working scientists.

LOCO AND THE BIRTH OF JOIDES

Instead, well before the death of Mohole, a new initiative, focused on oceanic sediments, moved forward. AMSOC itself, under its chairman Hollis Hedberg, had been increasingly inclined toward sediment drilling, partly as a prelude for Mohole and partly as an end in itself. There was much talk of a second ship for this purpose. Then Cesare Emiliani, of the University of Miami, seized the moment by proposing to NSF in 1962 a modest plan to use a small chartered drilling vessel to core sediments in the Caribbean Sea, a project labelled LOCO (LOng COres). The aim, wholly in keeping with his own special research interests, was to decipher and extend the paleoceanographic history of the Neogene when continental glaciers waxed and waned repeatedly in the northern hemisphere, causing major swings in global sea level. The swings could be monitored through the changing microfossil contents and stable isotopic compositions of cored calcareous pelagic sediments.

To help guide this work, Emiliani formed a LOCO advisory group comprising scientists from the major U.S. oceanographic institutions, which evolved into the Joint Oceanographic Institutions for Deep Earth Sampling (JOIDES) organization in 1964, with much encouragement from NSF, and membership of four U.S. oceanographic institutions—Miami, Lamont, Woods Hole, and Scripps. JOIDES was to plan and provide scientific advice; the actual management of projects was to be contracted by NSF to individual JOIDES institutions.

In the interim after the LOCO committee disbanded but before JOIDES formed, Emiliani drove ahead with his project and in 1963, after several attempts, successfully cored through about 55 m of pelagic sediments in 610 m of water off the coast of Jamaica from the small drill vessel *Submarex* (Emiliani and Jones, 1981). The LOCO program, driven by the ideas and persuasiveness of a single scientist and operated as a normal NSF grant, was a technical and scientific success and cost only about $100,000!

D/V *CALDRILL* ON THE BLAKE PLATEAU

JOIDES was now hard at work planning future drilling. Its first project, accomplished during 1965, was the drilling of a transect of holes across the Blake Plateau, a marginal submarine plateau at depths of 25-1,000 m off the Atlantic coast of Florida. The objective was to determine the history of relative sea level changes as an entree to the history of tectonic subsidence of a sector of the continental margin, which was known to have been a shallow-water reef area during the Late Cretaceous, some 70 million years ago. For this venture, NSF, on the advice of JOIDES, awarded the managerial contract to Lamont, which seized the offer of an oil company to allow use of its chartered vessel D/V *Caldrill* while it was in transit from Panama to Canada. In the spirit of JOIDES, the shipboard scientists came from several

JOIDES institutions and the U.S. Geological Survey. *Caldrill* maintained position by monitoring deviations from the vertical of a taut wire from the vessel to an anchor on the seafloor. The data went to a computer that controlled four large outboard motors on the four "corners" of the ship and kept the ship on station. The cores (recovery about 25-70 percent) from the six drill sites nicely documented the Cenozoic drowning history of the old Cretaceous carbonate platform, but everyone understood that the taut-wire station-keeping system would not be applicable to operations in the deep sea, the place everyone wanted to go. For this, a larger vessel with dynamic positioning was required.

DSDP AND D/V *GLOMAR CHALLENGER*

After the three coring ventures, Guadalupe Mohole, LOCO, and the Blake Plateau, sediment cores were now seen as fairly easy to recover. Microfossils in the cores were generally sufficiently abundant to determine the geologic age of samples. Coring could be extended at least into the upper part of basaltic oceanic basement and its age estimated from the paleontological age of the immediately overlying sediments. In principle, these two simple facts opened the possibility of working out not only the paleoceanographic history of the ocean basins over the past 100 million years (the age of the then-oldest known samples from the ocean floor), but also the age of oceanic crust in all the oceans. A heady vision!

At NSF, awareness was growing that coring of sediments was probably better done from a ship other than the Mohole ship. A sediment-coring ship would need be on station only for days or weeks, while the Mohole ship would be on station for years. The two programs were now being viewed as independent, and so NSF, in 1963, proposed to Congress an Ocean Sediment Coring Program, distinct from the Mohole Project. Funds were provided for the new program in fiscal year 1965.

The trigger for realizing an oceanic drilling project was the acquisition of a practical dynamic positioning system. This system, considerably refined from that deployed from *CUSS I*, comprised an acoustic transponder dropped from the ship onto the seafloor and an array of four hydrophones lowered a little below the ship's hull. The arrival-time differences of signals from the transponder were processed in a computer, which controlled the ship's main propulsion system and lateral tunnel thrusters, keeping the ship for weeks at a time generally well within a circle with a diameter less than 10 percent of water depth.

JOIDES panels had recommended to NSF the acquisition of a drilling vessel capable of coring in water as much as 6,000 m deep for periods of months in moderate sea conditions and of coring continuously into both sediments and basement rocks to subseafloor depths of several kilometers. NSF, in 1966, awarded to Scripps a prime contract for 18 months of drilling, with a first year's infusion of $7.4 million. Scripps immediately set about acquiring a suitable vessel and recruiting the managerial and technical staff required to operate the scientific parts of the program, now called the Deep Sea Drilling Project (DSDP). Scripps, in 1967, subcontracted with Global Marine, Inc., to supply the drill vessel, which Global Marine christened *Glomar Challenger*, in honor of the great exploring vessel of the nineteenth century. Scripps equipped it with laboratories for the preliminary study and curation of core samples. By the middle of 1968 the ship was ready to sail, staffed by Global Marine ship and drilling crews and by Scripps technicians. Scientific parties were recruited for each eight-week leg from the entire scientific community, foreign and domestic. The contractual terms required that Scripps take from JOIDES its scientific advice on the definition of objectives for each leg, the general track of the vessel, the shipboard measurements to be made, the curation of the cores, and downhole measurements. JOIDES also recommended shipboard scientists to DSDP. JOIDES advised, Scripps managed as prime contractor, and NSF monitored—and paid.

The First 18 Months: Validating the Promise

The Deep Sea Drilling Project would not have been possible had it not been that the main features of the bathymetry of the oceans were known from echo sounding during and after World War II. In addition, a near-global web of seismic reflection profiles had already been collected, mainly by Lamont ships; and magnetometer surveys showing anomaly patterns had been made. The new plate tectonic hypothesis was the hottest topic in Earth science. There were big ideas to put to the test and there was a way to pick good sites for the testing.

What was hoped for and what was accomplished during the first 18 months? A quick overview of the major scientific achievements shows why nobody wanted to stop drilling at the end of that time. JOIDES planners had by now devised a nine-leg plan of drilling: a beginning leg in the Gulf of Mexico, partly to explore one of the Sigsbee Knolls (a group of buried salt domes); then one leg each across the North and South Atlantic, mainly to date oceanic crust; a leg in the Caribbean; then five legs in the Pacific, including a north-south transect of the thick pile of pelagic sediments close to the Equator; and a long loop westward to explore the possibly very old crust farthest from the active East Pacific Rise spreading ridge. Then to the home port of Long Beach to end the project.

The ship went first to the Atlantic, mainly to test the seafloor-spreading hypothesis. The first leg, led by Ewing, drilled into the caprock of a salt dome in 3,572 m of water, where geophysical evidence suggested the crust might be oceanic rather than continental. The drilling did not settle the question of the depth of water during salt accumulation. The drilling did make JOIDES aware of the risk of encountering uncontrollable hydrocarbons, and so planners created a Safety and Pollution Panel to screen proposed sites for their risk potential.

The following leg, across the North Atlantic, ran into serious problems with hard chert layers in the lower Cenozoic sediments, and reached basaltic basement at only three sites. Calcareous sediments at depths below the present compensation depth for calcite (about 4,500 m) suggested subsidence of the seafloor. The age distribution of basement was shown to be consistent with spreading from the Mid-Atlantic Ridge, but could not be considered a good test of the hypothesis.

Leg 3, across the South Atlantic from Dakar to Rio, was a blockbuster. The main objective was no less than a rigorous test of the then-new hypothesis of seafloor spreading. J. Heirtzler had identified magnetic anomalies on both sides of the Mid-Atlantic Ridge along a transect across the South Atlantic and had estimated the ages of the anomalies by extrapolating back into the Late Cretaceous the radiometric ages of magnetic reversals in Neogene lava flows on land, assuming uniform spreading rates. Two geophysicists, Art Maxwell and Dick von Herzen, were designated as co-chief scientists. Drilling showed a near-perfect match between magnetically predicted basement ages and paleontologically determined ages of basal sediments, and for this reason alone the leg was a triumph. Seafloor spreading leaped from hypothesis to ruling theory at a single bound. But also among the scientific party were two geologists, Ken Hsü and Jim Andrews, who persuaded their co-chiefs to take lots of sediment cores on their way to the crucial contact between sediments and basement—the single core that some geophysicists wanted from a hole. In the long sequences of near-continuous cores, Hsü and Andrews recognized changes in the degree to which calcareous fossils were preserved from destruction by dissolution at the seafloor. Their data provided the basis for others to reconstruct the history for the South Atlantic of fluctuations in the depth of complete calcite dissolution, the calcite compensation depth (CCD). Quantitative paleoceanography was now a discipline, and drilling was the way toward writing a paleoceanographic history, back to about 180 million years ago and for all the world ocean.

The final Atlantic leg, in the western South Atlantic and in the Caribbean, reconnoitered a diverse array of problems, solving none of them, but whetting appetites for more focused work, especially in the Caribbean. Reconnaissance legs—and there were a number of them in the early part of DSDP—open up problems but don't generally solve them.

In the Pacific, two big questions lay open to the drill: What was the history of pelagic sedimentation in the equatorial high-productivity zone, and what was the age of oceanic lithosphere in the western Pacific, far from the active East Pacific Rise spreading ridge? In the eastern Pacific, planners laid out a three-leg, north-south transect, from about 41°N to 30°S, but the results of the first of these, from 41°N to 14°N, showed that dissolution on the seafloor had destroyed most microfossils at these latitudes. JOIDES therefore asked DSDP to change plans, substituting an east-west transect along the equator for the southern part of the original track. This was an early demonstration of the need to keep feeding new results into future planning and established an enduring *modus operandi.*

The results from these three early Pacific legs documented the strong dependence on latitude not only of pelagic sediment accumulation rates, but also of the depth profiles of carbonate dissolution for the past 35 million years. The drilling further opened up the problem of sedimentation on an oceanic plate that is moving not only east-west, as in the Atlantic, but also north-south, across the equatorial high-fertility zone. From the sediment-thickness data, epoch by epoch, a quantitative estimate of the rate of northward plate motion could be made and compared with independent estimates coming from the recently introduced fixed-hotspot model for the evolution of the linear Hawaiian volcanic chain. Recognition of the abundance of well-preserved microfossils and of the near-continuous record of sedimentation in the Pacific equatorial zone led to several later coring legs that provided us with the material basis for an extraordinarily detailed biostratigraphic time scale, combining all three major groups of planktonic microfossils: foraminifera, coccolithophorids (nannofossils), and radiolarians.

The results from the two-leg swing into the central and western Pacific, the region farthest from the actively spreading East Pacific Rise, could only hint at the history of the Mesozoic Pacific. On one of the legs, the ship was used more like a dredge than a drill and few cores were recovered during repeated attempts (36 holes!) to core the sediment-basement contact. Two western Pacific oceanic plateaus, Shatsky and Ontong Java, were drilled, and Lower Cretaceous strata were confirmed on Shatsky. Ontong Java (still the cynosure of many eyes) is covered by about 1 km of pelagic sediments, but chert layers, here as elsewhere, blocked penetration of the flat-faced diamond bits used in the early days of the project and interfered with recovery of more than a few chips of rock. Better technology was urgently needed.

Lessons Learned Early

A lesson learned from drilling during the first 18 months from both engineering and scientific perspectives was that coring should be continuous, and that vastly improved methods were required to get these continuous records in piles of sediment with widely varying physical properties (e.g., alternating chert and chalk layers). The JOIDES Planning Committee later ordained continuous coring as the norm and engineers developed better drill bits, a system of heave compensation to keep the drill bit from moving up and down with the motion of the vessel (ready for sea trials on Leg 33 in 1973), and a hydraulic piston corer that recovers long, undisturbed cores free of vessel motion (ready for Leg 64 in

1978). Pressure core barrels have been deployed to retrieve gas hydrate samples under *in situ* pressures.

Gradually and intermittently, then more regularly, downhole logging was instituted for most holes, and a fruitful collaboration was established with logging companies, to improve and widen the scope and effectiveness of the logging tools available. These tools have not only helped to fill in gaps in the cores, but enabled correlation of drill results with those of seismic reflection profiling, and establishment of heat flow values and other geophysical parameters.

From the very beginning, scientific panels advocated using the holes as "natural laboratories," but budgets and time constraints kept this activity at a slow pace. Nonetheless, over the years, the drilled holes have been increasingly used for measurement of such variables as heat flow and for experiments on fluid flow, seismic velocity, and earthquake monitoring, to name but a few.

What was needed, from the very first, was money to design, test, and put into action these technologic innovations. The reality was that there was never enough money. The contract with Global Marine was fixed and the remaining funds were for all the rest. If the planners asked for better bits or more logging, then the money had to come out of science operations. For example, until the very end of DSDP in 1983, NSF allowed expenditure for only one computer for word processing for a project that was publishing a 1000-page hard-back report on scientific results every two months.

The Long Haul

As easily predicted, before the ship had progressed more than part way along its planned nine-leg track, plans were already changing and new proposals were submitted to extend the project. So excited was the scientific community by the early results that NSF, after suitable review and by simple amendment to the initial contract, extended the project for another two years. Time and again extensions were granted, going on now for 30 years. Contractors have changed, the JOIDES organization has expanded, international partners have been recruited, funding sources have been added (and deleted), names have changed, the drill ship has been replaced and project management shifted, while the drilling goes on and new scientific results pour in.

Almost immediately after the formation of JOIDES, the University of Washington was added to the group and U.S. JOIDES institutions now number eleven. A U.S. corporate entity, Joint Oceanographic Institutions, Inc., was created to provide fiscal responsibility for JOIDES, so that NSF could sign contracts to support JOIDES activities. From the beginning of DSDP, many non-U.S. scientists had been members of the scientific parties aboard the ship, but in 1975, by requests from several countries and with the active encouragement of NSF, the project was formally internationalized as the International Program for Ocean Drilling (IPOD). Several partner nations (Germany, the USSR, France, United

Kingdom, Japan) joined JOIDES, each paying a share of project costs as annual "dues." The list of member countries and consortia of countries has fluctuated over the years, but their combined contribution to the drilling program now constitutes roughly 40 percent of the costs. They also contribute significantly to site surveys in preparation for drilling, pay the salaries and travel costs of their nationals, and fund their shipboard scientists for postcruise analysis of samples and data. After IPOD had been in existence for a few years, NSF came to realize that U.S. shipboard scientists were at a competitive disadvantage on funding and set up a system of support through a U.S. Scientific Advisory Committee (USSAC) that also gives grants (through NSF) to support other drilling-related science. Each partner nation is responsible for its own site survey expenditures, and NSF has responded quite generously to U.S. proposals for geophysical surveys in support of drilling.

A breakdown of the $1 billion expended by NSF and its international partners on the various ocean drilling projects, from Mohole to the present day, is shown in Table 1.

DETOUR: THE OCEAN MARGIN DRILLING PROGRAM

In the late 1970s and early 1980s, paced by improvements in seismic reflection systems available in academia, scientific interest in the JOIDES community began to focus very seriously on thickly sedimented continental margins such as the Atlantic margin of North America and the margins of Africa. To reach prime objectives at these places without risk of encountering oil or gas accumulations that could escape to the seafloor, a ship equipped with a riser system was needed (i.e., a system in which the drill pipe is within a surrounding pipe and drilling fluids pumped down the inner, rotating pipe are circulated back to the ship within the annular space between the two pipes). This system allows pressure controls (i.e., drilling muds and shut-off devices). NSF, with support from the Carter Administration, approached representatives of the U.S. oil industry with a suggestion that JOIDES and industry might form a kind of consortium to accomplish scientific drilling on margins, with industry supplying technical expertise, some geophysical survey data, and some financial support.

Industry went along for a time with this new Ocean Margin Drilling (OMD) program to the extent of sending delegates to the OMD scientific planning committee meetings and paying for a set of data synthesis albums. Some participants from industry were from the beginning hesitant not only about the potential costs of the program, but also about the possible presence of an "open-book" program operating in waters of economic interest to the companies. One requirement troublesome to most U.S. academic scientists was that non-U.S. participation was excluded. In 1984, on hearing the final cost estimates and with the Reagan Administration now at the helm, many industrial participants withdrew

from the project, which then collapsed. About $16 million had been spent, nearly all on administrative expenses and engineering studies. No steel was cut, no holes were drilled.

During the OMD effort, a search had been made for a suitable drill ship for the riser program. The daily costs for commercial vessels of this class were prohibitively high, and planners then turned to the famous *Glomar Explorer*, the ship that the Central Intelligence Agency had commissioned to recover the coding device from a Soviet submarine that sank in deep water northwest of the Hawaiian Islands. This recovery effort, thinly disguised as a manganese nodule hunt, in an area where nodules were not very abundant and compositionally of little commercial interest, was successful and the special ship, with its derrick, drawworks with immense lifting power, dynamic positioning, and very large spaces available for laboratories, was in mothballs near San Francisco. NSF, as part of the OMD program, contracted with engineers to draw plans for conversion of this government-owned ship for riser drilling. The cost estimates were huge, in fact unacceptable. The ship remained in mothballs until 1996, when it was at last converted to a deepwater drillship for Chevron and Texaco, at a cost of about $160 million.

JOIDES RESOLUTION AND ODP: A NEW SHIP AND A NEW MANAGEMENT

Owing to strong pressures from the scientific community, the DSDP drilling program was kept on course through all the OMD detour. At about this time, a crisis in industry sent daily rates for drill ships plummeting, and an alert NSF moved quickly to hire a particularly suitable ship, the D/V *Sedco 471*, owned and operated by British Petroleum and Schlumberger, at bargain rates. By November 1983, D/V *Glomar Challenger* had completed 96 consecutive legs of drilling. The acquisition of the larger and more capable ship coincided with a move of management of the project from the Deep Sea Drilling Project at Scripps Institution of Oceanography to the Ocean Drilling Program (ODP) at Texas A&M University. After a hiatus of only 14 months, drilling began again using D/V *Sedco 471*, known henceforth to the scientific community as *JOIDES Resolution*, a name not only honoring Cook's eighteenth-century exploring ship, but also resonating with notions of community accord, group determination, and scientific problem solving. Drilling began (ODP Leg 100) in January 1985. We are now (Leg 182) drilling along the Great Australian Bight and the system is performing well, given that budgets are now so tight that some scheduled scientific plans cannot be carried out for lack of proper tools being available on the ship.

SCIENTIFIC MILESTONES: RESULTS THAT CHANGED OUR WAY OF THINKING

In looking back over the past 30 years of scientific ocean drilling, certain milestones mark signal achievements, some

TABLE 1 NSF Expenditures for Scientific Ocean Drilling Programs, 1958-1998 (thousand dollars)

Year	Mohole	LOCO	Caldrill	DSDP U.S. Ship	DSDP U.S. Science[a]	DSDP International	OMD	ODP U.S. Ship	ODP U.S. Science[a]	ODP International	Annual Totals
1958	20										20
1959	20										20
1960	220										220
1961	1,520										1,520
1962	1,970										1,970
1963	3,820	20									3,840
1964	8,000	88									8,088
1965	24,700		250								24,950
1966	16,970			5,400							22,370
1967											0
1968				4,160	42						4,202
1969				2,430	20						2,450
1970				6,300	175						6,475
1971				7,010							7,010
1972				8,930	282						9,212
1973				9,250	269						9,519
1974				10,700	295	250					11,245
1975				10,220	185	1,250					11,655
1976				11,800		3,500					15,300
1977				12,800	100	3,880					16,780
1978				12,700	204	4,400	70				17,374
1979				9,260	1,800	8,000	1,000				20,060
1980				13,940	1,020	4,880	3,400				23,240
1981				14,790	2,158	5,600	4,888				27,436
1982				13,070	1,468	8,560	5,578				28,676
1983				15,500	1,145	6,460	1,313	3,640			28,058
1984				1,340	1,482	1,000		22,530	2,026	4,170	32,548
1985				1,070				22,760	3,695	7,600	35,125
1986				2,120				19,430	7,342	12,510	41,402
1987				210				20,220	8,600	15,000	44,030
1988								20,650	8,900	15,000	44,550
1989								21,250	9,800	15,000	46,050
1990								21,550	9,700	16,600	47,850
1991								23,460	10,100	16,500	50,060
1992								24,980	11,600	16,580	53,160
1993								25,400	10,000	17,800	53,200
1994								28,430	10,100	15,770	54,300
1995								27,550	11,500	16,720	55,770
1996								27,680	11,400	16,720	55,800
1997								27,440	12,450	16,960	56,850
1998								29,946	10,950	17,454	58,350
TOTALS	57,240	108	250	173,000	10,645	47,780	16,249	366,916	138,163	220,384	1,030,735

[a]The category U.S. Science, under DSDP and ODP, includes grants to U.S. scientists for drilling-related research and for U.S. site surveys in support of drilling.

SOURCE: Bruce Malfait, NSF, August 1998.

of which are set out in recent overviews (Malfait et al., 1993; Larson et al., 1997). Other achievements, just as significant, are the cumulative result of many legs of drilling. I review here a sampling—probably reflecting my own interests—of some of the most important results, findings that truly changed our way of thinking. I also mention a few problems that remain as very important but unresolved by drilling.

Time Scales

It has been said that the special philosophical contribution of the geological sciences is the establishment of the immensity of geologic time. The elaboration and refinement of time scales have developed apace with new methods to measure the passage of relative and absolute time: superposition of strata, cross-cutting relations among rock bodies, biostratigraphy, radiometric decay, magnetic reversals, variations in isotopic compositions of strata, and rhythmically deposited sediments. Because cores of pelagic sediments from ocean drilling are commonly exceptionally rich in the remains of the most important planktonic microfossils—radiolarians, foraminifers, and coccolithophorids—the cores provide the basis for very detailed biostratigraphic zonations, based on first and last appearances and joint occurrences of taxa. The web of drill sites in the different oceans enables the establishment of a virtually global biostratigraphic scheme for the past 150 million years. The continuous cores also provide material for determination of magnetic-reversal sequences, which can in turn be linked to the sequence of seafloor magnetic anomalies.

The direct radiometric dating of volcanic ash beds in the sediments and of drilled oceanic crust, using laser technology that can yield 0.1 million-years resolution, plus radiometric dating of biostratigraphically constrained ash beds and igneous rocks on the land, has improved resolution by an order magnitude since the Ocean Drilling Program began. In the last decade, these scales, with resolution of about 0.5-2 million years, have been further refined by an order of magnitude by the realization that rhythmic sedimentation in step with the rhythmic changes in the Earth's orbital parameters is a common feature of pelagic sediments. We are now close to the definition of a time scale for the last 30 million years with 20,000- to 100,000-year resolving power. The road is open to extend this precision back into the Jurassic via our drill cores. Having a time scale with such fine resolution makes it possible to address a host of rate problems: rates of sediment accumulation, rates of evolution, rates of change of environment. It makes possible the detailed ordering of related events on a global scale and the unraveling of cause-and-effect, chicken-and-egg problems.

Paleoceanography

The planktonic microfossils in pelagic sediments fall to the seafloor from overlying near-surface waters and thus re-flect prevailing environmental conditions in these waters, while benthic fossils reflect conditions at the seafloor. This simple picture is distorted by the effects of dissolution: calcareous fossils tend selectively to dissolve in cold deep waters, owing to the greater dissolved carbon dioxide content there. Thus, to make paleoceanography quantitative, we need an independent method of estimating paleodepth. Almost concurrent with the start of drilling, an empirical relation was established, using an early version of the magnetic anomaly time scale, for the age of oceanic crust and its depth below the sea surface. The empirical curve, with correction for isostatic loading by sediments, fits closely to a simple curve $D = D_0 + K(\text{Age}^{1/2})$, where D is the depth of oceanic crust, D_0 is the depth at the spreading center and K is a constant, generally about 350. The curve is applicable out to crust about 80 million years old, where it begins to flatten. The immediate payoff was the charting of the regional and temporal fluctuations in the depth where carbonate supply and dissolution rates balance, the calcite compensation depth (CCD). A first-order finding was that there was an abrupt deepening of the CCD by about 1,000 m near the beginning of the Oligocene, about 35 million years ago, at about the time of the earliest continental-scale Antarctic glaciation. Global paleodepth maps of the CCD now exist for many levels in the post-Jurassic.

A paleoceanographical surprise emerged with the coring of organic carbon-rich layers at several levels in the mid-Cretaceous in both the Atlantic and the Pacific oceans. Paleodepth estimates for these sediments yielded a broad range of depths, excluding the abyssal waters of the Pacific, suggesting that the anoxic conditions were associated with a broadening and intensification of the oxygen minimum, possibly owing to relatively strong density stratification of the oceans during these extreme "greenhouse" times of raised global sea level and warm ocean temperatures.

Determination of $^{16}O/^{18}O$ in precisely dated mid-Cretaceous-Recent planktonic and benthic foraminifers has allowed construction of a detailed history of oceanic surface- and bottom-water temperatures and an estimate of the changing volumes of continental ice. What the isotope record shows, besides the contrast between the generally warm "greenhouse" ocean climates of the Cretaceous and the colder (in high latitudes) "icehouse" climates of the Neogene, is a stepwise history of long periods of relatively stable conditions and abrupt transitions to new, but different, stable conditions. The record also shows that tropical sea-surface temperatures have been relatively stable; it has been the high-latitude oceans (and the deep waters derived from these latitudes), that have changed the most. What we do not understand are the "why's" of the stepwise history. One promising avenue was explored in the South Atlantic by coring the summit and flanks of Walvis Ridge in a highly successful attempt to document the history of bottom-water temperatures along an oceanic depth profile (Shackleton et al., 1984). The depth-profile approach has not since been much exploited, but holds great promise in mapping the paleotemperature structure of the oceans.

Beginning with DSDP Leg 27, attempts were made to drill in very high latitude waters, mainly for paleoceanographic objectives. In spite of daunting conditions, drilling around Antarctica and in the seas off northeast Greenland and in the Labrador Sea has elucidated the Paleogene beginnings of continental glaciation and clarified the plate tectonic events that opened a circum-Antarctic Ocean and led to the formation of Antarctic Bottom Water. In the north, drilling has enabled reconstruction of the history of formation of North Atlantic Deep Water.

Drilling on the two sides of the Isthmus of Panama has established the timing of the late Neogene closure of the isthmus, isolating Atlantic from Pacific marine biotas and forming a land bridge for terrestrial animals.

Catastrophes: The K/T Event and the Desiccate Mediterranean

Two spectacular events have captured public imaginations, the impact of the cosmic bolide that struck Earth at the end of the Cretaceous and the drying-up of the Mediterranean near the end of the Miocene. The K/T bolide story has depended as much on data obtained from land outcrops as from the ocean drill cores, which have served mainly to provide an especially detailed record of the sequence of events in regions relatively close to the impact site, on the Yucatan Peninsula. The discovery of the Mediterranean events, on the other hand, was almost purely the result of drilling on DSDP Legs 13 and 42A, which showed that the salt deposits that accumulated in shallow salt marshes and brine basins at the bottom of several Mediterranean depressions are both underlain and overlain directly by deep-sea biogenic sediments. Only small tectonic movements were required to isolate the Mediterranean from the Atlantic, and near-total evaporation, which may have been repeated many times, was likely very quick. These two catastrophes are now so well documented that, taken together with the evidence about very rapid shifts in ocean temperatures and the long-standing evidence of catastrophic floods on land (e.g., the rapid emptying of Lake Missoula to create the scablands of Washington), they are softening the rock-hard beliefs of the Earth science community in traditional gradualism. We must now admit the possibility of rare and powerful events, the amplifying effects of critically located small events, and the wide range of possible rates of change. James Hutton, the father of classical uniformitarianism and his disciple Charles Lyell may be uneasy in their graves.

Gas Hydrates and Living Bacteria at Depth

Solid hydrates of methane are stable in the pore spaces of sediments where the temperatures are cold or the confining pressures sufficient. Vast regions of the arctic tundra are underlain by sediments containing gas hydrates, and drilling has confirmed that continental margin sediments containing concentrations of biogenic methane also contain crystalline gas hydrates where temperatures and confining pressures are right. These concentrations are commonly visible on reflection seismic records as "bottom-simulating reflectors." Drilling has permitted preliminary estimates of the locations of these buried hydrates and an appreciation of the quantities of methane that might be released into the atmosphere if bottom-water temperatures were to rise significantly.

Although evidence of bacteria has been recovered in cores from oil exploration and from pores in volcanic glass under 400 m of mid-Atlantic sediments, ODP drilling in plant-rich layers in turbidites of the Amazon deep-sea fan in the Atlantic Ocean has recovered bacteria that are actively reproducing at subbottom depths of hundreds of meters. Taken together with the evidence of living bacteria from the high-temperature vents along spreading ridges, we can agree with Reiche (1945) that "The infernos envisioned by medieval theologians can [hold] only limited terrors for such creatures." We are still exploring for the outer limits of the biosphere.

Oceanic Lithosphere and Hydrothermal Activity

Early attempts to drill into very young oceanic lithosphere, close to the active spreading centers where postemplacement alteration of rocks should be minimal, were defeated. The brittle and fractured basaltic rocks broke up in front of the drill and stopped progress. Except in areas of strong hydrothermal alteration, we have still not been able to sample more than a few meters into "zero-age" oceanic crust. The most successful drilling has been at a site off Costa Rica on crust about 6 million years old, covered by about 275 m of sediment. Here coring was successful to a depth of 1,836 m into pillow basalts and sheeted dikes. Surprisingly, seismic velocities commonly associated with Layer 3, generally believed to be gabbro, are measured at this hole in part of the zone of sheeted dikes and basalt flows.

The deeper parts of the oceanic lithosphere can be reached only where spreading was very slow and magma supply so skimpy that basalts are thin or absent and spreading has allowed gabbro and mantle rocks to emerge at the seafloor. Gabbros were cored almost continuously at a site on the slow-spreading Southwest Indian Ridge, southeast of Africa, yielding virtually the full suite of oceanic plutonic rocks and partly validating models erected on the basis of scattered dredge samples and from studies of supposed oceanic lithosphere tectonically emplaced onto continents—the ophiolites. The excellent drilling conditions at this site suggest that, in principle, one might reach the dreamed-of Moho here.

The mantle itself has been cored at a few places, (e.g., in the Atlantic off Iberia in tectonically disturbed locales, where ultrabasic rocks have been serpentinized and uplifted in diapirs that reached the seafloor, and in magma-starved segments of the Mid-Atlantic Ridge). In the Pacific, serpentinized mantle rocks were recovered by drilling in a tectonic rift zone. Because seismic data indicate that the Moho is present at depth even where altered mantle rocks are close to the seafloor, we are left with the original AMSOC question: What is the nature of this seismic discontinuity? Is it an original petrologic boundary, a tectonic boundary, or a level in the lithosphere marking the downward limit of alteration by circulating seawater? Or any one of these, depending on where you are? Repeated measurements over a period of several years at the Costa Rica drill site show that cool ocean water is being drawn down into the upper parts of the oceanic crust. Heat flow measurements and direct observation from submersibles show that hot waters, charged with ions from crustal alteration, emerge elsewhere at oceanic spreading centers and from outcrops of crustal rocks on abyssal hills. The drill has successfully recovered hydrothermal spring deposits close to an active spreading center, deposits that include tall chimneys of sparkling metal sulfides. Gradually, we are building up quantitative estimates of the rates and depths of circulation of seawater through the oceanic lithosphere and of the extent to which this flow moderates the composition of seawater. Drilling on crust ranging in age back to the Middle Jurassic shows that most hydrothermal alteration takes place while the lithosphere is very young.

Beginning with the clean test of seafloor spreading on Leg 3, the determination of the age of oceanic lithosphere has been made at many of the 1100 sites drilled, giving us a set of ties between the biostratigraphic scale and magnetic anomalies, back to the mid-Jurassic, and enabling the interpretation of magnetic anomaly patterns in terms of plate tectonic evolution.

Mantle Plumes, Hotspots, and Early Cretaceous Volcanism

The ruling theory for the formation of linear seamount chains is that they result from motion of a plate over a fixed melting anomaly, or hotspot, in the underlying mantle. Drilling along the Emperor Seamount Chain in the North Pacific and the Ninety East Ridge in the Indian Ocean showed that these fitted the model. Other drilled chains (e.g., the Line and Marshall chains in the Pacific) have messy records of progressive volcanism, and some undrilled chains, sampled by dredge and by hammer, (e.g., the Australs and the Puka Puka chains), show a scrambling of ages, inconsistent with fixed hotspots. Linear seamount chains remain a problem.

Several oceanic plateaus—great deep-rooted (tens of kilometers to the Moho), smooth-backed leviathans that rise to levels 1-3 km above the surrounding deep ocean floor—have been drilled, primarily for the continuous stratigraphy of the mainly calcareous pelagic sediments that blanket the basaltic basement. The origin of many of the plateaus is ascribed to mantle plumes, arising mainly during a short interval during the mid-Cretaceous from unknown depths. Beyond limiting the times of formation, drilling has got us almost nowhere on the plateau problem so far.

Dating of the age of emplacement of several of the major oceanic plateaus (Ontong Java, Manihiki, Kerguelen), of scores of seamounts spread over a large part of the western Pacific and the great volumes of basalt on the deep Pacific seafloor, far from any contemporary spreading ridge, points to a highly unusual time of massive volcanism during a relatively short time of only about 20 million years in the Early Cretaceous. The volumes are comparable to those produced along the entire global spreading system and suggest some very deep rooted cause, a veritable revolution in the Earth's mantle. The near coincidence of these events with the beginning of the long period of normal polarity of the Earth's magnetic field and of the widespread deposition of organic-rich black sediments and evidence for warm climates, has stimulated a search for causal connections among these effects.

Passive Continental Margins

The Atlantic Ocean is bordered by passive continental margins, segments of which have been subsiding and receiving continent-derived and carbonate sediments since the Middle Jurassic. The North American margin is covered by a prism of sediments too thick for full penetration with *JOIDES Resolution*, but the European-African margin has a much thinner cover, and the early history of the margin is thus within reach of the drill. Cores documenting the early history of the Morocco margin show a beginning with a Late Triassic proto-Atlantic saline basin below sea level and progressive Mesozoic evolution from fluviatile to deeper and deeper waters.

Farther north, in the Norwegian Sea, which opened much later than the Central Atlantic, drilling penetrated and sampled huge wedges of mainly subaerial early Tertiary basalts that were extruded from both sides of the widening rift between Norway and Greenland. Such marginal basalts are imaged on seismic records from many other segments of passive margin around the world and may be related to voluminous mantle plumes that may localize and even initiate seafloor spreading.

Active Continental Margins, Island Arcs, and Backarc Basins

Concentrated drilling has been done on several active continental margins, where oceanic lithosphere is being subducted beneath a volcanic arc. These places are the loci of major seismicity, and understanding processes in them should contribute to public safety. The clearest results have been obtained from a transect off Barbados, where drilling was carried through the surface separating the two opposing plates. Here, measurements of *in situ* fluid pressures showed that the oceanic sediments in the lower plate are overpressured. Cores from sedimentary strata of the upper plate showed strong evidence of tectonic kneading of sediments in the accretionary prism and also the presence of fluid escape channels carrying waters squeezed from the deforming sediments upward to the seafloor.

Several transects across the entire active margin complex (trench, forearc, volcanic arc, backarc basin, and remnant arc) have documented not only the materials in this system, but the timing and rates of development in them as well as the contrasting deformational styles in zones with thickly sedimented compared to near-barren trenches.

THE FUTURE

Now the drilling program is approaching another crossroads. In 2003, unless something new happens, drilling may well cease or be replaced by a quite different program strongly resembling OMD in its scientific objectives. Planning continues for the five-year ODP time between now and then.

One drilling prospect has been opened by the Japanese announcement that they intend to construct, at their own expense, a large ship ("*Godzilla Maru*") fitted out for riser drilling. Some tens of millions of dollars are said to be in the pipeline for design studies for a ship that will cost upwards of $500 million to build and have daily operating costs of something like $130,000 (about three times the *JOIDES Resolution*). Drilling from this ship during the first few years is planned to be in waters not more than about 2.5 km deep (shallower than most of the spreading ridge system, let alone the main ocean floor), and the ship would work for much of this time close to Japanese home waters, where a number of problems in the structure, hydrology, and seismicity of thickly sedimented active margins are available. Proposals for specific riser drilling objectives are now being formulated.

As for nonriser, ODP-style drilling, NSF is said to be looking at the possibilities of funding a *Resolution*-type vessel for operations post-2003, in addition to paying its share of the Japanese riser ship daily costs. The U.S. COMPOST-II Committee on Post-2003 Scientific Ocean Drilling issued a report in 1996 endorsing a two-ship program. The active scientific community is busy writing proposals to be discussed at a planned international conference in 1999. Urgent messages are in the air that we should all be demonstrating support for and submitting proposals for work in an Integrated Ocean Drilling Program (IODP) to follow ODP, and using two ships, riser and nonriser. We appear still to lack concordance on major new scientific initiatives, initiatives of the scope and imagination of the original Mohole project, initiatives that can capture the attention of large segments of not only the scientific community but the public and Congress as well. To arms! Enlist now!

ACKNOWLEDGMENTS

I thank Bruce Malfait for the data on year-by-year expenditures for drilling projects. Both he and Walter Munk made constructive suggestions on an early draft of the paper. D.K. Van Keuren kindly allowed use of his unpublished paper on the history of the Mohole Project. W.W. Hay made constructive suggestions for improvement of an earlier draft of the manuscript.

REFERENCES

Emiliani, C., and J.I. Jones. 1981. A new global geology: Appendix III: report on Cruise LOCO 6301 with Drilling Vessel *Submarex* (a reprinting of the report made to NSF). Pp. 1721-1723 in C. Emiliani (ed.), *The Sea*, volume 7: *The Oceanic Lithosphere*. John Wiley, New York.

Greenberg, D.S. 1964. Mohole: The project that went awry. *Science* 143:115-119.

Larson, R.L., and 29 others. 1997. *ODP's Greatest Hits*. Brochure issued by Joint Oceanographic Institutions, Washington, D.C. 28 pp.

Malfait, B., and 51 others. 1993. 25 years of ocean drilling. *Oceanus* 36(4):5-133.

Reiche, P. 1945. *A Survey of Weathering Process and Products*. Univ. New Mexico Pubs. Geol. No. 1. 87 pp.

Shackleton, N.J., M.A. Hall, and A. Boersma. 1984. Pp. 599-612 in T.C. Moore, Jr., and Rabinowitz, P.D. et al., (eds). *Initial Reports of the Deep Sea Drilling Project, 74*. U.S. Government Printing Office, Washington, D.C.

Solow, H. 1963. *Fortune* (May):138-141, 198-199, 203-204, 208-209.

Van Keuren, D.K. 1995. Drilling to the mantle: Project Mohole and federal support for the Earth sciences after Sputnik. Unpublished text of paper delivered at Annual Meeting, History of Science Society, Minneapolis, Minn. 13 pp.

Technology Development for Ocean Sciences at NSF

H. LAWRENCE CLARK

Division of Ocean Sciences, National Science Foundation

ABSTRACT

Great advances in our understanding of global oceans and their interactions with the Earth and the atmosphere have been made under NSF sponsorship over the past 50 years. Many of these achievements were enabled, in part, by scientists' having the technical capabilities and other means to collect samples, run experiments, and make appropriate observations. The NSF, primarily through the Ocean Sciences Division (OCE), addresses the provision and development of technology for conducting ocean research in three ways: (1) by supporting a variety of shared-use facilities and technical services, (2) by developing techniques and instruments through the disciplinary research programs, and (3) through establishment of a unique technology development program that supports development of new capabilities for the overall ocean science community. The mechanisms through which OCE provides technological capabilities and develops new ones have evolved as the field has matured. OCE has effectively met community requirements for supporting facilities and projects necessary to advance the field. Provisions for funding long-term development of new instrumentation and technological capabilities should remain a priority for continued advancements. Just as past progress has benefited from collaborations with other agencies and endeavors, establishing and maintaining partnerships to develop new technological capabilities are going to be critical for future progress in ocean sciences as well.

He [Benjamin Franklin] thought the thermometer could become an important aid to navigation, particularly to ships sailing in or near the Gulf Stream. He convinced Capt. Truxtun that this novel idea was a good one, and for many years the Captain went about plunging thermometers into most of the seas of the world. (Ferguson, 1956)

This brief passage describes an interesting aspect of ocean science research—and it illustrates how sometimes one type of measurement will lead to basic new knowledge about a seemingly unrelated oceanographic feature. The passage describes how one of the earliest discoveries and descriptions of the Gulf Stream was brought about by associating the relative sailing time for trans-atlantic passages with seawater temperature. As Postmaster General for the newly formed United States, Benjamin Franklin received complaints about mail delivery. Eastbound ships from America to England made the passage in half the time of westbound ships. There were suspicions of a trading conspiracy.

After looking at ships' logs and talking with captains (including a relative who was captain of a Nantucket whaling ship), Franklin related rapid eastbound passages and slow westbound passages to unusually warm seawater. The fastest westbound passages followed a more southerly crossing in colder water. The notion of a flow pattern he developed was one of the first physical descriptions of the Gulf Stream. Based on this new knowledge, sailing orders were issued to avoid the warm water when sailing west to America from Europe, but to stay in the warmer water when sailing east to Europe.

Ocean science is in large part an observational science, so it follows that our knowledge of the oceans has increased as we have increasingly gained the ability to make measurements and observe natural processes on, within, and under the oceans. To do so, ocean scientists need appropriate tools and observational capabilities. It's not always obvious what tools will be needed to make the appropriate measurements. As described in other papers in this volume, great scientific achievements have been made under NSF sponsorship over the past 50 years. These achievements were enabled, in part, by scientists' being able to make appropriate observations, which in turn has depended on the availability of technological capabilities and the development of new technology, including that specifically for ocean science research.

Technology for ocean science research covers a wide spectrum, ranging from ships and satellites and underwater vehicles and buoys, to sophisticated laboratory instrumentation. Providing the appropriate technologies and developing tools and new technologies for research constitute a complex process. Initially, one needs to figure out how to measure what it is you're trying to measure. For example, it's figuring out how to make routine measurements of temperature and salinity from the surface to full ocean depths with enough precision, accuracy, and repeatability that one can describe the movement of water masses—when the physical differences between them are slight. How does one measure the heat content and heat distribution within these water masses and its exchange with the atmosphere in order to make predictions about climate variability? How does one measure the amount of material that sinks from the productive surface waters to the seafloor? How does one measure and describe the microbial processes in the water column and on the seafloor, as this sinking material decomposes and provides nutrients for other ecosystems? How does one measure the geological structure and properties of the seafloor so as to be able to understand the processes that gave origin to the Earth and that are continually shaping it? Once a decision is made as to what measurements are needed, then the issue is how does one get there and what does one use to make the necessary observations?

What we can learn about the oceans from direct observations with scuba tanks and surface measurements isn't particularly insightful and doesn't provide much new information on scales necessary to study basic processes at work in the oceans. It's when one goes to deeper water that things get interesting. First of all, investigators need to get out on the ocean with adequate tools and capabilities to handle whatever it is that was designed and built to make the measurements or collect the samples. Providing technological capabilities for the overall ocean science research community is where NSF has taken the lead and structured its programs to support these capabilities.

The NSF, primarily through the Division of Ocean Sciences, addresses the provision and development of technology in three ways: (1) by supporting a variety of shared-use facilities and technical services, (2) by developing new techniques and instruments through the disciplinary research programs, and (3) through establishment of a unique technology development program that supports development of new capabilities that might lead to enhanced capabilities for the overall ocean science community. The mechanisms through which OCE provides technological capabilities and develops new ones has evolved as the field has matured.

TECHNOLOGY VIA SUPPORT FOR MAJOR FACILITIES

Central to almost all oceanographic research endeavors in all disciplines is the research vessel. Research vessels and their equipment represent a major technological asset, and as such, they are critical to the advancement of ocean science research. Although the ships themselves have different owners and lineages, NSF has become the major source of support for providing, operating, coordinating, and maintaining this technological capability. This capability evolved over time and within some severe financial constraints, but it also evolved in response to some time-tested managerial decisions.

Because of their high costs of construction and operations, ships have always been the focus of special attention. It took the British Navy several years to come up with the resources in 1876 to provide the H.M.S. *Challenger* for the famous four-year expedition that initiated the field of ocean science research. Government ships provided the seagoing capability for civilian ocean science research in this country for decades.

Prior to World War II, there were four or five academic research ships in the country, each of which was operated and maintained by the few oceanographic laboratories at the time, for their own projects and personnel. During the rapid growth years of the 1960s, the Navy, primarily the Office of Naval Research (ONR), provided most of the support for ocean research and technology. The number of oceanographic research institutions grew and the number of ships grew. By 1970, the academic fleet totaled at least 24 ships—the operation of which had become big business. Also by 1970, the NSF had become the major source of support for ocean science research as major new programs, such as the International Decade of Ocean Exploration (IDOE), started up. Other agencies such as ONR and the Department of Energy (DOE) were major sponsors as well, but their relative support was diminishing.

The NSF took a different approach for funding research and facilities than did the Navy and other agencies supporting oceanographic research at that time. ONR research programs generally funded entire research projects inclusively—the research, the equipment, the technology, and the necessary ship time. NSF, on the other hand, separated research from seagoing logistics and facility support. In 1960, NSF established a separate office for the construction, conversion, and operation of research ships. Mary Johrde first headed this office, which went by different names with different reorganizations. But the Oceanographic Centers and Facilities Section (OCFS), as it is called today, has had responsibility for providing ship time and other facility support for Ocean Science Research Section (OSRS)-sponsored projects and other projects sponsored throughout the NSF.

The "NSF model" of separating ship and facility support from the research programs had some interesting consequences with respect to technology. The separation of facility support from research support enabled more focused attention to be given to improving technology as a community resource. The "ONR model" of inclusive project support worked well in the 1950s and early 1960s when institutions took on individual projects from start to finish. Having a research program buy ship time, technical services, and equipment was helpful to the successful completion of the individual project, but it did little to enhance research and technological capability for the community as a whole. During the 1950s and 1960s, institutions that operated ships did so primarily for their own scientists. Everything necessary for a study was taken on the ship at the start of a cruise, and off the ship at the end. There was little reason to think about what type of technologies or capabilities a ship required, other than the basic equipment-handling capabilities provided by winches and cranes.

A ship's technological capability became increasingly important as ocean science matured in the 1970s. As programs such as the IDOE progressed, ocean research became more expeditionary, multidisciplinary, multi-institutional, and much more complex. Scientists were increasingly making use of research vessels that were operated by an institution other than theirs. Ship scheduling and management plus the acquisition and management of technology became an important matter for the newly established University-National Oceanographic Laboratory System (UNOLS), which is the topic of an earlier paper in this volume.

At the very first UNOLS meeting in November 1971, the issue of providing technological assistance to science projects using UNOLS research vessels was identified as a matter that needed addressing. The NSF model of separating ship and facility operations from science support enabled the Office of Facilities Support to tackle the technology provision issue by establishing two new programs: the Shipboard Technician Program and the Oceanographic Instrumentation Program.

The Shipboard Technician Program was established in 1972 to provide technical assistance to users of the academic research vessel fleet. Technical services funded by NSF had an at-sea component and an onshore component. Technical support activities at sea involve maintenance and repair of shared-use scientific equipment, plus supervision and training of scientific personnel in the safe and effective use of this equipment. Activities ashore included the maintenance, calibration, and scheduling of the shared-use equipment that was made available to ship users. Additionally, the technical support activities provided a liaison between the scientific party and the ship's support personnel and crew. As the use of research vessels by visiting investigators increased and as the complexity of equipment on varying ships increased, this liaison function became increasingly important in making best use of time spent at sea.

UNOLS concerned itself with improving technological

capabilities as well. The Technical Assistance Committee (TAC) was established in 1974. It developed a set of standard technological capabilities for the different classes or sizes of academic research vessels and worked toward improving these capabilities. The NSF Technician Support Program, working with TAC, developed new capabilities for research vessels as well. One such new development was the installation of SAIL (serial-ASCII instrumentation loop) systems. SAIL systems onboard UNOLS ships allowed scientists to automatically display and record a number of environmental parameters, such as date and time, navigational coordinates, sea-surface temperature, and other meteorological data plus the project's experimental data. It's difficult to realize in these days of powerful personal computers and local area networks, that the ability to walk off a research vessel with a data tape from a just-completed cruise represented a new technological capability 20 years ago. This seemingly trivial advancement was an important step for conducting oceanographic observations, because it facilitated the integration and assimilation of multiple observations, which is the focus of much oceanographic research today.

EQUIPMENT AND TECHNOLOGY ACQUISITION

Until the mid-1970s, the acquisition of all facility equipment by NSF for use on ships and ashore was managed by a single equipment acquisition program. Ships' equipment, such as winches, cranes, echo-sounding gear, and other permanently affixed equipment, was proposed and reviewed along with pooled-use scientific instrumentation. Proposers and reviewers had a difficult time sorting out the relative priorities of robust ships' equipment versus precision scientific instrumentation, especially given the rapid evolution of seagoing scientific instrumentation and the intense competition for funds. Many people felt that the ability to make technological improvements through the acquisition of new instrumentation was being hampered by the ongoing need for permanent shipboard equipment. In response to this concern, a separate Oceanographic Instrumentation Program was established in 1974 to support the acquisition of shared-use scientific instrumentation. This newly acquired instrumentation was to be placed in a pool of equipment and made available to users of the facility, be it a research vessel or a shore-based laboratory. The overall research support capability of the institution and its ability to make effective use of the requested instrumentation for conducting NSF-sponsored research projects were main criteria for evaluating proposals.

Accelerator Mass Spectrometry (AMS) Facility

Although ships and their related activities have been the major focus for providing new community-wide technological capabilities for ocean science research, they have not had sole attention. In planning for the major global change research programs, the World Ocean Circulation Experiment (WOCE) and the Joint Global Ocean Flux Study (JGOFS), considerable attention was given to determine whether adequate facilities and capabilities were in place in order to do the ambitious programs. A particular shortcoming was identified in the research community's ability to analyze a very large number of radiocarbon and other tracer samples that were envisioned for WOCE and JGOFS. These chemical tracers, carbon-14 in particular, have become valuable tools for describing oceanographic processes. They provide information on long-term mixing and circulation in the deep ocean, on upwelling, and on air-sea carbon dioxide exchange processes. These processes have major implications for understanding the forces that affect climate variability and the chemical interaction of the carbon cycle and biological productivity. Given the large number of samples needed for WOCE, JGOFS, and other geosciences programs, it was recognized that available analytical and logistical capabilities were inadequate to meet scientific requirements. Plans called for the analysis of up to 4,000 carbon-14 samples annually with precision of 0.3 to 0.4 percent.

Following several workshops and advisory meetings, OCE issued an Announcement of Opportunity in 1987 to establish an ocean science Accelerator Mass Spectrometry Facility. Newly developed AMS technology could reduce the required sample size by a factor of 1,000, to 250 ml of seawater, for achieving the requisite level of precision. However, considerable effort would be necessary to develop automated sample preparation procedures and new instrumentation for a high level of throughput.

Funds for establishing the AMS facility were identified in the fiscal year 1989 NSF budget request to Congress. Approximately $1.8 million per year for three years was planned for construction, installation, and initial operation. Five institutions submitted proposals. The end result is the National Ocean Sciences Accelerator Mass Spectrometry (NOSAMS) Facility at the Woods Hole Oceanographic Institution. The facility's goal is to provide the oceanographic community with a large number (up to 4,300 per year) of high-precision radiocarbon analyses. This includes rapid dissemination of the results of these analyses to the user and scientific communities. A commitment to automation has been made throughout the facility, including sample preparation, analysis, and data reduction, and a comprehensive relational database and bar-coding system tracks every sample and every process performed at the facility.

INCREMENTAL ADVANCES IN TECHNOLOGY

An overall characteristic of ocean science research is the fact that scientific advances and improved technological capabilities are incremental. With few exceptions, such as the hydrothermal vent discoveries from *Alvin* that are dis-

cussed elsewhere in this report, advances in our knowledge of the oceans are measured in small steps. The great advances that have been made in our knowledge of the oceans in the past 50 years are not so much in response to great technological advances, such as in space expeditions sponsored by the National Aeronautics and Space Administration (NASA), but rather from the continuous application of technologies and incremental new developments arising from scientific investigations. Essential to scientific advancement are the provision of technology to accomplish the research and providing mechanisms for developing and applying new technologies.

As a relative newcomer to the study of the oceans, compared to naval and fisheries interests, NSF has been a beneficiary of a long history of focused technological and scientific research. During this century and especially since World War II, the major provider of technological capabilities has been the U.S. Navy. There is a long and distinguished list of scientific accomplishments derived from Navy-developed instruments and technologies. These include

- SWATH bathymetric sonar,
- laser line scan optical sensors,
- global positioning satellite system,
- ocean bottom seismometers,
- seagoing flux gate total field magnetometer,
- *Alvin* and *Flip*,
- acoustic Doppler current meters,
- bioluminescence sensors, and
- long-term mooring technologies.

NSF's research requirements are oftentimes compatible with capabilities provided for the Navy interests, but there have been issues of accessibility, further refinement, adaptation, and cost-effective usage. Many requirements are unique to ocean science research and therefore require a focused and specific effort to make the right type of measurements at the right scale and with the needed precision and accuracy.

Throughout the period of the IDOE and subsequent reorganizations, nearly all NSF-sponsored technology developments were funded through individual research projects. Observational and measurement capabilities were developed by scientists in direct response to the progression of their scientific inquiry. One example, among thousands, is the successive development of plankton nets and other devices for enumerating and describing the distribution of plankton. Traditional conical nets gave way to multiple opening and closing nets, to which sensors were added to relate physical factors to the abundance of collected plankton. Nets in turn gave way to optical and acoustic sensing systems that work on varying time and space scales. No single device or capability is necessarily an objective. Differing research objectives call for differing research capabilities. Developing new capabilities oftentimes had to be accomplished over a succession of different proposals, reviews, and awards under sponsorship of different agencies and programs.

TECHNOLOGY DEVELOPMENT PROGRAM

In 1981, the Office of Technology Assessment (OTA) reported that technology development across the federal ocean programs was poorly coordinated and was provided mainly through specific objectives of mission-oriented agencies such as the Navy, the National Oceanic and Atmospheric Administration (NOAA), and NASA (OTA, 1981). NSF was shown to have a minimal role in ocean instrumentation and technology development. Research programs were attributed to whatever technology support was provided on an ad hoc basis.

About this same time, observers of the NSF ocean science peer-review process noted that in matching available resources to highly rated proposal budgets, instrumentation development was one of the first items to be eliminated. The focus was on research, more than on new tools to accomplish it. This was especially true of multidisciplinary instrumentation. Funding pressures and the conservative nature of the peer-review process required that NSF-sponsored technology development for basic ocean research either be essential for the accomplishment of the highest-rated research projects or be done at no cost to NSF.

Given these somewhat subjective observations, an experimental program area was established in fiscal year 1982 to consider proposals for developing new instrumentation and new technological capabilities that would have broad applicability. The Oceanographic Technology (OT) program was established within OCFS as part of an overall reorganization of OCE. The OT Program also assumed responsibility for supporting shipboard technicians, the acquisition of commercially available shared-use research instrumentation, and the development of new instrumentation and technology by individual investigators. In keeping with the multiuser facility responsibilities of OCFS, initial proposal submission guidelines for technology development emphasized data collection and general-use instrumentation.

Since this was a new program area and the first of its type for ocean science at any agency, there was a lot of latitude in the scope of the original proposals. Ocean science instrumentation development proposals had to satisfy two major proposal requirements: technological or engineering quality and ocean science relevance. Bimodal ratings occasionally resulted when scientists were enthusiastic about a proposed new measurement capability, but engineering reviewers judged that the proposal was technically flawed. The opposite also occurred when a proposed new development was well reviewed from the technical side, but the science reviewers found the scientific relevance or utility of the new device to be lacking.

From its inception in fiscal year 1982 through fiscal year 1998, slightly more than $55.5 million has been awarded for supporting more than 150 ocean science instrument development projects. Three general categories of projects have been supported, reflecting different community requirements: (1) demonstration projects that typically seek part-time support for a technician or engineer, plus supplies to test an idea for enhancing existing instrumentation; (2) implementation projects that span a range of activities for further developing or modifying existing instrumentation for general ocean science research applications; and (3) instrumentation systems development, which involve major projects, represented by cooperative efforts between scientists and engineers to integrate several instruments and technologies into an observational system. Parallel advances in theory and instrumentation are usually necessitated. Bioacoustic and satellite remote sensing, long-term moorings, tomography, autonomous underwater vehicles, conditional sampling devices built around knowledge-based systems, and fiber-optic sensors are examples of this complex category of development project. A long-term effort is required at relatively high annual cost, and risk of failure is a further consideration.

The peer-review system does not lend itself well to long-term, forward-looking projects with a significant risk of failure. However, to develop new capabilities that are driven by scientific needs, risk can be reviewed and managed. A case in point is the development and establishment of long-term seafloor observatories. The scientific need to make long-term measurements, in both the coastal zone and the deep sea, coupled with newly developed sensors and other technologies, has set the stage for a new way of conducting certain types of ocean science research. The nature of these observatories suggests that they will have to be a new type of facility. However, as with other facilities, their long-term support and viability will depend on their ability to provide the technological capabilities that will be needed to support ongoing ocean science research.

If one considers the phenomenal advances that the academic ocean science research community has made in the past several decades, sponsored primarily by NSF, one may conclude that the provision of technology and the development of new capabilities have been appropriately addressed. An adequate mix of ships and facilities has been provided to the community, research projects have been underpinned by a growing technological base, and OCE has provided funds to lay the groundwork and develop new capabilities for research envisioned in the future. OCE has effectively met a community requirement for supporting projects to enhance and upgrade existing observational and analytical research capabilities. The availability of significant levels of funding for long-term development of new instrumentation and technology should remain a priority for growth. Just as past progress has been based on collaborations with other agencies and endeavors, establishing partnerships and maintaining them are going to be critical for future progress.

REFERENCES

Ferguson, E.S. 1956. *Truxtun of the CONSTELLATION.* Johns Hopkins Press, Baltimore, Maryland.

Office of Technology Assessment. 1981. *Technology and Oceanography: An Assessment of Federal Technologies for Oceanographic Research and Monitoring.* U.S. Government Printing Office, Washington, D.C. 161 pp.

Large and Small Science Programs: A Delicate Balance

The Great Importance of "Small" Science Programs

G. Michael Purdy

Division of Ocean Sciences, National Science Foundation

Any discussion of the merits of "large" versus "small" science programs (as alternative mechanisms for the organization and funding of basic research) must begin with a description of the factors that govern progress in research.

The core of basic research in the natural sciences is the generation of new ideas that explain natural phenomena in useful ways. Therefore, one essential goal of any organizational structure designed to support basic research must be the creation of new ideas. This is not a simple matter! Ideas are created by individuals. They are not arrived at by consensus, they are not directly the result of any formal process, and the best ideas cannot be produced according to any predetermined schedule. It is not always possible to predict which area of science will produce the best new ideas or, indeed in what direction these new concepts will lead. Ideas require stimulation beyond simply the curiosity of a bright mind, and the source of this stimulation can vary widely. Unexpected observations, new theoretical approaches, other investigators' ideas, or even the discovery of an error or an oversight in some previous work—all can play the catalytic role that converts a long period of unsatisfying bewilderment into a joyful flash of insight and understanding.

It is wrong, however, to represent basic research as *nothing but* idea generation. Progress in research depends on many other less abstract factors. If models and hypotheses are to be verified, appropriate data and observations are needed. If complex data sets are to be understood and made useful, data analysis tools are essential. If new fields are to be explored, the necessary measurement technologies must be developed. The design of any structure to support basic research must take into account these factors and many others.

Large science programs, such as those described later in this volume, involve many investigators in their planning and implementation and necessarily depend on the development of a consensus among the participating researchers concerning investigative strategies and plans. The process of developing this consensus allows participants to share ideas and opinions, and produces, most often, the optimal set of compromises required to match objectives with available capabilities and resources. The successes of this planning process for large programs are well documented in the articles cited later in this paper. A range of models for the management of large research programs have been developed and implemented, and their strengths are clear. They have achieved their objectives of developing global strategies for coordinated data collection, of building new cross disciplinary connections in the community, and of efficiently directing substantial resources toward focused research problems of particular significance to society. The strengths of community consensus-based planning for large research programs are well established. The shortcomings are not so obvious, but they are precisely the strengths of the individual-investigator, small science approach to the support of research and are most effectively described in these terms.

By far the single most important attribute of the individual-investigator, small science approach to the support of basic research is its superior ability to recognize, select, and support the best new ideas, new approaches, new investigators, who often challenge existing dogma and take the research in unpredicted directions. Small science projects are built around the single most important resource: the individual investigator. As emphasized earlier, ideas come from individuals and ideas are the foundation for all research progress. An effective system for the support of research must be as open as possible to all investigators and all ideas, so there is the richest possible field of opportunities from which to select when the harsh reality of prioritization and resource allocation is faced.

The Division of Ocean Sciences at the National Science Foundation (NSF) supports five core disciplinary programs in biological, chemical, and physical oceanography; oceanographic technology; and marine geology and geophysics. These five programs are the engines that generate the ideas that drive ocean discovery. The rich diversity of the topics that these programs support makes it impossible to summarize their effectiveness or their contributions. This can be achieved only by example. It was with this goal in mind that four leading researchers were invited to present at the symposium their perspectives on the role of small programs in the progress of basic ocean research. Susan Lozier, Cynthia Jones, Miguel Goñi, and Maureen Raymo—physical, biological, chemical, and geological oceanographers, respectively—used different approaches to present convincing evidence of the importance of small research programs to the health of the field.

The theme developed earlier, concerning the importance of the individual investigator, was emphasized by Susan Lozier with an eloquent quotation from the great oceanographer Hank Stommel (1989, p. 50):

> Breaking new ground in science is such a difficult process that it can only be done by an individual mind.

Lozier described clearly a number of specific contributions by individuals that have shaped our understanding of ocean dynamics today and showed how each contribution constituted one more step toward understanding—each successive investigator standing on the shoulders of his or her predecessors to gain a deeper understanding of the ocean's complex processes. The earliest beginnings of physical oceanography lie in the first recorded temperature measurements of the deep ocean by British sea captain Henry Ellis in 1751, resulting in the first suggestion of a generally global feature of our oceans—the thermocline—that has proven surprisingly difficult to understand quantitatively. Lozier chronicled the ideas and approaches of Iselin (1939) and Montgomery (1938), and the progress of Welander (1959, 1971), but explained that it was not until the work of Luyten, Pedlosky, and Stommel (1983) that a theory was developed that could be used to predict the vertical and horizontal structure of the ocean's density field. Other examples of the stepwise nature of progress toward understanding the physics of the oceans were described with continuing emphasis on the importance of the contribution of the individual, and with a particular plea that everyone in the field make the *individual* effort to teach, to mentor, and to support students and younger colleagues. Lozier described the rewards of progressing through the often lonely and frustrating process of problem solving (e.g., Lozier et al., 1994; Lozier, 1997) to that special moment of insight and first understanding, as "the lightness of discovery"—that special and unique moment of satisfaction and clarity. This precious reward is a strangely powerful motivator and is to be experienced only by the individual investigator. The important theme of Lozier's presentation was effectively summarized in her closing words: " . . . as we collectively progress toward that elusive ocean of truth, we would do well to remember that we do so with many individual steps."

A different, but comparably compelling perspective, was provided in Cynthia Jones' paper on fisheries ecology, which served also to emphasize the important role that tech-

nology (in this case the development of inductively coupled plasma mass spectrometry [ICP-MS]) plays in enabling breakthroughs in research. Unlike many other marine organisms, fish provide clues to understand the processes that affect population dynamics because they contain a dated record of important life history events encoded in their bones. The most reliable bones that serve as data loggers in fish are the earbones or *otoliths*. Fish encode a history of their age and growth as the result of periodic rings that are visible in a cross section of an otolith, in a pattern similar to that found in trees (e.g., Jones, 1986, 1992, 1995). The elemental composition of the annual bands in the otolith reflects to some degree the environmental characteristics of the water in which the fish lives. Since the physical and chemical composition of the water varies spatially, otolith microchemistry records the water mass characteristics specific to a particular area and thus provides a possible technique for defining population associations and providing insight into population dynamics. The commercial availability of ICP-MS has enabled the development of techniques to read the chemical composition of the otolith and reveal a retrospective datable history of migration contained within the otolith bands. This research has been carried out over the past eight years or so, supported by a series of modest grants to individual researchers. This science was not part of a major initiative developed from the consensus of leading researchers, but rather was developed by a few independent investigators proposing to extend the frontiers of knowledge in understanding the ecology of marine fish with a novel and (at least in the early days of the research) high-risk approach.

Miguel Goñi's presentation provided examples of the critical contributions of individual-investigator research to the field of biogeochemistry, a field within which the large international program known as the Joint Global Ocean Flux Study (JGOFS) plays a dominant role. Goñi made the point that although continents have long been identified as key suppliers of dissolved and particulate matter to the oceans, and oceans and continents are (obviously) intimately connected by rivers, groundwater, and wind, much of the ocean biogeochemistry research of the past several decades has focused on internal ocean processes. Major ocean programs have almost exclusively investigated the marine carbon and nutrient cycles in the context of ocean productivity and indeed have led to considerable increases in the understanding of internal carbon and nutrient dynamics in the upper ocean. In contrast, the efforts to further investigate the role of land-derived materials in ocean chemistry have been led predominantly by individual investigators working on small independent grants (e.g., Goñi et al., 1997). Their findings in recent years represent important breakthroughs in the understanding of ocean biogeochemistry. Three examples that were well developed in Goñi's talk were the importance of terrigenous organic carbon in marine sediments, the role of mineral surfaces in the preservation of organic matter in marine sediments, and the importance of groundwater inputs to the ocean. The complexity of ocean processes essentially guarantees that there are always fertile areas away from the focused efforts of the major programs that, as Goñi points out, can yield important and fundamental results.

The core of Maureen Raymo's talk on paleoclimatology and paleoceanography was the description of two excellent examples of exciting progress that has been made in this field during the past decade. In the early 1980s, there were two main views as to why climate changed on tectonic and millennial timescales. In the first, it was suggested that critical sills or gateways opened or closed, perturbing ocean and atmospheric heat transport to the degree that Earth's albedo, and hence global climate, changed. The second view, championed by Walter Pitman, Jim Hayes, Jim Walker, and Bob Berner, was that changes in the rates of seafloor spreading, and hence mantle degassing, changed the amount of carbon dioxide, a greenhouse gas, in the atmosphere. Although this second idea was intriguing to Raymo, the mismatch in timing between when seafloor spreading rates slowed down (in the late Cretaceous) and when Cenozoic cooling occurred (post-Eocene), caused her to develop an alternative hypothesis whereby the late Cenozoic cooling was caused instead by enhanced chemical weathering and consumption of atmospheric carbon dioxide in the mountainous regions of the world, in particular the Himalayas. This controversial hypothesis remains unproven, but it stimulates much valuable debate among scientists working not only in marine geology, but also in tectonics, geomorphology, river chemistry, weathering reactions, climate, and carbon cycle modeling. Importantly, all of these ideas are attributable to individual scientists' questioning, testing, and refuting or confirming the ideas of colleagues.

The second example quoted by Maureen Raymo is of particular interest to this debate because it is concerned with the interaction of big and small science. In the early 1990s, researchers first realized that the dramatic and rapid air temperature changes observed in Greenland ice cores could also be seen in records of sea-surface temperature variability recorded in North Atlantic sediments. It is now recognized that changes in the chemistry of the deep and intermediate ocean also occur on these time scales, suggesting that such climatic cycles are global in extent and potentially involve reorganizations of ocean thermohaline circulation on timescales as short as decades to centuries. To investigate this phenomenon Raymo and her colleague Delia Oppo determined that they needed to recover deep-ocean sediment cores containing millennial-resolution sequences extending far back in time, into periods warmer than today. In this way the physical behavior of the climate system could be studied under a number of different climate regimes. However, the only way that such sediment cores could be recovered was by using a deep-ocean drillship. This challenge was overcome by submitting a successful proposal to the Ocean Drilling Program, which subsequently scheduled the drilling vessel *JOIDES* [Joint Oceanographic Institutions for Deep Earth Sampling] *Resolution* on Leg 162 with Maureen Raymo as co-chief scientist to collect the samples required (Raymo et al., 1999). It was six years or less after they had received their Ph.D. degrees that Raymo and Oppo, through their intellect and originality, were able to steer a major international resource—*JOIDES Resolution*—to attack *their* problem, and investigate *their* idea. This is an excellent example of how big science, when well managed, *can* be responsive to the best ideas of individual scientists.

The subjects are varied, but all four of these presentations were uncompromising in their praise of the value and effectiveness of individual-investigator research projects. Later in this volume, a similarly compelling case is made concerning the essential contributions of large organized programs. Both mechanisms—small and large programs—contribute in important ways to the overall research endeavor. In fact, a strong case can be made that the success of the U.S. basic research enterprise is due in large part to the diversity of management approaches and funding mechanisms that are available to U.S. academic researchers. It is not a meaningful or useful quest to search for the "one best way" to support basic research. There is no such thing. It is appropriate to end these brief comments with a quotation from a 1995 National Academy of Sciences (NAS, 1995) report that eloquently states a fundamental truth:

> . . . in reality pluralism is a great source of strength, an advantage over the ways research and development are organized in many other countries. The diversity of performers fosters creativity and innovation. It increases the number of perspectives on a problem. It makes competition among proposals richer, and it induces competition to support the best work . . . diverse funding alternatives give original ideas a better chance to find support than would a more centralized system. A pluralistic research and development system thus enhances quality and our national capacity to respond to new opportunities and changing national needs. (p. 29)

REFERENCES

Goñi, M.A., K.C. Ruttenberg, and T.I. Eglington. 1997. Sources and contribution of terrigenous organic carbon to surface sediments in the Gulf of Mexico. *Nature* 389:275-278.

Iselin, C.O'D. 1939. The influence of vertical and lateral turbulence on the characteristics of waters at mid-depths. *Trans. Amer. Geophys. Union* 20:414-417.

Jones, C.M. 1986. Determining age of larval fish with the otolith increment technique. *Fishery Bulletin* 84:91-102.

Jones, C.M. 1992. Development and application of the otolith increment technique. Pp. 1-11 in D.K. Stevenson and S.E. Campana (eds.), *Otolith Microstructure Examination and Analysis. Canadian Special Publication of Fisheries and Aquatic Sciences,* Volume 117.

Jones, C.M. 1995. Summary of current research in chemical tags and otolith composition. Pp. 633-635 in D.H. Secor, J. Dean, and S.E. Campana (eds.), *Recent Developments in Fish Otolith Research.* University of South Carolina Press, Columbia, South Carolina. 764 pp.

Lozier, M.S. 1997. Evidence for large-scale eddy-driven gyres in the North Atlantic. *Science* 277:361-364.

Lozier, M.S., M.S. McCartney, and W.B. Owens. 1994. Anomalous anomalies in averaged hydrographic data. *J. Phys. Oceanogr.* 24:2624-2638.

Luyten, J.R., J. Pedlosky, and H. Stommel. 1982. The ventilated thermocline. *J. Phys. Oceanogr.* 13:292-309.

Montgomery, R.B. 1938. Circulation in upper waters of southern North Atlantic deduced with the use of isentropic analysis. *Papers Phys. Ocean. and Meteorology.* Massachusetts Institute of Technology, Cambridge, Massachusetts. 55 pp.

National Academy of Sciences. 1995. *Allocating Federal Funds for Science and Technology.* National Academy Press, Washington D.C.

Raymo, M.E., E. Jansen, P. Blum, and T. Herbert (eds.). 1999. *Proceedings of the Ocean Drilling Program, Scientific Results*, Vol 162. Ocean Drilling Program, College Station, Texas.

Stommel, H. 1989. Why we are oceanographers. *Oceanography* 2:48-54.

Welander, P. 1959. An advective model of the ocean thermohaline. *Tellus* 11:309-318.

Welander, P. 1971. Some exact solutions to the equations describing an ideal fluid thermocline. *J. Mar. Res.* 29:60-68.

The Role of NSF in "Big" Ocean Science: 1950-1980

Director, Office of University Research, Texas A&M University (ret.)

ABSTRACT

Between 1950 and 1980 the National Science Foundation (NSF) was assigned administrative funding responsibility for three major programs involving ocean sciences. The first of these was the International Geophysical Year (IGY), 1956-1959, which included all of the geosciences. Less than 5 percent of the funds were available to ocean sciences, but this was a big boost in the amount NSF had for oceanography. The second was the International Indian Ocean Expedition (IIOE), 1962-1967, during which almost $13 million was spent, primarily at the nation's academic institutions. The third was the International Decade of Ocean Exploration (IDOE), 1971-1980, during which more than 200 million dollars were spent on oceanographic research, including ship operating costs, at U.S. academic institutions. All of these programs were "big science," in that they involved multiscience, multi-investigator, and multi-institutional projects. The process by which NSF, ocean scientists, and the academic institutions learned how to administer and carry out these large programs is discussed. That they were successful in the learning process is evidenced by the large-scale ocean sciences research programs that are still an integral part of the NSF ocean science program.

The purpose of this paper is to briefly discuss the three programs that marked the beginning and growth of NSF's role in the administration and management of "big science oceanography." The IGY provided the first significant funding for ocean sciences in NSF. Following IGY, IIOE and, subsequently, IDOE each contributed to the growth of funding for ocean sciences in NSF. Both IGY and IIOE raised the level of ocean sciences support by NSF only for the period of these programs. After they were completed the funding level fell back almost to that which existed beforehand. IDOE differed in the fact that the support for large-scale ocean research continued, but not under the IDOE banner.

INTERNATIONAL GEOPHYSICAL YEAR (1956-1959)

The IGY was initially proposed as the Third Polar Year by Lloyd Berkner of Brookhaven National Laboratory and Sidney Chapman of the University of Alaska and was adopted by the International Council of Scientific Unions (ICSU), a non-governmental organization, founded in 1931 to bring together natural scientists in international endeavors. Later in 1952 the program was broadened to include scientific study of the whole Earth. The program was to be "the common study of our planet by all nations for the benefit of all" (Chapman, 1959).

By 1956, 14 scientists were named to coordinate and lead separate parts of the IGY program (Box 1). Dr. G. Laclavére of France had the responsibility for oceanography. He met with working groups of scientists to develop the international program in oceanography. Altogether 67 nations took part in IGY.

At the urging of the National Academy of Sciences (NAS), NSF was selected as the lead agency for planning and managing U.S. participation in the IGY. A special coordinating office was set up in the Office of the Director because the multidisciplinary nature of the program prevented it from fitting in either of the research divisions which, at the time, were the Division for Mathematics, Physical, and Engineering Sciences, and the Division of Biological and Medi-

BOX 1
IGY International Reporters

For the IGY, fourteen scientists (called reporters) had special duties, namely to coordinate and lead the development of separate parts of the enterprise. Two reporters dealt with parts that affected more than one of the scientific branches.

1. World Days and Communications: A.H. Shapely
2. Rockets and Satellites: L.V. Berkner
3. Meteorology: J. Van Mieghem
4. Geomagnetism: V. Laursen
5. Aurora and Airglow: S. Chapman
6. Ionosphere: W.J.G. Beynon
7. Solar Activity: H. Spencer Jones; Y. Öhman; M.A. Ellerson (in succession)
8. Cosmic Rays: J.A. Simpson
9. Longitudes and Latitudes: A. Danjon
10. Glaciology: J.M. Wordie
11. Oceanography: G. Laclavére
12. Seismology: V.V. Beloussov
13. Gravity Measurements: P. Lejoy; P. Tardi (in succession)
14. Nuclear Radiation: M. Nicolet

Source: Chapman (1959).

TABLE 1A IGY Oceanography Funding by Institution (dollars)

Institution	Number of Awards	FY 1956	FY 1957	FY 1958	FY 1959
Columbia University	3	146,180	299,070	138,475	43,000
DOI	1	47,000	11,300	0	0
U.S. Navy	1	0	0	30,421	0
Scripps	3	86,920	561,570	68,005	12,000
TAMU	2	23,070	71,055	16,000	3,000
University of Washington	3	23,350	97,075	45,425	0
WHOI	3	51,180	205,995	49,700	5,000
Total	16	378,700	1,246,065	348,026	63,000

NOTE: DOI = Department of the Interior; TAMU = Texas A&M University; WHOI = Woods Hole Oceanographic Institution

TABLE 1B IGY Oceanography Funding by Scientific Category (dollars)

Category	Number of Awards	FY 1956	FY 1957	FY 1958	FY 1959
CO$_2$	5	112,000	174,465	90,292	
Island Observations	3	132,600	234,225	56,405	7,000
Currents	6	82,100	786,375	193,004	56,000
Arctic	2	52,000	51,000	8,325	

SOURCE: Lambert (1998b).

cal Sciences. The budget for U.S. participation in the 18 months of field operations totaled $43.5 million. The funds for IGY were entirely "new money"—appropriations over and above those for ongoing NSF programs. The ocean sciences component was a small part of the total IGY funding totaling $2,035,791, but it was far in excess of any previous support for ocean research in NSF's Research Division. Its impact on the ocean sciences budget during 1956 through 1959 is shown in Table 1A.

The oceanographic program was carried out by five U.S. academic institutions: Columbia, Scripps, Texas A&M, University of Washington, and Woods Hole, and by the Department of the Interior and Department of the Navy. The funding during the four years 1956-1959 is shown in Table 1A, by institution and in Table 1B by scientific category. According to Thomas F. Malone (1997), Lloyd Berkner was quoted as noting the IGY was a program "operated by scientists, with consent, cooperation, and aid, but not the direction of the governments."

INTERNATIONAL INDIAN OCEAN EXPEDITION (1962-1967)

Even before the IGY was completed, the International Council of Scientific Unions asked Roger Revelle (Director

of Scripps Institution of Oceanography) to appoint a special Committee on Oceanic Research, (eventually changed to Scientific Committee on Oceanic Research—SCOR) so that oceanographers could play a major role in affairs of ICSU. The 15-member SCOR, at its first meeting in Woods Hole in August 1957, decided to plan an international expedition to the Indian Ocean. The Indian Ocean was the least understood ocean, physically and biologically, although there were indications that it might have a biological productivity higher than either the Atlantic or Pacific. The seasonal reversal of monsoon winds made it an ideal natural laboratory for observing the effects of wind stress on oceanic currents.

On the basis of input from 40 scientists, national and international, invited by SCOR, representing different disciplines in oceanography, a prospectus for exploration of the Indian Ocean was prepared and finalized in August 1960 by a group of three eminent scientists, namely: Roger Revelle, United States; George Deacon, United Kingdom; and Anton Bruun, Denmark (Lambert, 1998a).

In 1961, NSF awarded a grant to the National Academy of Sciences for "Support of Coordinator, IOE" (Lambert,

TABLE 2   International Indian Ocean Expedition Oceanography Funding by U.S. Institutions (thousand dollars)

Institution	1962	1963	1964	1965	1966	1967	Total
LDGO	150	544	1,296	1,940	300	230	4,460
SIO	150	680	285	150			1,265
WHOI		150	2,178	1,560	110	280	4,278
Stanford University				529			529
University of Washington			122	282	42		446
University of Hawaii			250	229	433		912
WXBUR			201				201
URI	100						100
USC			50	5		23	78
Smithsonian				76			76
University of Michigan			22	73	83		178
USAF				50			50
NAS-NRC	44	19					63
Others				48	7		55
Total	444	1,393	4,933	4,413	975	533	12,691

NOTE: LDGO = Lamont-Doherty Geological Observatory; NAS-NRC = National Academy of Sciences-National Research Council; URI = University of Rhode Island; USAF = United States Air Force; USC = University of Southern California; WXBUR = U.S. Weather Bureau

1998b).   From a U.S. perspective, this was the beginning of the International Indian Ocean Expedition. During the first three years of the program, including the 1961 grant, the planning and direction were accomplished by a contract with the National Academy of Sciences. For the remaining four years, 1964-1967, the NSF funded grants on the basis of proposals from the institutions. The overall direction of the program came from the academic scientific community. But within the Foundation, the NSF Coordinating Group on Oceanography (CGO) was established and specifically tasked with the coordination of oceanographic facilities, conversion, construction of ships, and the International Indian Ocean Expedition.

The expenditures for the six years of the IIOE are listed in Table 2, which shows the level of funding for each of the participating institutions. The major participating institutions were: Lamont-Doherty Geological Observatory (LDGO), Woods Hole Oceanographic Institution (WHOI), and Scripps Institution of Oceanography (SIO). The reason for this is not only because these were by far the largest oceanographic institutions, but also because they were also the laboratories with ships large enough to travel to and carry out research in the Indian Ocean. Ship operation costs are included in Table 2.

The IIOE was a very interdisciplinary program, but the major expenditures were for marine geology and geophysics (gravity and magnetics, rock analyses, bathymetry, and sediments), atmospheric circulation and air-sea interaction, oceanic circulation, marine biology, and geochemistry.

The relatively independent nature of the IIOE cruises is highlighted by Edmund (1980) who, in discussing the IIOE geochemical efforts, states, "Data from different cruises could not be contoured together. Hence, the intended division of labor—different areas of the ocean assigned to different groups—led to a database of little use." The same statement does not hold for the extensive work in marine geology and geophysics, which was to prove very useful in the *Geological and Geophysical Atlas of the Indian Ocean* published in 1975 by the Academy of Sciences and Main Administration of Geodesy and Cartography of the USSR.

INTERNATIONAL DECADE OF OCEAN EXPLORATION (1971-1980)

Origin—The International Decade of Ocean Exploration was carried out during the 10-year period, 1971 to 1980. Unlike IGY and IIOE, which were initiated by the academic scientific community, IDOE was the brainchild of the National Council of Marine Resources and Engineering. The council was established by Congress in the Marine Sciences Act of 1966.

The act instructed the President, through the council, to advance marine initiatives that would contribute to cooperation with other nations and international organizations. The President (Lyndon B. Johnson) stressed the need for cooperation of all maritime nations. According to Ed Wenk (1980), who served as Executive Secretary of the Council during the Johnson and Nixon administrations, Johnson's philosophy went well beyond an abstraction of scientific interchange. It was driven by a quest for a stable, lasting peace, despite the paradox of a growing commitment to Vietnam.

Mindful of the international emphasis of the Marine Sciences Act and the President's pronouncements, the Marine Council under the leadership of Vice-President Humphrey generated, among other marine policy initiatives, an initiative in international marine activities. This was approved in December 1966. This initiative evolved into the IDOE, and in December 1967 the Vice-President recommended it to the President "arguing the case in terms of food for expanding world population, maritime threats to world order, waterfront deterioration in coastal cities, increased pollution at the shoreline, expanding requirements for seabed oil, gas, and minerals, and expanding ocean shipping" (Wenk, 1980). The full blessing of the White House was given in March 1968 in the President's conservation message as an International Decade of Ocean Exploration for the 1970s.

International support for the program by other nations and international marine organizations was actively sought by the Marine Council, with the result that on June 13, 1968, the Intergovernmental Oceanographic Commission (IOC) of the United Nations Educational, Scientific, and Cultural Organization (UNESCO) recommended support for IDOE. United Nation's support for the program was obtained in proposition 3 of the General Assembly Resolution 2467(XXIII) cosponsored by 28 nations. This ensured government-to-government endorsement for the program.

Participation of the U.S. marine scientific community in the planning of the IDOE was not ensured until a contract was signed in July 1968 between the Marine Council and the National Academy of Sciences and National Academy of Engineering to elicit the ideas of scientists and engineers relative to the broad goals developed by the Council. The Academies completed their studies and presented their findings and recommendations in a joint report entitled *An Oceanic Quest: The International Decade of Ocean Exploration* (NAS, 1969).

The program became official in October 1969 when President Nixon announced five initiatives in marine affairs including a commitment of $25 million for IDOE. The National Science Foundation was given lead responsibility for the program.

Goals and Objectives—The goals of IDOE identified by the Marine Council in its January 1970 report were

- preserve the ocean environment,
- improve environmental forecasting,
- expand seabed assessment activities,
- develop ocean monitoring systems,
- improve worldwide data exchange, and
- increase opportunities for international sharing of responsibilities and costs for ocean exploration.

The NAS (1969) *Quest* report identified the science and engineering programs, and the resources needed to address the Marine Council goals. The report included a broad statement of the basic objectives as follows:

To achieve more comprehensive knowledge of ocean characteristics and their changes and more profound understanding of oceanic processes for the purpose of more effective utilization of the ocean and its resources.

The report went on to state that the emphasis on utilization was considered of primary importance and that the primary focus of IDOE activities would be on exploration efforts in support of such objectives as:

- increased net yield from ocean resources,
- prediction and enhanced control of natural phenomena, and
- improved quality of the marine environment.

Thus, IDOE investigations should be identifiably relevant to some aspect of ocean utilization.

Distinguishing features of IDOE programs should include (1) ocean investigations involving cooperation among investigators in this country and abroad; (2) long-term and continuing nature requiring the facilities of several groups; (3) programs within the United States to be cooperatively implemented by government agencies (federal and state) and private facilities (academic and industrial); and (4) international cooperation.

In describing the kind of research and exploration needed to address the objectives of the IDOE, the Academy report identified four major topics:

- geology and non-living resources,
- biology and living resources,
- physics and environmental forecasting, and
- geochemistry and environmental change.

Within these major topical areas, specific programs and studies were described. Most of them required further study and development, but some like the Geochemical Ocean-Section Study (GEOSECS), the Mid-Ocean Dynamics Experiment (MODE), and Climate: Long-range Investigation, Mapping and Prediction (CLIMAP) already were formulated.

Implementation—Responsibility for the planning, management, and funding of IDOE activities was assigned to the National Science Foundation by the Administration. Funding of $15 million for the first year of the program was included in the fiscal year 1971 federal budget.

IDOE was initially established as an office reporting directly to the assistant director of NSF responsible for national and international programs in company with other programs such as the Office for Oceanographic Facilities and Support and the Office of Polar Programs, as shown in Figure 1. In 1975, another internal reorganization subsumed IDOE within a new Division of Ocean Sciences, one layer more remote from the assistant director level.

Although both the Marine Council report and the NAS (1969) report envisaged significant participation by federal agencies in IDOE, it became evident in the first year of the program that such an arrangement was unworkable. Each of the agencies had its own mission, which did not necessarily coincide with the kinds of projects identified for emphasis by the IDOE program managers.

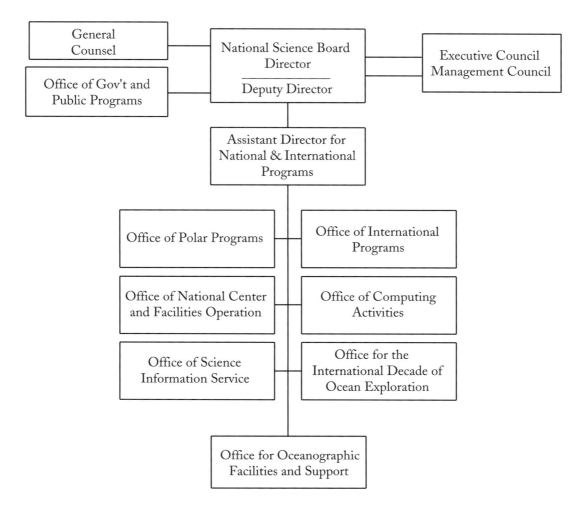

FIGURE 1   Simplified National Science Foundation organizational chart for the period 1969-1975 (see Appendix F for complete chart).

Furthermore, proposals from the agency scientists, most of whom were unfamiliar with the procedures and requirements for submitting research proposals to NSF, did not receive very good reviews from the traditional mail reviews utilized by NSF. Proposals that might have been suitable for gaining support within the agencies were not favorably received by the academic reviewers.

The final obstacle to agency participation in IDOE lay in the fact that even when the agency mission coincided with a particular IDOE project, there remained insurmountable problems resulting from differences in management style, funding procedures, and long-range research objectives. These barriers tended to discourage any significant participation by other federal agencies in the IDOE.

As a result, after the first year of the program, during which half of the IDOE funds were essentially passed through to those agencies having marine responsibilities, agency participation in the program was minimal. The one exception was the North Pacific Experiment (NORPAX), which addressed problems of direct interest to the Office of

Naval Research's (ONR's) oceanographic research mission. NORPAX became a jointly funded program in which ONR and IDOE each supported research carried out by the academic oceanographic institutions. The research was closely coordinated by the program managers from each agency.

International Participation—Another area in which the IDOE was unable to carry out the concepts envisaged by both the Marine Council and the National Academy of Sciences was the extent of international cooperation. While the U.S. marine science community was quickly able to design large research projects responsive to purpose of more effective utilization of the ocean and its resources, other maritime nations were not able to organize themselves quickly enough for meaningful participation.

In each year of the program, the U.S. IDOE submitted its plans and programs to the IOC and received the endorsement of member states. But procedures followed by scientists of the member countries in obtaining financial support from their own governments for the participation were agonizingly slow, and in most cases, these governments were unwilling or unable to fund these projects. The IOC itself had very little funding for research and was unable to support the projects, and the U.S. IDOE could not use its funds to support scientists of other nations.

Two notable exceptions to this state of affairs were FAMOUS and POLYMODE. FAMOUS was a joint French-U.S. study of the Mid-Atlantic Ridge, which was in planning stages when IDOE was established and which was carried out during the first few years of the program. POLYMODE was a joint USSR-U.S. study of midocean dynamics in the Central/North Atlantic. This project was carried out during the last half of IDOE and was a truly cooperative effort in planning and execution. Scientists from both countries designed the experiments, planned the logistics, and carried out the research, and the governments of both countries supported the operations.

Focus Areas—Of the four major topics of study identified in the NAS (1969) report, IDOE was prohibited from funding "biology and living resources" during the first year of the program. We in the IDOE office were informed that the prohibition was imposed by the Bureau of the Budget because of arguments regarding fisheries. At the time, we thought the disagreement was between the budget bureau and the Bureau of Commercial Fisheries. However, according to Wenk (1980) the problem lay in the Department of State, which wanted to exclude fisheries from IDOE's scope. The Department of State's traditional roles in multi- and bilateral fishery policy might be endangered. The issue was settled after IDOE's first year, and biology and living resources became an integral part of the program beginning in the second year of IDOE.

Program Development—Although, for reasons stated above, it was not possible to achieve the kind of international and government agency participation envisaged, NSF was very successful in implementing a program with the remaining distinguishing characteristics identified in the NAS (1969) report. We in the IDOE office required IDOE projects to be identifiably relevant to the more effective utilization of the ocean and its resources, the major goal of IDOE. Further, the project had to be large-scale, long-term research, drawing on the expertise and skill of scientists from all applicable disciplines. The result was "big science" projects involving key research scientists from the major U.S. academic institutions and having a duration of three to ten years. Table 3 illustrates some of these features (Jennings and King, 1980). Throughout its 10-year history, IDOE supported 21 major projects, totaling approximately $189 million. This did not include ship operating costs; which were included in budgets of the Office for Oceanographic Facilities and Support.

One other characteristic of IDOE was that all data would have to be submitted to the appropriate national or international data centers. The cost to each project for adhering to this policy was included in the project proposals and funded by IDOE as an integral part of the projects. Safeguards were established to protect the proprietary interests of the researchers. IDOE also provided special funds to the data centers to ensure their capabilities to manage the influx of additional data.

After the first frustrating year of dealing with required pass-through funds to other government agencies and less than satisfactory proposals from academic scientists, which failed to adequately address the goals of the program, IDOE managed to develop an operating philosophy that served the program well for the remainder of the decade.

In order for a project to become part of IDOE it would have to address some aspect of ocean utilization and would have to be comprehensive enough to hold the promise of a significant advance toward solving the problem under study. The importance of the project and its design would need to have the consensus of those individuals most knowledgeable about the issue to be studied. Some of the projects undertaken during the early days of IDOE had already reached this stage of development by the time IDOE was established: GEOSECS, MODE, CLIMAP, and Nazca Plate are examples.

For projects that had not reached this stage of development we were to become dependent on a series of planning workshops, each addressing its own project. The academic scientific leaders wishing to establish an IDOE project were called on to organize a planning workshop and to invite all research scientists who were knowledgeable about the subject and who might ultimately become important research members of the final project. If the workshop was successful, the leader or leaders of the project organized and submitted to IDOE a complex proposal describing the administration of the project, the scientific approach, and the role of each individual investigator in the project including a proposal from each investigator.

We came to refer to the leaders of these projects as the "Heroes" and the philosophy as the "Hero Principle." The Heroes were responsible for all administrative aspects of the project including budgets, logistics, planning, and so forth. In almost every project, the Hero was really an executive committee. For example, in GEOSECS, the executive committee included a scientist from each of five institutions: Lamont, Scripps, Woods Hole, Miami, and Yale. IDOE funds were granted to the home institution of each participating scientist.

The planning workshops were very successful and were carried out without difficulty during the first half of the decade. However, as the program matured, parts of the scientific community became concerned that the workshop organizers were inviting participants on the basis of "old boy" networks and excluding some scientists who could make meaningful contributions. Thereafter, it became necessary to publicize, well in advance, our intention to sponsor these

TABLE 3 Major U.S. IDOE Projects

Programs/Projects	Number of Institutions	Number of Scientific Investigators	Year Begun	Expected Year of Completion	Estimated Total Cost ($M)	U.S. Agencies Providing Funds
Environmental Forecasting						
NORPAX[a]	28	45	1971	1982	29.7	ONR
CLIMAP	8	22	1971	1980	8.0	
MODE	16	45	1971	1974	8.0	ONR, NOAA
ISOS	9	16	1974	1981	10.2	NSF
POLYMODE	12	35	1975	1982	15.5	ONR, NOAA
Environmental Quality						
GEOSECS	14	28	1971	1980	23.5	ERDA
Pollutant Baseline	17	30	1971	1978	2.3	
Pollutant Transfer	9	10	1972	1979	10.0	
Biological Effects						
Field (CEPEX)	5	10	1973	1980	6.5	
Laboratory	6	8	1973	1979	10.0	
SEAREX	9	15	1977	1983	4.6	
PRIMA	5	6	1978	1984	2.4	
Seabed Assessment						
South Atlantic Margins	2	15	1971	1975	4.0	
Nazca Plate	3	25	1971	1977	6.0	
FAMOUS	4	10	1972	1975	2.0	NSF, ONR, NOAA
Manganese Nodules	10	18	1972	1977	4.0	
MANOP	11	21	1977	1984	8.0	
Galapagos	3	9	1976	1979	1.4	
RISE	5	7	1977	1980	1.3	
SEATAR	7	15	1975	1980	5.4	
CENOP	11	15	1978	1982	2.8	
Living Resources						
CUEA	13	11	1972	1979	16.1	
SES	10	11	1974	1981	7.0	

[a]See Appendix H for the definitions of acronyms.

workshops. Although there was surely some merit in the concerns expressed by those scientists who felt neglected, I do not believe the early projects themselves failed to address any important significant aspects of the scientific research needed to achieve the objectives of the projects.

Once the projects resulting from the workshops had been identified as appropriate for consideration by IDOE, and the proposals submitted, the well-established NSF peer-review process played a critical role in the final selection of projects for funding. Like most NSF proposals, IDOE proposals were subjected to peer review. In the case of IDOE, these were mail reviews, and the mail reviews for each project were then carefully considered by a panel of specialists that made its own recommendations.

One of the major difficulties in reviewing IDOE projects was that traditionally NSF reviewers were accustomed to reviewing only individual projects and the reviews focused on the question of scientific excellence and receiving ratings accordingly. But IDOE projects included *all* of the tasks necessary to achieve success, and while not all of

these tasks were the type to receive excellent ratings, each of them was essential to the success of the project. Mail reviewers were quick to point out the deficiencies in these proposals, to note the routine character of certain tasks, and to give them only fair ratings. In NSF, the administration was accustomed to funding only those individual projects receiving excellent ratings by the reviewers. Early on, we in the IDOE office were able to explain to the NSF chain of command, without too much difficulty, that these routine tasks were essential to the projects even though they did not receive high marks from the reviewers. At the time, we reported directly to the Assistant Director for National and International Programs, whose office understood the problem and fortunately proposals did not receive heavy scrutiny above that level.

Later on, in response to pressure from Congress, a review board was established in each directorate, plus a review board for special items requiring approval by the Foundation's governing body, the National Science Board. The review boards compounded the prospects for delay and frustrations in the movement of funds to the researchers. Further, they elevated the importance of procedures to the point where administrative form became as important as scientific judgment as criteria for moving grant actions through NSF.

We were still able to move IDOE proposals through the system but with more delays and a great amount of bureaucratic effort. One successful procedure for avoiding the final hurdle of a National Science Board review was to break the IDOE projects into small enough segments to stay below the million-dollar level that would cause it to become a special item. This was done to avoid the additional delay, not because we were concerned about the merits of the projects or about final approval by the Board.

As pressure from congressional scrutiny of NSF management practices continued, the Foundation established a set of guidelines that made it even more difficult for the IDOE program. The new guidelines for the selection of reviewers were excessive in their zeal to avoid all biased judgment. They included a restriction against using reviewers from any university involved in the proposal, even though the reviewers were from different fields or in different parts of the university. Under these conditions, most of the scientists from major institutions were prevented from assisting in the review of large IDOE projects because most of their institutions were participants.

In spite of these bureaucratic hurdles, the decision to give responsibility for IDOE to the National Science Foundation was the right one. The Foundation had in place the review and granting mechanisms and the experience in dealing with the academic research community that would ultimately be responsible for carrying out the work of the program. Aside from the scientific results, which are not the subject of this paper, IDOE provided very important lessons to both NSF and the academic community in organizing, reviewing, and archiving the results of large-scale, cooperative research projects.

In 1977, NSF asked that the National Academy of Sciences and the National Academy of Engineering continue to provide advice and guidance on the nature of programs to follow the IDOE. In response, the Ocean Sciences Board of the Academy appointed a post-IDOE Planning Steering Committee, which organized a series of five planning workshops. These workshops formed the basis for a report issued by the NAS (1979) titled *The Continuing Quest: Large Scale Ocean Research for the Future.*

The 1979 NAS report concluded that, "the IDOE was a watershed in the history of Ocean Research. By providing the structure and resources for large scale, long term coordinated projects, the program gave a powerful impetus to the transformation of marine science from a descriptive effort to one increasingly driven by experimental and theoretical concerns."

The report recommended that a program of cooperative ocean research should follow and evolve from IDOE and that it should be sponsored by NSF as a major component of its overall efforts in fundamental ocean research. It listed 28 principal conclusions and recommendations regarding the future program and identified oceanographic opportunities for the 1980s. The Foundation has followed this advice and continues to fund the type of large-scale, long-term, cooperative projects that were the heart and soul of IDOE.

REFERENCES

Chapman, S. 1959. *IGY: Year of Discovery.* University of Michigan Press, Ann Arbor.

Edmund, J.M. 1980. GEOSECS is like the Yankees: Everybody hates it and it always wins. *Oceanus* 23(1).

Jennings, F.D., and L.R. King. 1980. Bureaucracy and science: The IDOE in the National Science Foundation. *Oceanus* 23(1):12-19.

Lambert, R.B. 1998a. Management of Large Ocean Programs. Unpublished manuscript.

Lambert, R.B. 1998b. The Emergence of Ocean Science Research in NSF, 1951-1980. *Marine Technology Society Journal* 32(3):68-73.

Malone, T. 1997. Building on the legacies of the International Geophysical Year. *EOS* 78(18, May 6).

National Academy of Sciences (NAS). 1969. *Oceanic Quest: The International Decade of Ocean Exploration.* National Academy Press, Washington, D.C.

National Academy of Sciences (NAS). 1979. *The Continuing Quest: Large-Scale Ocean Science for the Future.* National Academy Press, Washington, D.C.

Wenk, E., Jr. 1980. Genesis of a marine policy: The IDOE. *Oceanus* 28(1):2-11.

Major Physical Oceanography Programs at NSF: IDOE Through Global Change

Richard B. Lambert, Jr.

National Science Foundation (ret.)

ABSTRACT

The transition from the major coordinated research programs of the International Decade of Ocean Exploration to the global projects of the U.S. Global Change Research Program is described. Special emphasis is placed on physical oceanography, in which global programs include the World Ocean Circulation Experiment and the Tropical Ocean and Global Atmosphere Program. Contrasting management structures are described, and speculation is made as to the challenges of future global programs.

Following on from the paper by Feenan Jennings, I would like to describe briefly how some of the major physical oceanography programs developed subsequent to the International Decade of Ocean Exploration (IDOE) through the rise of the global change programs. My emphasis is primarily on the major physical oceanography programs, WOCE (the World Ocean Circulation Experiment), which was primarily an oceanographic program, and TOGA (the Tropical Ocean and Global Atmosphere program), which was a truly interdisciplinary program involving both the atmospheric and the oceanic communities.

Before I proceed, I would like to acknowledge one person who, more than any of the rest of us, defined what the physical oceanography program is today—Curt Collins. Curt sent me an e-mail a few weeks ago expressing his regrets that he could not be here this week. However, he is where he has always felt most at home—at sea! Curt began his career at the National Science Foundation (NSF) in the early days of IDOE, kept the helm through the transition from IDOE to disciplinary programs, and was one of the major proponents of both the WOCE and the TOGA programs. Ocean sciences in general, and physical oceanography in particular, owe him a great debt of gratitude.

Below, I give a brief historical sketch, showing some of the legacies of IDOE, from both scientific and management perspectives, including the rise of collaborative research and the growth of the community. Then I compare and contrast what we refer to as midsize programs with the truly major global programs of the U.S. Global Change Research Pro-

gram (USGCRP). I then describe some of the major management issues with interagency, international programs by comparing TOGA and WOCE, and finally attempt to draw some conclusions, from the point of view of a program manager, as to where we go next.

You have already heard about the NSF reorganization of 1975—right in the middle of the IDOE—that created the Division of Ocean Sciences essentially as it exists now. Five years later, the IDOE ended, and the programs merged with the existing research section, essentially along disciplinary lines. The early 1980s was a time of consolidation, with *midsize* programs paving the way for dealing with problems on a scale that would lead to programs with a more truly *global* outlook. It was a time of planning, and the beginnings of TOGA and long-lead-time activities for WOCE. Subsequently, the Geosciences Directorate was created under the leadership of Bill Merrell, and in 1989, the USGCRP was formally initiated, with its first budget called out in 1990. At the same time, the WOCE field program funding began.

However, although the international WOCE program had begun, the U.S. WOCE field program did not really begin until nearly 1992, due to a variety of delays, primarily with the availability of adequate ships. In 1995, the TOGA program officially ended. Soon after that, the WOCE field program, due to end in the same year, was extended through 1998, and with agreement among most international participants, a period of analysis, interpretation, modeling, and synthesis was initiated, with an expected lifetime through at least 2002, and perhaps longer. In the meantime, a follow-on program, under the direction of the World Climate Research Program was begun. Called CLImate VARiability, (CLIVAR), it is now viewed as the next major atmosphere-ocean program, with the goal of greatly improving our ability to forecast variations in climate on very long time scales. CLIVAR field programs are expected to last at least through the next decade and perhaps provide at last scientific motivation for the long-awaited Global Ocean Observing System, perhaps even a Global Climate Observing System.

In looking back at the IDOE, recall that one of the major characteristics of programs during this time was the initiation of major programs and then their dissection into smaller but still collaborative programs for the sake, primarily, of simplifying the management required. Recall the four major components of IDOE: Environmental Forecasting (EF), Environmental Quality (EQ), Living Resources (LR), and Nonliving Resources, or Sea Bed Assessment (SBA). Consider for a moment the breakdown, or dissection, of the EF program, which consisted largely of physical oceanography programs. The major examples are MODE (Mid-Ocean Dynamics Experiment), POLYMODE (the U.S.-Soviet follow-on to MODE), NORPAX (North Pacific Experiment), ISOS (International Southern Ocean Studies), CLIMAP (Climate Long-range Investigation, Mapping, and Prediction Study), and CUEA (Coastal Upwelling Ecosystems Analysis). The latter two showed the way to truly interdisciplinary work, with CLIMAP studying physical phenomena in the distant past using paleoceanographic techniques and CUEA showing the way to investigating the physical impacts on fisheries, or "living resources."

Another legacy of the IDOE was the start-up of a number of midsize programs during the last year of the decade. These were clearly multiyear projects, with a requirement that funding continue in order to maintain them. Whether they were started in order to guarantee funding continuity or whether the continuity was already planned is not clear to me. In any case, they were logical follow-ons, but also led the way into the large global programs to follow. Some examples are the Coastal Ocean Dynamics Experiment (CODE); Tropic Heat (TH), a study of the Eastern Pacific Cold Tongue; Pacific Equatorial Ocean Dynamics Experiment (PEQUOD); the Western Equatorial Pacific Ocean Circulation Study (WEPOCS); and Transient Tracers in the Ocean (TTO), which was a follow on to the Geochemical Ocean Sections (GEOSECS) study, and a precursor to the tracer work to be done in WOCE and other survey experiments. This is another example of two disciplines coming together to study common problems.

Perhaps a better way of looking at the transition is shown in Table 1. MODE, which Walter Munk describes briefly, was the first comprehensive look at the mesoscale eddy field. Followed by the joint Russian-U.S. POLYMODE, it paved the way for the World Ocean Circulation Experiment. In a similar fashion, the GEOSECS program, leading into the

TABLE 1   Time Line Summary Illustrating the Development from the Coordinated Programs of IDOE Through Mid-size Programs of the Transition Period (1980–1985) into the Global Programs of the USGCRP (WOCE and TOGA)

1970-1980	1980-1985	1985-1990
IDOE	Midsize Follow-ons	Global change
MODE	POLYMODE Transient Tracers	WOCE
NORPAX	NORPAX (cont.) (Hawaii - Tahiti Shuttle) PEQUOD Tropic Heat WEPOCS	TOGA

study of Transient Tracers, also paved the way for the high-precision tracer work during WOCE.

Similarly, NORPAX, expanding its range with the Hawaii to Tahiti Shuttle, largely a survey using expendable bathythermographs with frequent crossings of the equator, was one of many midsize programs that paved the way for TOGA. Others include Tropic Heat, PEQUOD, WEPOCS, the National Oceanic and Atmospheric Administration's EPOCS (Eastern Pacific Ocean Climate Study), and others.

Major characteristics of these midsize programs include the following:

1. They are needed to address problems too big for one or two principal investigators (PIs).
2. They are usually regional, not basin-wide or global.
3. They require several, but usually a small number of PIs.
4. They usually involve coordinated field work.
5. They are usually fully collaborative.
6. The cost averages approximately $1 million to $3 million per year.
7. There is little need for international or interagency coordination.

Some of the parallel characteristics of global change programs are the following:

1. The studies are usually long-term (several years) and large-scale (global or at least basin-wide).
2. They require a large number of PIs, although funding may be accomplished through individual grants.
3. They require a collective review process that may differ from the normal review of individual proposals.
4. The cost may average $5 million to $10 million per year or more for any given program.
5. They are usually fully inter-agency (national programs).

6. They are also usually fully international: a characteristic that dictates international coordination and management.

Table 2 illustrates some of the differences between the management structures of WOCE and TOGA. In this table, national steering groups are indicated by SSC (Scientific Steering Committee), whereas international steering groups are denoted by SSG (Scientific Steering Group). There are pros and cons to both mechanisms, which will be the subject of a future paper. For the purpose of this paper, it suffices simply to state the facts.

In conclusion, I would like to simply state the obvious and list some of the strengths of the major physical oceanography programs. Then I indicate what, to me, are some of the challenges remaining as we move into the CLIVAR era.

First of all, both large and small programs are needed to make scientific progress, even though more extensive review procedures are required for the large programs. This time and effort, however, seem warranted, since large programs usually add resources to the community. New major programs are much more likely to be interdisciplinary, and midsize programs are needed as bridges and to deal with pieces of the bigger puzzle.

Nevertheless, some challenges remain. More manpower is needed (especially strong leadership) in order to realize the potential for new programs already conceived. It is particularly incumbent on academic institutions to develop ways in which their faculty are recognized for sometimes thankless and onerous tasks. There is a need for community consensus in order to ensure community support for implementation of these major programs, and there is a special need for new ideas as old ways of doing things become obsolete. New approaches and new agreements are also needed in the issue of data collection and sharing.

From the agency side, the securing and allocation of adequate funding are crucial, without compromising agency missions, but exploiting the differences to accomplish a wide

TABLE 2 Contrast of Management Structures and Institutions Involved in WOCE and TOGA. Both International and National Components are Listed.

	WOCE	TOGA
Sponsors		
International	IOC/SCOR (WCRP)	WMO/ICSU (WCRP)
National	NSF (ONR, NOAA, DOE, NASA)	NOAA (NSF, ONR, NASA, DOE)
Project Offices		
International	IOS/SOC (UK)	NOAA (Boulder) $\rightarrow$ Geneva
National	University (TAMU)	Government (NOAA)
Science Steering		
International	SSG report to JSC	SSG report to JSC
National	SSC report to IAG	NAS panel report to BASC
Government Oversight		
International	IWP (IOC and SSG)	ITB (WMO and SSG)
National	IAG (Agency PDs)	Formal part of the USTPO
Panels and WGs	Established by SSC, SSG	Established by USTPO & Panel
Data Management	Distributed system	Centralized system

NOTE: BASC = Board on Atmospheric Sciences and Climate; DOE = Department of Energy; IAG = Inter-Agency Group; ICSU = International Council for Science; IOC = Intergovernmental Oceanographic Commission; IOS = Institute of Ocean Sciences/Southampton Oceanography Center, UK; ITB = Intergovernmental TOGA Board; JSC = Joint Scientific Committee for the WCRP; NAS = National Academy of Sciences; NASA = National Aeronautics and Space Administration; ONR = Office of Naval Research; PD = program director; SCOR = Scientific Committee on Oceanic Research; TAMU = Texas A&M University; USTPO = U.S. TOGA Project Office; WCRP = World Climate Research Program; WG = working groups; WMO = World Meteorological Organization.

diversity of tasks. This would help in reducing inertia in a system that is somewhat resistant to new approaches and would help in such specific issues as the costs of data sharing and dissemination.

Major International Programs in Ocean Sciences: Ocean Chemistry

Peter G. Brewer

Monterey Bay Aquarium Research Institute

INTRODUCTION

As I searched through the correspondence for this meeting I discovered that the topic I have been asked to speak on appeared to change. It appeared variously as the topic of ocean chemistry within "large and small science programs," "large oceanographic programs," and "major international programs." This does send a message; after all the so-called "large programs" are identifiable as discrete from "small science" aren't they? And if the programs were large, well then they were probably international, and vice versa. Is all this true? I cannot give an objective answer, for I have been so intimately involved in the continuum of large and small, national and international programs in ocean chemistry of the National Science Foundation (NSF) over the last 35 years (indeed I am an alumnus of the famed NSF "rotator" program) that subjective perception rules. It is my thesis that the large ocean chemistry programs sponsored by NSF have evolved enormously in method and style over time. Two key program officers, Neil Andersen and Rodger Baier, together with a steady stream of visiting rotators, have had enormous impact on the field over this period.

Please then permit me to give a purely subjective personal account of this period and my own recollections of human effort and scientific achievement. Although some programs may therefore escape comment here, I suspect that my experiences are typical. In my view we need a mixture of large and small—large simply because of the immense scale of the processes we seek to observe, and because human beings are fundamentally captured by the grandeur of the ocean enterprise. And because the best large programs are simply the true ensemble effort of many creative individuals, who have their roots in a first-class small laboratory with theory, and experiment, and all the ferment, creativity, and loyalty that such groups and programs stimulate within their circle.

A companion paper on the overall progress of ocean chemistry is provided by John Farrington earlier in this volume.

EARLY DAYS

I first went to sea as a graduate student from Liverpool University, in 1962, on a small boat in the Irish Sea. There I heard rumors of an expedition to the Indian Ocean, which seemed wonderfully exotic and unknown. With a grant from the Royal Society I was drafted to serve on the RRS *Discovery* on this two-year effort. With some encouragement from John Riley and training in salinity-nutrient-oxygen analyses, I helped perform thousands of measurements on the 1963-1964 International Indian Ocean Expedition. I heard legends of the International Geophysical Year (IGY) while at sea, and it was there that I first made contact of sorts with the National Science Foundation, in the Seychelles, when we met with the RV *Anton Bruun*. David Menzel grabbed me and, in the midst of a very boisterous party, grilled me about the controls on nutrient ratios in different monsoon seasons. I suspected much later that he was taking an early look at denitrification in the enormous suboxic regions, which we now know proceeds to an extraordinary extent. I was curious at what might happen to this vast collection of hydrographic data; no one told a graduate student, and I began to suspect that some of the senior scientists just didn't know. Years later I had in my hands a beautiful atlas compiled by Klaus Wyrtki, published, of course, by NSF. Capturing the power of large data sets is something that is done infinitely better today.

Although the expedition was dubbed International, and it clearly was, I was very struck by the fact that each nation was locked in its own ship and that at-sea contact with other nations' scientists, in the form of direct joint experiments, appeared to be small. We appeared to be "national by facility," but international by concept and desire. Indeed it was clear to me that a large and charming international guild existed and that information was traded here with extraordinary speed and generosity, possibly more so on the social occasions wrapped around the formal meetings. This is still true today. The International Indian Ocean Expedition was one of the last of the old-style efforts, where the expedition was coherent in place, but consisted of a mixture of many unrelated scientific activities, from net hauls to seismology.

On September 11, 1964, on the trip home, the *Discovery* stopped in the middle of the Red Sea for one last hydrographic station. John Swallow (UK) and Rocky Miller (U.S.) who was funded by NSF, had independently noted a warm, saline anomaly and acoustic reflecting layers in the deep water of the Red Sea, and had shared the data. John wanted to track it down. I was assigned the task of (pencil) plotting the data called out from the echo sounder, and we found a definite depression at the site. I then drove the steam winch, hung the bottles, lowered the hydro cast as close to the bottom as we could, carried the samples, and logged the data. The results were extraordinary, with a bottom temperature of 44°C and saturation with salt at >300 grams per kilogram seawater! I was proud of my hard work in the hot sun, after 8 long months at sea, and was very surprised to see a large fraction of the sample being taken and stored in a big plastic bottle that I knew nothing of. "What is that for?" I asked, and was told it was for Harmon Craig at Scripps. I knew nothing of Harmon then, but I did have a first glimpse of how far one could push this international thing.

THE PRE-GEOSECS PERIOD

In 1966 I was offered a job at Woods Hole by John Hunt, who liked the work we had done on the Red Sea brines, and by Derek Spencer, who was building an energetic new chemistry initiative. I knew nothing of the way U.S. science was conducted, but on my first venture into work I observed a plaque on the building commemorating its construction with funds provided by NSF. That summer I went to sea on the *Atlantis II* and found a similar plaque there too. This seemed to everyone to be quite normal, but such generosity made a great impression on me.

The results of the Red Sea hot brines discovery, soon extended by the group on *Atlantis II* finding a collection of still hotter, more chemically extreme solutions, with vast metal deposits, occupied much of the ocean chemistry community in the late 1960s, and it received strong NSF support. Dave Ross and Egon Degens led the effort and produced a very fine book (Degens and Ross, 1969). The total saturation with halite made these hot brine pools completely sterile, and it was not until much later, in 1979, that exotic animal communities were discovered in association with hot vents, near the Galapagos, on the East Pacific Rise. Interestingly, if one takes the earlier Red Sea data and simply strips off the NaCl component, the residual chemistry is almost identical to the midocean ridge venting fluids, that is, in showing a dramatic loss of magnesium and sulfate, and strong enrichment in iron and manganese, from water-rock interaction.

The fundamental science of what drove the fluids, and the mechanism by which seawater could be altered so dramatically, were very much on our minds. For the first time I began to appreciate why my early sample had been shipped to Scripps so quickly, when I saw the stable isotope data from Craig's lab and realized the constraints it provided.

The Red Sea cruises and the Black Sea cruise in 1969 (Degens and Ross, 1974) were excellent examples of medium-scale ocean science. They had finite goals and a well-constrained geographic area, but they were clearly much larger in scope than a single-investigator laboratory could handle. The papers from these efforts were first class and had many international contributions.

I was advised that I needed to branch out, and in 1967 I wrote my first proposal to NSF, requesting funds to adapt a new fluoride electrode into deep-ocean instrumentation. Funds were awarded, and we quickly wrote a paper on $(Mg-F)^+$ ion-pairing that provided the first experimental test of theoretical models. In some small way we were making a contribution to the building of the elegant thermodynamic model of seawater that we now take for granted. Almost all of this fundamental work on solution physical chemistry was supported by NSF, and all models of the ocean uptake of fossil fuel CO_2 today depend critically on this important thermodynamic framework.

A cruise was scheduled, which yielded no unusual results, but it was on this trip, on a warm night on the fantail of the RV *Chain* in harbor in the Azores, that I first heard, from Derek Spencer, of the plans for a global set of geochemical sections. It seemed to be a very attractive idea. Hank Stommel (Stommel and Arons, 1960) had been persuaded by the early evidence (see Broecker et al., 1961, for the earliest discussion) from Wally Broecker and colleagues, for ^{14}C dating of water masses and of the possibility of putting new time constraints on the mean rate of global ocean circulation by use of the radioactive clock. The new International Decade of Ocean Exploration (IDOE) initiative was arising and would be housed within NSF. This presented a fresh and important opportunity to make a case, and the excitement we felt at revealing the picture of the circulation of the greatest fluid on Earth, painted anew in chemical colors, was palpable.

THE GEOSECS PERIOD: 1968-1978

Building the Program

The object of GEOSECS, the Geochemical Ocean Sections Program, was to trace the picture of the abyssal circulation, using the power of radiotracers to accomplish this, with major cruises in the Atlantic, Pacific, and Indian Oceans. But only just beyond this goal lay much uncertainty. There were many potential tracers in addition to natural ^{14}C; the suite of ^{226}Ra and ^{228}Ra; ^{222}Rn for gas exchange rates and bottom boundary layers; ^3H and ^{14}C from the nuclear tests; and the elusive potential of ^{32}Si. Did they *all* have to be measured? What were the potential gains? What sampling pattern and density were required? How might the results be incorporated into physical models? These questions, first posed within the GEOSECS context (Craig, 1972), have only been answered with any rigor in the last few years. They are fundamentally hard topics, and it is a tribute to the NSF of those days that it ventured into (literally) such uncharted waters.

Moreover each of these tracers had chemical reactivity as well as radioisotopic decay. This had to be constrained too, and for ^{14}C it meant attacking the full ocean CO_2 system, while for the radium isotopes the chemical analogue of barium was selected. Each of these efforts had its advocates, and once the science case had been made within the GEOSECS steering committee, it would petition the NSF-IDOE for funds.

It was plain at the outset that the GEOSECS program would be fundamentally different in style and scale than anything before. It was also very confusing. At least three major institutions were big players: Scripps, where Harmon Craig had persuaded Arnold Bainbridge to set up the GEOSECS Operations Group that was to craft the advanced instrumentation and staff the technical support activity; Lamont, where Wally Broecker had pioneered many of the radiochemical tracer techniques and gas exchange rate concepts; and Woods Hole, which was to provide the RV *Knorr* for the first, Atlantic, expedition and where Derek Spencer created the coordinating center. Karl Turekian at Yale and Gote Ostlund at Miami provided wisdom and refereed the sometimes amazing disputes that arose. Incidents involving fire extinguishers, epoxy, and roller derby are best not mentioned here. At each institute there were young scientists eager to be involved, but all had different views on what would be needed and on how to make a personal scientific effort within this large enterprise.

The problems were typical. Big programs need to be staffed with first-class scientists who will remain with the program for years. First-class scientists cannot be cogs in a big wheel, but are very inventive people of rapidly evolving interests who need to create their own identity and establish their own careers. How to balance these conflicting needs often lies at the heart of ambivalent feelings about large programs. Although the decade was dubbed International, it was not clear what this meant for a particular program, but in the case of GEOSECS several individuals in other countries (Yoshio Horibe, Devendra Lal, Wolfgang Roether, Brian Clarke, and Roger Chesselet) made extraordinary personal efforts.

So far as I can tell, the IDOE programs were not successful in making use of SCOR (the Scientific Committee for Oceanic Research) or the IOC (Intergovernmental Oceanographic Commission), the formal ICSU (International Council for Science) and UN-affiliated international bodies, respectively, to carry out their planning or execution, in spite of a strong effort to do so. Lou Brown was brought into NSF to serve as the internationalist, and he remains at NSF today. In practice the dominant new factor was the forcing by NSF of domestic interinstitutional, not international, expeditions and programs. This radical intrusion into the sovereignty over their ships enjoyed by the major institutions was a source of great discomfort to traditionalists, but it opened the door wide to young and ambitious scientists.

The Expeditions

The Atlantic Ocean—The GEOSECS Atlantic Expedition in 1972, the Pacific Ocean Expedition in 1974, and the Indian Ocean Expedition in 1978, all presented unique challenges. The Atlantic cruise was preceded by at least two test stations, or cruises, that showed somewhat alarming results (Craig and Weiss, 1970). Measurements of the CO_2 system properties made by different techniques gave discordant results, calling into question the basis for using the ^{14}C tracer. It was shown that the precision of the Δ^{14}C measurement, achieved by Gote Ostlund and Minze Stuiver, would be ± 4 per mille, giving an age resolution $\cong 30$ years. The product of total CO_2 and ^{14}C was required, and the supposedly easier total CO_2 measurement should not degrade the signal. But it did, and the confusion was to last for several years. It is a tribute to the drive of the leaders, and the courage of NSF, that the expedition went forward.

This courage was soon tested. The *Knorr* left Woods Hole for a nine-month cruise on July 18, 1972—and lost the entire, horribly expensive, conductivity-temperature-depth probe/profiler (CTD)-rosette sampling package on the very first station. Apparently a locking pin had not been set in place. In spite of heroic efforts by Arnold Bainbridge, not all the advanced analytical systems worked, and it would be months before the CO_2 system was fully operational. Not everyone trusted the new systems to work at all; Joe Reid had insisted that a separate conventional Nansen bottle cast be done as backup at every station, and this was laboriously carried out. The cruise tracks are shown in Figure 1.

Tensions soon arose. The work was long and hard, the cost of supporting the expedition was high, and in contrast to small- or medium-scale science, there was no natural break to stop and write papers. It began, to some hostile critics, to look like a large, expensive, general data gathering hydrographic exercise, possibly similar to the International Indian Ocean Expedition a decade before. A review was held at NSF, and the case for going ahead with the Pacific cruise was made. "What," Admiral Owens asked, "would be the consequence of *not* funding the Pacific cruise?" Amid the uproar he had made his point: this was NSF research, managed by the Foundation which did have the last word. And there were expectations for individual scientific accomplishment.

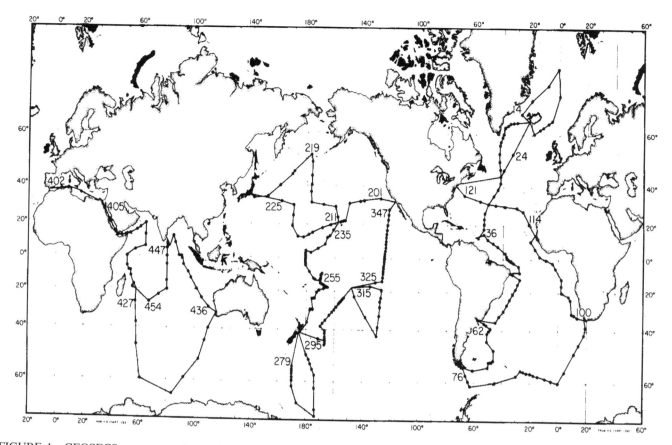

FIGURE 1   GEOSECS program cruise tracks, 1972-1978.   Reprinted from Craig and Turekian (1980) with permission from Elsevier Science.

Hank Stommel noted "a profound sense of beauty" in seeing for the first time the tracer signal of the North Atlantic deep water overflows as they began the abyssal tour, and an intense flurry of activity resulted in a strong collection of papers (Craig, 1974). Feenan Jennings at NSF made sure that the program proceeded.

The Pacific Ocean—The GEOSECS Pacific expedition, from August 1973 to June 1974, on the RV *Melville*, continued the pattern. The Nansen bottle cast requirement was, thankfully, dropped, but the discordant CO_2 data problem, latent in the Atlantic cruise results, was now much worse. Nonetheless the classic picture of the chemical response to "aging" of our global circulating fluid was emerging beautifully, and the first glimpse of a global CO_2 picture was tantalizingly close.

A Damoclean list appeared above the chief scientist's bunk of the cruise legs on which major equipment was lost; "Bomber" Takahashi led the list since he had had the bad weather legs, for which we were all grateful. The effort to measure the cosmogenic isotope ^{32}Si, requiring the processing of a thousand liters of seawater through smelly manganese-loaded fibers, was particularly messy. And the early

results were showing very little signal. We were to find much later that the half-life had been in error by a factor of four!

NSF realized at some point that this was an enterprise of historic scale and decided to memorialize it on film. A contract was awarded, and a very new cinematographer was flown to Tahiti. The movie is still fun to watch, but it was his personal comment afterwards that shook me. "My God!" he said, "I didn't know the work was that hard!" A plane crash in Samoa sadly resulted in death and injury for the team.

The Indian Ocean—The strains of multiyear devotion to such an all-consuming effort were beginning to show, and by the end of the Pacific cruise, time was needed to regroup, analyze samples, upgrade equipment, and repair relationships with NSF, which, through an evolving stream of program managers, had kept close watch on progress. The urge to focus on showing success and building scientific knowledge, through work on the Atlantic and Pacific results, was getting in the way of creating the Indian Ocean expedition. This illustrates a common problem of large programs—the balance between keeping the technical skills and facility in readiness, and taking definite individual time for research.

In my view it is very much to their credit that NSF program managers have always been wise and pragmatic about this.

The four Indian Ocean legs, from December 1977 to April 1978, were thus a much smaller effort. Taro Takahashi found that the CO_2 problem grew yet worse, so that the measured pCO_2, and that calculated from the measured alkalinity and total CO_2, differed by well over 30 ppm (parts per million). The urge to tackle directly the growing fossil fuel CO_2 problem was now very strong, and this stood in the way. The elegant CO_2 model by Hans Oeschger and colleagues in Switzerland had appeared two years earlier and had stimulated renewed interest.

New interests were also arising; the "particle reactive tracers" such as thorium and ^{210}Pb were proving more tractable than anyone had thought—not for the original problem of the abyssal circulation rate, but for insights into how the ocean biogeochemical cycle worked. The program was evolving, and important breakthroughs in trace-metal geochemistry, organic geochemistry, and observing the rain of particles to the seafloor were occurring.

GEOSECS Synthesis

The general release of data from the shipboard program was keenly sought, but those close to the measurements were always aware that things could be improved. More problematic still were the results from the shore-based laboratories. These were closely held by the principal investigators (PIs) so as to maximize their advantage in publication; yet for such a conspicuous program there was widespread desire for full disclosure. The data release problem is commonly dealt with by NSF today, but it was the pattern that was created during the GEOSECS era that laid the rules. Pressure, official (the purse string) and peer, was brought to bear on PIs, and the results emerged.

The GEOSECS period resulted in all manner of fundamental insights. I wrote a paper on modifying the equation of state, thus eliminating an ambiguity in connecting the interocean abyssal flow (Brewer and Bradshaw, 1975). Wally Broecker carried on an amazing and sustained attack on the use of the fundamental nutrient relationships to decode water masses, mixing, and chemistry (beginning with Broecker, 1974; see Broecker and Peng, 1982, for a masterful analysis). And eventually the long haul of collecting, stripping, and measuring the radiocarbon signal was completed. Stuiver et al. (1983) published a remarkably simple and elegant paper compiling the ^{14}C results (Figure 2). They reported that "the mean replacement times for the Pacific, Indian, and Atlantic ocean deep waters (more than 1500 meters deep) [are] approximately 510, 250, and 275 years respectively. The deep waters of the entire world ocean are replaced on average every 500 years." These ages were much shorter than first expected: the promise of a radiocarbon solution to the "age" problem was fulfilled; the task of shedding more light on geophysical fluid dynamics by the

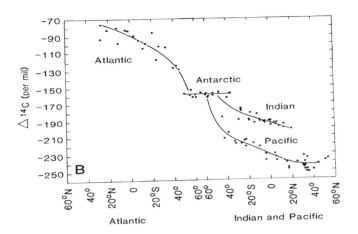

FIGURE 2   The $\Delta^{14}C$ values of the cores of North Atlantic, Pacific, and Indian Ocean deep waters. The oldest waters are encountered near 40°N in the Pacific Ocean. Reprinted from Stuiver et al. (1983) with permission from the American Association for the Advancement of Science.

tracer approach proved to be far more complex. Of course such a paper was principally a necessary and welcome formality. Thanks to the wise NSF data release policy, the GEOSECS results had already been in use around the world, for all kinds of innovative uses, for many years.

Follow On

In 1978, John Steele called me down to his office at the Woods Hole Oceanographic Institution (WHOI), to meet with John Ryther and Hank Stommel. He wanted to see some fresh starts, and he was concerned about the ending of GEOSECS. He particularly wished to see WHOI tackle the CO_2 problem in some way. Hank referred him to a short NAS report, written with Jules Charney, on the anticipated thermal changes; I volunteered to look at the GEOSECS data to find the oceanic chemical signal. Since I had served as co-chief scientist on GEOSECS Atlantic Leg 6, I simply went to those data and wrote a provocative paper on the procedure for detecting the fossil fuel CO_2 signal above the very large natural background.

At the same time, Gote Ostlund in Miami was fretting about the lack of a GEOSECS follow-on. The classic dilemma with large programs is the problem of continuity versus innovation; a superb observing system had been created and refined, and a talented team of people, particularly the Operations Group under Arnold Bainbridge at Scripps, existed. I had heard that Gote was to hold a meeting, with Department of Energy support, to discuss this and I called him to ask if I could attend. I gave the fossil fuel CO_2 paper and pointed out that, due to the early GEOSECS technical problems, we had no Atlantic data north of 20°N in the critical deep water formation regions. Others pointed out the new information from the chemical tracers in the region, and we conceived of a program to measure intensively the invading wave of chemical tracers from the industrial activities of man. The project was soon dubbed "Transient Tracers in the Ocean" (TTO), and of course, we went to NSF for support.

THE TTO PROGRAM

While the TTO program evolved from the GEOSECS experience, it had marked differences, driven both by PI desire for hands-on research and NSF desire for the accountability of individual components to peer review. It was smaller in space and time, the support from the Operations Group was cut in half, and a much more efficient set of observing protocols was adopted. The unseen hand of the scientific marketplace was at work, and our big program follow-on was now to be staffed by an ensemble cast. Within NSF the latent problem of whether this was chemistry (the technique) or physics (a major application) had to be dealt with, and an impasse occurred. Physical Oceanography program manager Curtis Collins at NSF was to rise to the challenge and ably represent the program.

The program was national, but as with GEOSECS, a very strong informal international flavor was simply assumed to exist. I recall driving to Lamont for one meeting with Canadian, German, Japanese, and English participants, which seemed quite normal.

The planning ran into two problems very quickly: facing up to the undiagnosed error in the GEOSECS CO_2 results, now nine years old, could no longer be postponed. NSF, quite properly, would not let a new program go ahead without it. And the design of a cruise track that would attempt to cover a very large area of the North Atlantic in one snapshot, proved challenging. In the midst of this, Arnold Bainbridge, the talented, gracious hero of GEOSECS, suddenly died. He was only 48 years old. Years of stress and failure to take care of a chronic health problem had taken a dreadful toll. The shock was enormous.

The death of Arnold Bainbridge left a huge hole and much confusion. When his team went through his office, to put affairs in order and recover original files, they found a drawer full of carefully labeled tapes archiving all the programs that we were using. The problem was that all the labels simply read, "Test"! We were lost.

A meeting was held at Lamont to review the CO_2 measurement and data recovery problem, and Bob Williams kindly loaned me a very large binder of FORTRAN printout, which probably contained the answer somewhere. I digested it on a plane flight to Seattle, and by the end of the trip, red eyed, I had found the few lines of code that seemed to count. Arnold had been creative with his chemistry coding and had not told any of us! Al Bradshaw and I painstakingly pulled things apart, and ran some tests (Bradshaw et al., 1981); yes, we could rewrite the equations, and yes, a coding error had occurred during the Atlantic to Pacific transition. We were learning hard lessons—that big programs can be vulnerable.

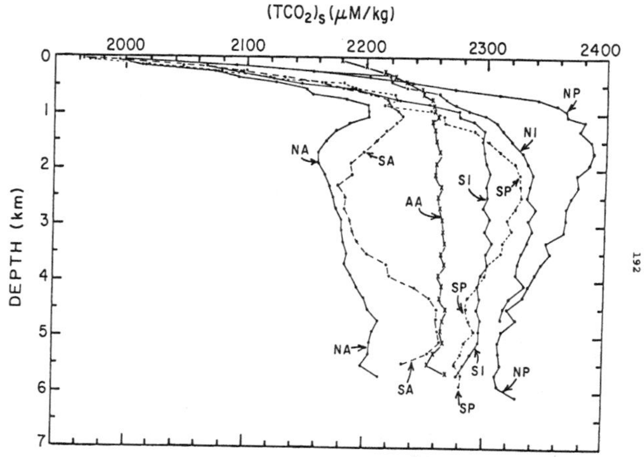

FIGURE 3 Total CO_2 results from the GEOSECS program, showing the progressive enrichment due to respiration and carbonate dissolution accompanying the deep circulation. Reprinted from Takahaski et al. (1981) in SCOPE 16, *Carbon Cycle Monitoring*, edited by Bert Bolin, with permission from SCOPE, John Wiley & Sons, Ltd., UK.

But we could put the problem to rest and advise NSF that publication of the GEOSECS atlases, long stalled by this problem, could proceed. It fell to Taro Takahashi to compile the data (Takahashi et al., 1981), and the classic picture that resulted is shown in Figure 3.

A test cruise in 1980, and a wonderful year in 1981, saw a large-scale attack on the tracer chemistry of the North Atlantic Ocean. Some 250 stations were occupied (Figure 4), and no equipment was lost. Richard Gammon made the first, exciting measurements of the chlorofluorocarbon tracers. A marked freshening of the North Atlantic was found, which became part of the "Great Salinity Anomaly." The tracer signals showed beautifully the evolution of ocean water masses in the nine years since GEOSECS. Tritium-helium dating of water mass ventilation came of age, thanks to the superb efforts of Bill Jenkins. And chemical coherence within the CO_2 system was attained, thanks in large part to the (on-shore) presence of Dave Keeling who provided a limited data set of unassailable integrity. The program provided a superb benchmark for carbon-cycle science.

INSIDE THE FOUNDATION

In the fall of 1981 I took a two-year leave from Woods Hole to serve as program director for Marine Chemistry at NSF. Neil Andersen had been appointed to the Intergovernmental Oceanographic Commission in Paris, and I was his "rotator" replacement. Most of my colleagues were shocked at the move; I had a wonderful time—eventually, and thanks above all to my program colleague Rodger Baier. I learned from Rodger that NSF program managers were not dull; they could play the piano cross-handed while lying down and

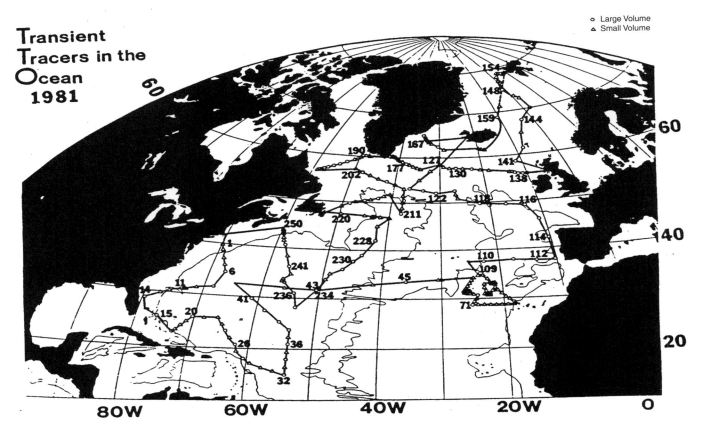

FIGURE 4   Cruise tracks for the Transient Tracers in the Ocean (TTO) North Atlantic Program, 1981.

balance a $10 million budget perfectly on a tiny calculator bought with "global change" in an airport in Tunisia.

My first impression was one of a strange trading market, where almost nothing was done by one program alone, but I saw that I had nonetheless inherited a superbly balanced program. Next I realized how fuzzy our scientist's picture was of how decisions were made. And I learned that nothing counted for more than a clear scientific question to address an unsolved problem; this made a confusing job easy. There were big programs I had never heard of and small programs of all kinds. The folding of the former IDOE big programs into the Division of Ocean Sciences had just occurred, and I began to realize that creating a new big program in this environment would be very difficult indeed. It was the new funds and separateness of IDOE that had allowed the big programs that had shaped our scientific lives to succeed. They also provided the organizational framework to make full use of our ships and yielded dense, well-populated data sets that constrained the ocean in ways not possible by other means.

I attended hearings and learned more about the need to address scientific problems fundamentally important for society. I signed off with pleasure for the publication of the long-awaited GEOSECS atlases. And I served, with Bill Nierenberg and Roger Revelle, on a very special NAS study

of "Changing Climate." For 30 years Roger had kept his focus on the CO_2-climate problem. He had educated Al Gore at Harvard, and the political world was beginning to catch up to the issue.

We had at NSF an excellent group of dedicated people: Grant Gross, Bob Wall, Curt Collins, Don Heinrichs, Bruce Malfait, Larry Clark, and rotators Mike Reeve and Rana Fine. All three rotators were to return to academia at about the same time in 1983. We met for lunch to discuss the inevitable exit interview, which we decided to do as a team. It was clear to us that the pattern we had observed, of frequent emergency requests from on high on a Tuesday afternoon for a new long-range plan by Thursday morning ("latest"), barely tolerated by the veterans, was unsustainable. Bob Wall listened, and soon after, a much more vigorous, NSF-initiated planning process took shape in several forms. In some ways it was a natural response to the vacuum created by the demise of a separate IDOE program.

THE JOINT GLOBAL OCEAN FLUX STUDY

The program era that followed was to prove to be fundamentally different. NSF was about to lead, with considerable courage, the new "global change" programs: larger in scale and complexity, longer in planning, international in scope, and far more visible to policy makers. None of us anticipated the work involved.

The Global Ocean Flux Study (GOFS)

I returned to WHOI in late 1983. The World Ocean Circulation Experiment (WOCE) program was then taking shape under Carl Wunsch's leadership. In many ways it seemed a natural ally to the TTO program that had proceeded to the Equatorial, and South, Atlantic. But the latent conflicts between physics and chemistry had not disappeared: how many tracers were really necessary? Could a few sparsely placed samples yield adequate constraints? How solid were the boundary conditions? A rather fierce debate took place. The NAS Ocean Studies Board, with NSF support, provided the forum for this.

John Steele observed the emergence of WOCE, with roots in a desire to use an altimetric satellite combined with a global hydrographic survey, in order to study the global circulation. He was concerned that no program of similar scale existed to constrain the biogeochemical cycles of the ocean and that the promise of an ocean color satellite, hinted at by the Coastal Zone Color Scanner (CZCS) on Nimbus-7, might not be realized unless we took action. The descriptions of the major ocean biogeochemical cycles, which support life in the sea, rested on largely untested ground. He used the Ocean Studies Board to organize a major meeting, held in September 1984, at the Woods Hole Study Center.

The meeting (NRC, 1984) itself was plainly important, but confusing. Mixed together were satellites, primary productivity, higher organisms, sediment traps, radioisotopes, benthic instruments, and the sediment record. Linkages of this kind had been drawn in "horrendograms" by Francis Bretherton as visual drama but executing a coherent study was another matter alltogether. The CO_2 story was barely mentioned—would WOCE take care of this? John Steele, Jim Baker, Wally Broecker, Jim McCarthy, and Carl Wunsch kept a careful eye on proceedings. Ken Bruland had been selected to head the Planning Committee. The individual papers were good, but nothing seemed to gel—the topic was so broad, and almost no one had experience dealing with the National Aeronautics and Space Administration (NASA) and satellites. What to do next?

It was Neil Andersen who stepped forward and quietly asked a small group (Ken Bruland, Peter Jumars, Jim McCarthy, and me) to attend a meeting in Washington, ostensibly to edit the report. We met at the NAS building a week or two later. There Jim Baker walked us through the problem; getting a new start in NASA for an ocean color observing satellite would not be easy at all. And the political scene was fundamentally changed by the privatization passion of the current administration, so that commercial possibilities must be factored in. We did edit the GOFS report, and I urged that a far more prominent role for directly observing the controls on the oceanic carbon cycle be included.

A most difficult period then followed. Many participants had naively assumed that simply issuing a report would guarantee funds! The breadth of subject matter left room for a very large number of potential participants. And the review nature of the report, without any early crafting of tactics, left no road map with which to proceed. I was asked to chair the group and instantly felt these problems. A proposal to proceed was submitted through the National Research Council (NRC), and the first "pitch" was made to Burt Edelson at NASA. It went well.

The first step had to be consensus building, and a tense set of small working meetings followed through 1985, touching on each of the subthemes in the GOFS report. I became the sole member of a "Planning Office," helped enormously by the astonishing rise of electronic mail (pioneered by Omnet for the ocean science community). Great credit at this point must go to Neil Andersen, who saw the end point of a powerful program through the clutter of early discussions and carefully guided science along. Again, the Ocean Studies Board meetings served as the debating ground.

In October 1986, plans were more advanced, and I attended a major WOCE meeting, again at NAS. There I addressed the science behind the ocean carbon cycle and received a very enthusiastic audience response (my presentation materials were promptly "borrowed" by a complete stranger!). We had by then conceived within GOFS planning of a three-part attack on the problem: establishment of time-series stations at Bermuda and Hawaii to observe seasonal cycles and secular trends, a set of carefully crafted process studies to illuminate the controlling functions, and a global survey of the CO_2 field. It was this latter component that we wished to see accomplished as a collateral program with the WOCE global hydrographic survey, for it would be the critical glue that would scientifically link the two principal ocean observing programs. WOCE was measuring ^{14}C distributions, building on the GEOSECS legacy, and our point was that the full CO_2 system, with its embedded biogeochemical content, naturally followed. We could not afford two global surveys.

But enthusiasm and practicality do clash. There were basic problems of space and funds, let alone the interdisciplinary science. A blunt compromise was quickly reached; WOCE would provide bunk and laboratory space and access to samples. And GOFS would provide trained people, instruments, funds, and data, and would represent the program to appropriate bodies. It was a deal.

It was at once clear that this forced some new steps. WOCE and the global survey were now formally international. GOFS was still national. Within 24 hours a proposal was drafted to the Scientific Committee on Oceanic Research to request its attention to this program and to propose a true international effort. It was immediately hand carried to the SCOR General Meeting in Tasmania and well received.

The first SCOR-sponsored international meeting was held in Paris, at ICSU headquarters, in February 1987. It was chaired by Jim Baker and carefully observed by Gerold Siedler (Kiel) as president of SCOR. Two key pieces of information came to light at this critical time. We had received the challenge to produce a satellite global chlorophyll image from compositing the CZCS fragments. Gene Feldman rose to the occasion and produced a beautiful image; we saw it for the first time in the luggage area at De Gaul airport in a gray dawn light. It made an enormous impression.

We had earlier received a scientific challenge for the proposed global CO_2 survey. In essence it was, "Show us that you can treat ocean CO_2 data in the same rigorous manner that we treat the transport of heat." This was very reasonable. By placing ocean CO_2 and heat transport in the same observational and theoretical framework, we could link the climate and greenhouse gas signals much more directly. But, as we have seen, the history of such measurement was fraught with difficulty. David Dyrssen (Goteborg) and I had drafted a position paper for the meeting on this very topic and had computed the CO_2 and nutrient fluxes across $25°N$ in the Atlantic Ocean. The problem was far more tractable than we had believed, and new concepts of constraining the mass balance by incorporating some adventurous organic carbon measurements had to be called upon.

It was Jim Baker who suggested that the program henceforth be called JGOFS (not Japanese, but Joint, he quipped). It took. Bernt Zeitschel became the first chair and the first JGOFS International Expedition took place in the North Atlantic in 1988.

This was my first experience at attempting a truly international effort, and I think we all found that it wasn't easy. Neil Andersen's courage, tenacity, and international experience were to serve us well throughout this period. Special mention must also go to Elizabeth Tidmarsh (now Elizabeth Gross) as executive secretary of SCOR for superb efforts in implementing the international form of the program.

Understanding the role of the ocean in the global CO_2 equation is not easy, even for many chemists. And here we had a diverse international collection of scientists of several different disciplines, many of whom were now being asked by their government agencies for informed comment on this topic as interest in greenhouse gas policy grew. We soon found that huge differences of opinion occurred. The ocean uptake of fossil fuel CO_2 from the atmosphere is not controlled by biological activity, but is an inorganic phenomenon. But the background ocean CO_2 level, which the rising trend is imprinted on, is. We are not writing the industrial signal on a blank ocean page. This detail was lost on many, and several highly contentious meetings took place. Years later, we find that a large international population of ocean scientists is now fluent in these issues, and this is a very good thing. The JGOFS program is still in place today, and the results are superb.

The transition from observation and diagnosis of the carbon cycle to active intervention by changing industrial policy, and ocean CO_2 manipulation by disposal and/or fertilization, is about to occur and I have no doubt that NSF will provide the leadership for the scientific underpinnings needed.

IMPRESSIONS TODAY

What can we learn from the big program versus small program theme of this session? Firstly, the style of so-called big programs has changed enormously over the years—from the miscellany of the Indian Ocean expedition, to the large dedicated staff and many-year theme of GEOSECS, to the ensemble cast of TTO, to the remarkable coalescence of individual efforts within JGOFS to attack a very broad problem in a structured way. At each step NSF has shown leadership and creativity in crafting these efforts. And it has enabled the discovery of the fundamental pattern and time scale of ocean circulation, the invading chemical signal of the twentieth century, the chemistry of strange seas, and the fundamental basis for biogeochemical balance as we approach a warmer world. Big programs are not impersonal, but are unusually intense experiences for dozens of small groups. They are a critical part of our ocean science community.

While I have concentrated here on personal experiences, I suspect that others, in parallel programs, have similar tales to tell. The balance of small and large programs comes naturally; theory, instruments, methods, all typically come from small efforts, and the big programs cannot do without this. The best large programs embrace theory, create "small" initiatives, and provide superb opportunities for a very large number of scientists. The future is a bit more worrisome. With the desire to detect global change there is a call for very large scale operational programs, with data continuity and massive modeling as the goal, rather than a set of evolving questions. Fortunately, there is a new class of medium-scale projects emerging, there are satellites for global observations, and there are exciting new possibilities of sensing and manipulating ocean chemistry in entirely novel ways. NSF is not a mission agency, and I hope that over the next 50 years of ocean discovery the Foundation will keep the healthy large-small ocean program balance in place.

ACKNOWLEDGEMENTS

This paper is dedicated to all my friends at NSF, over many years. It is supported by a grant to MBARI from the David and Lucille Packard Foundation.

REFERENCES

Bradshaw, A.L., P.G. Brewer, D.K. Shafer, and R.T. Williams. 1981. Measurements of total carbon dioxide and alkalinity by potentiometric titration in the GEOSECS program. *Earth Planet. Sci. Lett.* 55: 99-115.

Brewer, P.G., and A. Bradshaw. 1975. The effect of the non-ideal composition of sea water on salinity and density. *J. Mar. Res.* 33:157-175.

Broecker, W.S. 1974. "NO," A conservative water-mass tracer. *Earth Planet. Sci. Lett.* 23:100-107.

Broecker, W.S., and T.-H. Peng. 1982. *Tracers in the Sea.* Eldigio Press. 690 pp.

Broecker, W.S., R.D. Gerard, M. Ewing, and B.C. Heezen. 1961. Geochemistry and physics of ocean circulation. Pp. 301-322 in M. Sears (ed.), *Oceanography.* American Association for the Advancement of Science, Publication #67.

Craig, H. 1972. The GEOSECS program: 1970-1971. *Earth Planet. Sci. Lett.* 16:47-49.

Craig, H. 1974. The GEOSECS program: 1972-1973. *Earth Planet. Sci. Lett.* 23:63-64.

Craig, H., and K.K. Turekian. 1980. The GEOSECS program: 1976-1979. *Earth Planet. Sci. Lett.* 49:263-265.

Craig, H., and R.F. Weiss. 1970. The GEOSECS 1969 intercalibration station: Introduction and hydrographic features, and total CO_2–O_2 relationships, with 9 supporting papers. *J. Geophys. Res.* 75:7641.

Degens, E.T., and D.A. Ross (eds.). 1969. *Hot Brines and Heavy Metal Deposits in the Red Sea.* Springer-Verlag, New York. 600 pp.

Degens, E.T., and D.A. Ross (eds.). 1974. *The Black Sea—Geology, Chemistry, and Biology.* American Association of Petroleum Geologists, Memoir 20. Tulsa, Oklahoma. 633 pp.

National Research Council. 1984. *Global Ocean Flux Study.* National Academy Press, Washington, D.C. 360 pp.

Stommel, H., and A.B. Arons. 1960. On the abyssal circulation of the world ocean, II, An idealized model of the circulation pattern and amplitude in oceanic basins. *Deep-Sea Res.* 6:217-233.

Stuiver, M., P.D. Quay, and H.G. Ostlund. 1983. Abyssal water carbon-14 distribution and the age of the world oceans. *Science* 219:849-851.

Takahashi, T., W.S. Broecker, and A.E. Bainbridge. 1981. The alkalinity and total carbon dioxide concentration in the world oceans. In J. Bolin (ed.), *Carbon Cycle Modelling,* SCOPE Volume 16, J. Wiley & Sons, New York.

Ocean Sciences
Today and Tomorrow

The Future of Physical Oceanography[1]

WILLIAM R. YOUNG

Scripps Institution of Oceanography, University of California, San Diego

SUMMARY

The National Science Foundation (NSF) tasked the U.S. physical oceanographic community in 1997 to evaluate the current status of research in physical oceanography and to identify future opportunities and infrastructure needs. A workshop was held in Monterey, California from December 15-17, 1997 and was attended by 46 scientists representing the community of NSF-supported investigators. A subtheme of the meeting was the role and effectiveness of the NSF's core program in physical oceanography. Input via electronic mail from the wider scientific community was sought both before and after the meeting.

The community was asked to consider advances in physical oceanography over the last twenty years. The following items were widely hailed as significant recent achievements: a revolutionary understanding of the coupling of the tropical ocean and atmosphere and the development of predictive El Niño models; estimation of the global distribution of mesoscale variability in the world ocean and theories and models of this geostrophic turbulence; completion of the World Ocean Circulation Experiment and improved estimates of the pathways and timescales of the circulation; and quantitative measurements of the strength of small-scale ocean mixing and the dependence of this mixing on the strength of the internal wave field and other environmental conditions.

[1]Excerpted from *The Future of Physical Oceanography: Report of the APROPOS Workshop.* http://www.joss.ucar.edu/joss_psg/project/oce_workshop/apropos/report.html, 9/11/99. The APROPOS committee was co-chaired by William Young (Scripps Institution of Oceanography) and Thomas Royer (Old Dominion University). Other members of the steering committee included John Barth (Oregon State University), Eric Chassignet (University of Miami), James Ledwell (Woods Hole Oceanographic Institution), Susan Lozier (Duke University), Stephen Monismith (Stanford University), Peter Rhines (University of Washington), and Peter Schlosser (Lamont-Doherty Earth Observatory). William Young presented a summary of the APROPOS activity at the symposium.

The community was also asked to look into the future and forecast advances for the next twenty years. Great excitement was expressed at the prospect of new tools that might solve the problem of observing the global ocean. Already the TOPEX/POSEIDON satellite mission has measured the topography of the sea surface to 3 cm accuracy at 7 km spacing for 5 years. Future developments in satellite oceanography promise global measurements of sea surface salinity and precipitation. These measurements are crucial if we are to understand the climate system and the hydrologic cycle. Yet sea-truth is essential and *in situ* water-column observations made by an unprecedented class of autonomous instruments are anticipated. Integrating measurements, such as tomography, and the installation of cheap and easy-to-use probes on ships-of-opportunity, hold great promise.

Even with present technology, a description and an understanding of the spatial distribution of turbulent processes in the global ocean is achievable in the next decade. Our present conception of ocean dynamics is largely ignorant of processes with relatively short horizontal length scales (say 100 m to 50 km). Yet biological variability is concentrated on these short scales. It is the dynamics on these same scales that is parameterized by eddy-resolving circulation models. Further, in the coastal zones, cross-shelf exchanges are likely mediated by instabilities and topographic influences whose horizontal scales are much less than those of the well-studied alongshore flows. Exploring these largely unvisited scales is a new frontier for physical oceanography.

Several problems facing physical oceanography were identified at the meeting. These are: (1) large sea-going groups are retrenching and there is a consequent loss of technicians, engineers, and the hardware that these people maintain; (2) sustaining the funding of long time series observations is difficult; (3) physical oceanography is not visible to undergraduate mathematics, physics, and engineering majors, and so does not attract many graduate applicants from that population; (4) the organization of NSF physical oceanography makes it difficult to fund projects of intermediate size and this difficulty is compounded if the project is interdisciplinary.

Despite these problems, there was consensus that the National Science Foundation's core program is an invaluable asset of the field. The peer-review system maintains a balance between scientific rigor and responsiveness and ensures continuing support for innovative and fundamental science.

FUTURE DIRECTIONS

Challenging as it may be to make progress on any scientific problem, it is even more difficult to predict the future course of scientific progress. One might say that every important discovery in science is, almost by definition, unpredictable and so it is futile to guess at future triumphs. Indeed, it is worse than futile if these guesses are used to "manage" the direction and content of science. It is our belief that basic research, independent of any practical concerns, is critical to the advance of science and the development of technology. Science is the most serendipitous of human enterprises and the ability of physical oceanography to solve problems of social concern depends on a healthy commitment of resources to basic research on fundamental scientific issues.

Climate

The economic benefits of understanding the role of the ocean in the climate system are enormous. And accumulating evidence of man-made climate change has brought these issues to the attention of the public. These concerns coincide with recent successes in long-term weather forecasting associated with El Niño, and with advances that enable detailed measurement of climate variables. (For instance, in the last ten years, the errors in surface heat fluxes obtained from moorings have been reduced by a factor of forty so that the present uncertainty is 5 Watts per square meter.) These factors imply that climate studies will be a significant path for future research in oceanography.

The development of long-term forecasting skill raises challenging scientific problems. These include: understanding and quantifying turbulent mixing, convection, and water-mass formation and destruction; the thermohaline circulation and its coupling to the wind-driven circulation; the generation, maintenance, and destruction of climatic anomalies; climatic oscillations and the extratropical coupling of the ocean and atmosphere on seasonal, decadal, and interdecadal timescales; and the physics of exchange processes between the ocean and the atmosphere. All these problems are of fundamental scientific and practical importance.

Will there be substantial progress on these issues during the next decade? Many physical oceanographers have already begun an enthusiastic frontal assault under the banner of CLIVAR. It is likely that the economic issues that surround global change and climate prediction will motivate continued financial support from society. If people and money are what counts, then we have every reason to be optimistic.

The problem of global climate prediction is the most difficult that our field has encountered. Unlike equatorial oceanography and El Niño, there is not going to be a theory based on linear waveguide dynamics that decisively identifies timescales and cohesively binds oceanography and meteorology. Further, the decadal timescale of extratropical dynamics means that scientists see only a few realizations of the system within their own lifetime. This is bad for morale, but even worse, we cannot wait to gather enough data to reliably verify the different predictions of climate models. Could meteorologists have developed daily weather prediction models if these scientists saw only three or four independent realizations of the system in a lifetime? The only way around this statistical problem is to expand our data base and frame hypotheses about past climate change and ocean circulation using paleo-oceanographic studies. An important challenge is to test the dynamical consistency of these hypotheses.

The Hydrologic Cycle

An emerging theme, which is strongly related to climate, is the ocean's role in the hydrologic cycle. New satellite technologies promise to measure sea surface salinity and precipitation. These, coupled with improvements in the computation of evaporation via indirect methods, will improve our picture of the freshwater flux in the oceans. The freshwater sphere is an encompassing topic that spans oceanography, the atmospheric sciences, polar ice dynamics, and hydrology. Our knowledge of the oceanic freshwater source-sink distribution is far poorer than our knowledge of the source-sink distribution of heat. Yet salinity and temperature contend in their joint effect on the density of seawater and in their influence on the ocean circulation, and the climate system. Knowledge of freshwater input from continents, precipitation, and sea-ice is poor. Observational techniques addressing these issues (for example, the use of oxygen isotopes, and tritium/helium to diagnose freshwater sources) herald progress.

Coupled with improved estimates of the freshwater sources at the surface, will be an increased understanding of water-mass dynamics and transformations. We can look for advancement on such fundamental issues as the causes of the temperature-salinity relationship, thermocline maintenance, and interhemispheric water-mass exchanges.

Observing the Ocean

We will see explosive development of new observational tools, such as those used by the TOPEX/POSEIDON satellite mission. Future developments in satellite oceanography promise more of the same at ever-increasing accuracy, coupled with the deployment of new satellite-borne instruments. Yet sea-truth is essential and we envisage *in situ* observations that will be made by an unprecedented class of autonomous instruments and probes. The ability to manipulate these tools in mid-mission is developing. While we are making enormous strides in sampling the global ocean better, we still have far to go for truly adequate spatial and temporal sampling, though the era of grossly undersampling the global ocean is dead.

A national effort to support sustained high-quality global observations over decades is needed. Measurements of air-sea fluxes of heat, fresh water, and gases, of surface and sub-surface temperature, salinity, and velocity, are all necessary to meet new scientific challenges and practical needs. Looking beyond the equatorial TOGA-TAO array, long-term subsurface measurements spanning the global ocean are required.

Given the rapid increase in Lagrangian measurements by drifting and profiling floats, and the parallel increase in geochemical tracer data, an intense approach to Lagrangian analysis of advection and diffusion is warranted; our existing base of theoretical tools and concepts is not worthy of the observations that we are about to receive.

Global and Regional Connections

Many emerging physical oceanographic issues concern connections between large-scale and small-scale motions; for example, the relation between small-scale turbulent mixing and the large-scale meridional overturning circulation. Analogous connections and interactions between scales are arising in issues of societal concern, often centered around the increasing recognition that many issues previously regarded as regional now require a global perspective. Anthropogenic pollutants have reached the open ocean and are known to be transported far from their sources. A better understanding is needed of small-scale processes and small-scale aqueous systems (estuaries, wetlands, coral reefs) and their impacts on global issues. For example, the growth of plankton populations, which affect carbon dioxide levels and thus may be important in global warming scenarios, is dependent on details of circulation at fronts, sea-ice, and mixed-layer boundaries.

Cross-Shelf Transports

In most coastal regions, the strongest persistent gradients in properties (for example, salinity, temperature, nutrients or suspended materials) are found in the cross-shelf direction. This is because cross-shelf flow is often inhibited by topography and because the coastal ocean is the contact zone between terrestrial influences, such as river runoffs, and oceanic influences characterized by nonlinear physical

dynamics and oligotrophic biological conditions. Progress has certainly been made on some aspects of the flows that determine cross-shelf transports, especially those related to surface and bottom boundary layer processes. A good deal more has yet to be learned about exchanges that occur in the interior of the water column. The problem is difficult because it often appears that the processes that are relevant for the dominant alongshore flows do not apply to cross-shelf flows. For example, it is likely that instabilities and topographic influences may dominate the exchange process. The exchange itself needs to be understood if we are to address issues such as the control of biological productivity in the coastal ocean, or the removal of contaminants from the near-shore zone.

In addition to cross-shelf exchange processes themselves, there is the question of how the coastal ocean couples to its surroundings on both the landward and seaward sides. Estuarine processes are important for determining the quantity and quality of terrestrial materials that reach the open shelves. The oceanic setting, including eddies, filaments, and boundary currents, in turn determines how effectively coastal influences can spread offshore, or how the oceanic reservoir will affect shelf conditions. Consequently, the study of the continental shelf demands consideration of both offshore and near-shore (estuarine and surf zone) dynamics.

Inland Waters and Environmental Fluid Dynamics

Our understanding of inland waters, such as estuaries, wetlands, tide flats, and lakes, will be aided by the same observational and computational technologies that promise progress on the general circulation problem. This work will afford exciting opportunities for interdisciplinary research blending physical oceanography with biology, geochemistry, and ecology. Examples are tidal flushing through the root system of a wetland, and the physical oceanography of coral reefs.

Lakes can be useful analogs of the ocean, with wind and thermally driven circulations, developing coastal fronts, and topographically steered currents. Lakes are important as model ecosystems that are simpler and more accessible than ocean ecosystems. Significant progress can be foreseen in the coming decades in limnology, helped by the tools and ideas developed for the ocean.

The expertise of the physical oceanography community should make possible substantial advances in the understanding of all these shallow systems. Because of the major roles played by turbulence and complex topography, these systems pose impressive and fascinating challenges to physical oceanography.

Turbulent Mixing and Unexplored Scales

Past achievements in quantifying small-scale turbulent mixing in the main thermocline, coupled with exciting recent measurements in the deep ocean, suggest that a description and an understanding of the spatial distribution of turbulent mixing in the global ocean is achievable in the next decade. Unraveling the possible connections between the spatial and temporal distribution of mixing, the large-scale meridional overturning circulation, and climate variability are important aspects of this research.

Knowledge of the horizontal structure of the ocean on scales between the mesoscale (roughly 50 km) and the microscale (roughly less than 10 m) will be radically advanced and altered. The growing use of towed and autonomous vehicles, in combination with acoustic Doppler current profilers, will revolutionize our view of the ocean by exploring and mapping these almost unvisited scales throughout the global ocean. While this research is driven by interdisciplinary forces (biological processes and variability are active on these relatively small horizontal scales) it is also a new frontier for physical oceanography, and one in which even present technology enables ocean observers to obtain impressive data sets.

Numerical Modeling as an Integrative Tool

Large-scale numerical models of the ocean, and of the coupled ocean-atmosphere, are becoming the centerpiece of our science. This is not to say that numerical models dominate our science, but rather that results of theory and observational data are often cast into the form of numerical models. This happens either through data assimilation or through process-model explorations of theoretical ideas. Yet the fundamental difficulty of computer modeling remains: the ocean has, in its balanced circulation, energy-containing eddies of such small scale (less than 100 km) that explicit resolution of these dominant elements is marginally possible. Compounding this difficulty are the unbalanced, three-dimensional turbulent motions that are known to be important in select areas, such as the sites of open ocean convection.

We now have a well-acknowledged list of subregions of general circulation models that are greatly in need of improvement. These include: deep convection; boundary currents and benthic boundary layers; the representation of the dynamics and thermohaline variability of the upper mixed layer; fluxes across the air-sea interface; diapycnal mixing; and topographic effects. Progress in all of these areas is likely as our capacity for modeling smaller scale features increases, and as physically-based parameterizations are developed.

The Future of Ocean Chemistry in the United States[1]

FOCUS STEERING COMMITTEE

INTRODUCTION

As part of the long-range planning process within the NSF Geosciences Directorate, the chemical oceanography community completed an examination entitled "Future of Ocean Chemistry in the U.S. (FOCUS)." There was strong consensus that the field has advanced dramatically during the past 20 years, and is poised to make fundamental new contributions in the coming decades. Chemical and biogeochemical processes in the ocean have profound implications for atmospheric chemistry, ocean biology, and (via hydrothermal processes) the evolution of the ocean crust. Chemical tracer studies underlie much of our understanding of ocean circulation, while the chemical and isotopic composition of microfossils provides a record of past ocean circulation, temperature, and climate. Chemical processes in the marine realm thus have a profound impact on key aspects of the environment while providing the tools to study other fundamental properties.

A steering committee of nine scientists commissioned a series of Progress Reports, organized a workshop of forty participants held in South Carolina in January 1998, and involved the broader oceanographic community via mailings, a town meeting at the 1998 American Geophysical Union-American Society for Limnology and Oceanography Ocean Sciences meeting in San Diego, and a participatory web-site.

This process summarized the status of the field via a review of recent progress, identification of major questions and future research trajectories, and assessment of the status of people and facilities in the field. The full report is published by the University Corporation for Atmospheric Research (UCAR) and available via the Consortium for Oceanographic Research and Education (CORE). This chapter summarizes the findings of the community as described in the report.

RECENT PROGRESS

Over the past two to three decades, research accomplishments by individuals and groups have led to major advances in understanding chemical processes in the ocean. These advances deal with processes ranging in scale from molecules to ocean basins. This chemical perspective has provided critical information and insight to many central oceanographic questions as well as to issues in the companion fields of biological, physical, and geological oceanography. It has also spun off many chemical tools with which other fields could make their own progress.

The FOCUS report assesses this headway in detail. Progress Reports, written by experts in the various fields of ocean chemistry, highlight these significant research accomplishments in ten areas: biogeochemical cycles, oceanic sources and sinks, gases, ocean paleochemistry, physical chemistry of seawater, sedimentary processes, organic matter, anthropogenic effects on the oceans, chemical tracers of ocean ventilation, and chemical analyses and approaches. Among the many advances during this period are:

- learning most of what we now know about internal cycling of materials within the ocean, such as vertical and horizontal fluxes. Previously, we had only basic understanding of chemical fluxes between the oceans and neighboring Earth compartments (e.g., land, sediments, atmosphere). New sub-fields have been developed, addressing areas such

[1]Excerpted and adapted from *The Future of Ocean Chemistry in the U.S.: Report of a Workshop*. http://www.joss. ucar.edu/joss_psg/project/oce_workshop/focus. The FOCUS committee was co-chaired by Ellen Druffel and Lawrence Mayer. Its members are Lawrence Mayer (University of Maine), Ellen Druffel (University of California, Irvine), Cindy Lee (State University of New York, Stony Brook), Michael Bender (Princeton University), Ed Boyle (Massachusetts Institute of Technology), Richard Jahnke (Skidaway Institute of Oceanography), William Jenkins (Woods Hole Oceanographic Institution), George Luther (University of Delaware), and Willard Moore (University of California). Cindy Lee presented a summary of the FOCUS activity at the symposium.

as photochemistry, vertical particle flux and scavenging, and organic complexation of metals.

• producing trustworthy measurements of concentrations of most of the elements in seawater and realizing that the concentrations generally followed patterns that are predictable based on a combination of elemental chemistry, biological processes, and global water circulation patterns. Previously we had either zero or erroneous values for concentrations of a majority of the elements in the ocean.

• uncovering the relationships between biological processing and chemical diagenesis in sediments, critical to areas such as the coupling of water column and sediment processes, mineral formation, and paleoceanographic interpretations.

• assessing the crucial role that hydrothermal processes play in chemical budgets of the oceans, as well as the role of chemical processes in the very dynamic biological and geological phenomena found in rift zones.

• determining the importance of coastal margins as locations of very high productivity, high carbon flux, and high sedimentary carbon storage as compared to the rest of the open ocean.

FUTURE RESEARCH TRAJECTORIES

We used our review of past progress and our sense of impending issues to guess at the future of the science. Our time horizon forward is on the order of two decades, roughly that used for our backward look. Forecasting progress in a field as broad as ocean chemistry requires that we break it into smaller chunks amenable to handling. Organizing this effort along the lines of our review of progress, however, allows the past to rule the future. To avoid the smaller tribalisms that exist in our (or any) field, it was necessary to keep the participants from breaking into their natural caucuses. We needed a means to shake up the traditional thinking patterns.

We therefore chose to divide oceanic processes (not necessarily chemical processes) along the lines of the time scales within which their characteristic patterns emerge. We focused on processes occurring at (1) seasonal and shorter time scales, (2) seasonal to annual time scales, (3) annual to millennial time scales and (4) greater than millennial time scales. Of course, this choice forces its own structure onto the field, so we also considered possible omissions and overlaps. This time-scale approach reinforces the role of ocean chemistry in the solution of a variety of interdisciplinary oceanic problems.

Synthesizing questions raised using the time-scale approach required identification of major themes that the field of ocean chemistry will address over the next few decades. We attempted to balance a desire to identify exciting problems, apparent at this time, with the need to provide umbrellas likely to contain the unexpected discoveries of coming

decades. The results of our deliberations can be grouped into eight themes.

1. Major and minor plant nutrients—how they are transported to the euphotic zone and affect community structure, and how these processes are influenced by natural and anthropogenic changes. The ocean's ability to support life and the role of life in maintaining the chemical constitution of the ocean are strongly affected by the transport and redistribution of nutrients. Despite exciting progress over many decades, it is clear that unknown processes are controlling the patterns of these mutual controls. Rapid progress will show how subtleties in nutrient dynamics affect end states of great importance, such as fisheries and harmful algal blooms.

2. Land-sea exchange at the ocean margins. Margins influence biogeochemical cycles to an extent much more than their areal extent might imply, while being especially susceptible to anthropogenic influence. Processes that occur disproportionately in margin environments, such as organic matter burial, mineral formation, and denitrification affect the oceanic balances of many elements. Unraveling the highly variable complex of chemical, physical, geological, and biological linkages in margins will provide needed context for human colonization of the coastline.

3. Organic matter assemblies, at molecular to supramolecular scales, their reactivity and interactions with other materials. Organic matter must be characterized at scales including, but also greater than, its molecular constituents, to enable understanding its preservation, transport, and interactions with inorganic materials. The "micro-architecture" with which constituents are assembled controls reactivity with important implications for primary and secondary production, photochemical processes, mineral formation, and trace metal dynamics.

4. Advective chemical transport through the ocean ridge system (ridges and flanks), ocean margin sediments, and coastal aquifers. Fluid flow through these environments appears to have greater importance than previously appreciated, and may strongly influence many oceanic chemical cycles. Greater understanding of the magnitude and variability of these advective transports will improve budgeting of chemicals in the ocean and provide explanations for many regional processes affected by the flow, such as mineral formation and nutrient inputs.

5. Forecasting and characterization of anthropogenic changes in ocean chemistry: consequences at local and global scales. Climatic as well as chemical changes to the oceans will affect many different biogeochemical cycles. Assessing natural variability will be critical to determination of anthropogenic effects. Linkage to other oceanographic variables, such as biological and physical processes, will enable better assessment of the role of the oceans in global environmental change.

6. Air-sea exchange rates of gases that directly influ-

ence global ecosystems. Carbon dioxide and other greenhouse gases, halocarbons that affect stratospheric ozone, and sulfur gases that create sulfate aerosol all have important source and/or sink terms in the oceans. More accurate determination of air-sea fluxes of these gases, of both natural and anthropogenic origin, are critical to assess processes affected by these gases.

7. Relationships among photosynthesis, internal cycling, and material export from the upper water column. Most production and remineralization of organic matter occurs in the shallow euphotic zone. Our understanding of processes such as CO_2 and N_2 sequestration from the atmosphere and pelagic-benthic coupling are thus critically dependent on improving our understanding of euphotic zone recycling.

8. Controls on the accumulation of sedimentary phases and their chemical and isotopic compositions. Further development of paleoenvironmental indicators will enable better understanding of past climatic and carbon cycle variations. Earth historical records provide an invaluable guide to natural variability of the chemistry/climate system, including natural "experiments" in which the whole system has responded to a perturbation.

Synthesizing these eight topics, three major areas appear especially fertile for future discovery. The first is boundary interactions between major reservoirs, including gas exchange between air and sea and advective flows through ridge systems and coastal aquifers, which promise resolution of important mass balances for the surface of the Earth. The second area where we are on the verge of making sizable discoveries is the ocean's ability to support life, its effect on the cycling of elements in the upper ocean, and the forms of organic matter that fuel various life forms. Last, and perhaps most important are the links between environmental changes (e.g., anthropogenically induced impacts) and the chemistry of the ocean—links that have both local and global significance.

RESEARCH INFRASTRUCTURE

Future advances in ocean chemistry will require new approaches to infrastructure to support the science. Emerging technologies, and access to them, will be critical for the next advances. These include methods to sample, analyze, and visualize chemical distributions in the oceans at vastly wider ranges of time and space scales than heretofore possible. As we focus more strongly on variability in the ocean, higher data densities over longer time scales will be required. Sensor technology that can be used at sea is particularly well-poised to enable new insights into the functioning of the oceans. There is need for some tuning in the funding approaches to certain kinds of research, such as new opportunities for mid-size research groups or long-time series.

Shifts in approach to recruitment, training, and career guidance are needed to provide the human resources for growth of this field. Because ocean chemistry is among the most interdisciplinary of marine sciences, greater linkage is recommended to other oceanographic and materials science disciplines. For example, shifts in training from chemical to oceanographic programs must not lead to atrophy of our connections to the chemical sciences. Examples of such connections include more active recruiting of chemistry undergraduates and involvement of ocean chemists in their environmental science program areas.

CONCLUSION

The picture that emerges from this self-assessment is one of a field with a record of impressive recent gains and a prospect of imminent further advance. These advances should benefit not only the field of ocean chemistry narrowly defined, but will be central to a variety of other fields of Earth science.

The Future of Marine Geology and Geophysics: A Summary[1]

Marcia K. McNutt

Monterey Bay Aquarium Research Institute

The Marine Geology and Geophysics (MG&G) and Ocean Drilling programs at the National Science Foundation together spend approximately $29 million on science annually in the United States (not including ship costs). Although this is an impressive figure, increases to the program budget have barely kept up with inflation, the pool of potential investigators has been ever increasing, and other federal agencies that traditionally supported research and technology development in MG&G have had their budgets cut drastically. The net effect is a stressed research community in which innovative science, novel collaborations, and the next generation of technology development are more difficult to justify and support. Given the realities of the U.S. federal budget deficit, the financial situation is unlikely to improve in the near term. In this type of environment, there is the danger that creativity, the lifeblood of science, will become stifled in an overly conservative peer-review system unless there is some community coordination and consensus that allows bold, new ideas to be pursued. Thus the questions asked of the participants at an Ashland Hills workshop held in December 1996, were:

- What are the most promising and exciting directions for future research in marine geology and geophysics?
- What research strategies will best address these problems?

The workshop participants divided into four thematic groups in order to formulate their vision of where their fields should be heading in the next two decades so they could address the most pressing issues. Each subgroup identified a set of major research questions that need to be answered in order to make progress in understanding critical issues in marine geology and geophysics. Some common issues that appeared in several of the reports are:

- The societal imperative of making rapid progress in scientific understanding of complicated, nonlinear systems. Many of the research topics central to marine geology and geophysics address issues of societal concern, such as changing climate, coastal pollution and erosion, and earthquake hazards. In some cases, there has been pressure to implement solutions to these problems without a complete understanding of these complicated systems. Even worse, some of these systems are now demonstrated to be highly nonlinear, such that input at one frequency can produce a response at very different frequencies. Human forcing may in fact lead to very unpredictable and undesirable consequences. An important area of future research will be in characterizing and modeling systems in which the input forcing is known or can be measured, and the system response can be inferred from the geologic record (geologic time scales) or from direct observation (human time scales).

- The central role of focused fluids in producing volcanic, tectonic, and thermal modification of the planet. Geologic modification of Earth is controlled by its fluids, whether it be water in fault zones, magma erupting on a midocean ridge or island arc, plumes rising from the deep Earth, hydrothermal circulation in ocean crust and sediments, or methane deposits on continental margins. These fluids determine the locus of geologic activity and are the agents for geochemical cycling between the solid Earth and the hydrosphere and atmosphere. Quantitative understanding of the physical and chemical processes that lead to concentrations and focusing of these fluids through the lithosphere,

[1]Exerpted from *Future of Marine Geology and Geophysics (FUMAGES)*. 1998. P. Baker and M. McNutt, (eds.), in Proceedings of a Workshop December 5-7, 1996, in Ashland Hills, Oregon, under sponsorship of a grant to the Consortium for Oceanographic Research and Education (CORE) from the National Science Foundation, OCE-MG&G and ODP, 264 pp. Additional copies are available from the Consortium on Oceanographic Research and Education, 1755 Massachusetts Ave., NW, Washington, DC, 20036. A complete list of the contributors is appended to the end of this summary. Marcia McNutt presented a summary of the FUMAGES activity at the symposium.

igneous crust, and sediments until their eventual expulsion into the water column or atmosphere, however, is in its infancy. We need to better understand the physical properties of the medium through which the fluids flow, the stresses acting on the systems, and their chemical, mechanical, and thermal interaction with their host rock.

• The recognition that present-day conditions may not be representative of the whole of geologic history. A glance at the recent past shows a climate system principally forced by the eccentricity of Earth's orbit. Present-day nearshore sedimentary sequences reflect flooding of the world's shelves following the melting of large continental ice sheets, and today's seafloor volcanic activity is completely dominated by steady-state formation of new crust at the midocean ridge. However, with the benefit of the geologic record, we see that just one million years ago variations in Earth's tilt were more important than eccentricity in modulating climate. During glacial maxima, sediments bypassed many continental margins through a series of canyons. In the Cretaceous, plume-type volcanism was far more important than it is today in the mass and energy transfer between the deep Earth and the surface. While in some cases, the causes of the changes in the geologic record are easily identified (e.g., rising sea level), in other cases they are not. More emphasis in the future will be directed toward documenting the various different stable states of Earth's systems, discovering what events trigger evolution from one stable state to another, and identifying the linkages between the states of very different systems (e.g., climate and tectonics).

• The importance of explicit incorporation of effects of and on the biosphere into marine geology and geophysics. Investigators in MG&G are extremely comfortable with introducing a fair amount of physical and chemical sophistication in their science. Many have their primary professional training in these allied physical sciences. The links to biology, in comparison, are weaker and must be shored up to make progress on a number of fronts. Just as ocean chemistry cannot be understood using the principles of chemical equilibrium without taking into account biochemical cycling of nutrients, the solid Earth is modified by biologic activity from the scale of bacteria to humans. Submarine ecosystems harbor some of the most unusual and extreme examples of life on Earth, and the implications of understanding how these systems have adapted to and how they modify their environments have implications for the origin of life itself.

• The appreciation that we must move beyond steady-state models to study geologic events as they happen. The geologic record contains evidence of many catastrophic events: earthquakes, landslides, volcanic eruptions, etc. Most of our models, however, smooth these events over time to create steady-state representations for what are really discontinuous processes such as erosion of headlands, glacial meltwater pulses, creation of oceanic crust, and filling of flexural moats. Such steady-state models distort the true impact of these events on human timescales and are useless for any hazard mitigation. Given the current lack of understanding of the temporal and spatial pattern of most geologic events, we require the technology to install undersea observatories and event-detection systems to catch geologic events in action.

• The limitations of present funding structures and technology for problems that span the shoreline. From the standpoint of many problems in geology and geophysics, the division between the Ocean Science and Earth Science divisions at NSF is somewhat artificial. Although most of the midocean ridge system is under water, sometimes it is easiest to map it where it lies above sea level (e.g., Iceland). Fluids vented along coastal margins may originate from terrestrial aquifers. Variations in sea level shift the shoreline position laterally for distances of kilometers over timescales of millennia. Ice core data from subaerial drilling can complement deep sea cores. Most efficient use of future resources will require close collaborations between land and marine geoscientists and their corresponding program officers. Even more of an impediment to working across the shoreline is lack of equipment to work near the shoreline, in shallow-water, high-energy environments. No amount of community interest in geologic processes at the oceanic margins will lead to progress unless improved technology is available for imaging, sampling, and monitoring the nearshore region.

Overall, the thematic reports, briefly summarized below, give the impression of anything but "business as usual." The community is enthusiastic about the opportunities to build new collaborations and apply new technology and expertise to find answers to the most intellectually challenging problems in marine geology and geophysics.

GROUP #1: THE SOLID EARTH

The solid Earth is continually in movement, and this movement reflects the processes of energy and mass exchange between the Earth's interior and exterior reservoirs. The current manifestations of these movements are represented by the diverse plate tectonic settings of the Earth, many of which are depicted schematically in Figure 1. This snapshot of the plate tectonic physical and geochemical circulation can also be considered conceptually as a cycle, as shown in Figure 2. The cycle begins with the formation of a new rift, followed by the opening of an ocean basin as new oceanic plate is created by spreading at ocean ridges. The aging oceanic plate is acted on by a variety of mid-plate processes such as hotspots (including seamount and island volcanism), sedimentation, subsidence, and deformation. As the plate approaches a convergent margin, it enters the "subduction factory," leading to the generation of earthquakes, release of fluids, varieties of volcanism, back-arc spreading, and ultimate recycling of residual peridotite, basalt, and sedi-

THE SOLID EARTH CYCLE

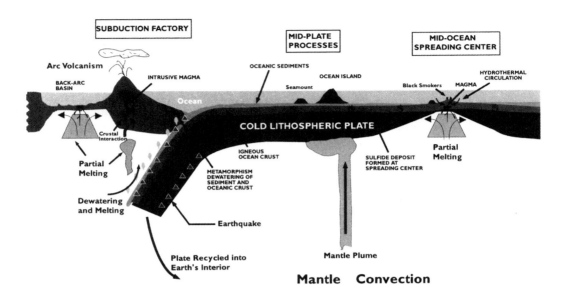

FIGURE 1   Cartoon of the solid Earth plate tectonic cycle, divided into three major tectonic components: plate creation at mid-ocean spreading centers, modification of the plate as it traverses the mantle, and subduction of the plate and creation of new crust at convergent margins. The initiation of an ocean basin through continental rifting is not shown in the figure. SOURCE: FUMAGES (1998), p. 8.

ment into the mantle. The subduction factory is the current mode of continental growth and modification, and over time may have been the principal process leading to the creation of the continents. Ultimately the continental crust again will rift, leading to the formation of a new ocean basin, and the cycle repeats.

Each aspect of the plate tectonic cycle has a host of scientific questions that remain to be answered. These questions naturally divide according to the diverse provinces of the cycle—extending from creation of crust at ridges, to the mid-plate region, to the convergent margin. Figures 1 and 2 present these domains and some of the processes that take place in them, and the specific sections of the FUMAGES report discuss some of the outstanding questions that provide a basis for fruitful new directions for research.

There is an important additional aspect of the evolution of the solid Earth, however, that is not fully represented by Figure 1. Figure 1 is as an instantaneous view of the overall process; Figure 2 conveys the notion that not only is this view one of a continuing and repeating cycle, but also that this cycle may lead to and be influenced by the long-term geochemical and tectonic evolution of the solid Earth system. The Earth's current state is not necessarily typical of all tectonic regimes in the past. One obvious aspect of these changes is reflected in the very different apparent state of the

ocean floor during the Cretaceous, when large igneous plateaus were present over much of the seafloor. There may have been major changes that have yet to be discovered, such as, perhaps, changes in the mode of mantle convection. Therefore, the study of the evolution of the solid Earth cycle through time—in all settings—emerges as one of the clear frontiers of the science over the next decade. Old ocean floor contains one of the best records of this history.

The solid Earth cycle and its evolution through time are driven by fundamental processes that result in mass transport across the boundaries between the asthenosphere, lithosphere, hydrosphere, and atmosphere. These fundamental processes include the generation and segregation of magma, brittle and ductile lithospheric deformation, the scales and patterns of asthenospheric flow, and the influence of fluid flow on rheology and chemical exchange between the solid and fluid Earth. In addition to its intrinsic scientific interest, investigation of these processes can eventually lead to an understanding of the causes of great earthquakes or volcanic explosions, of the generation and concentration of mineral resources, of cataclysmic events in Earth's history that have modified Earth's climate, and even of the origins of life itself.

An aim in investigations of these processes is the development of quantitative, unifying principles that govern the

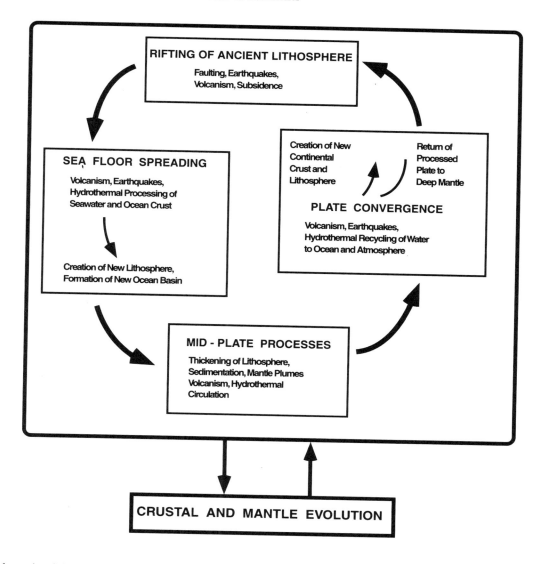

FIGURE 2   Schematic of the solid Earth cycle, beginning with rifting of ancient lithosphere, through creation of a new ocean basin by seafloor spreading, plate modification during passage over the mantle, subduction and continental addition, and continental collision, followed ultimately by another episode of continental rifting. This process through time leads to evolution and differentiation of the mantle, and mantle processes in turn influence all components of the cycle. SOURCE: FUMAGES (1998), p. 9.

formation and destruction of the crust and mantle lithosphere, and its interactions with the hydrosphere and biosphere. Recent advances in geophysical and geochemical observation techniques, combined with the computational capability to evaluate the effects of nonlinear, open systems, have led to the development of predictive models of the mid-ocean ridge system based on simple, geodynamic parameters. Based on these advances, it seems likely that a general theory of mantle differentiation and lithospheric genesis will emerge in the next ten to twenty years. To produce this theory, we will have to concentrate research on the upper and lower boundaries of the lithosphere—the regions of interaction with the underlying convective mantle and the over-

lying hydrosphere and biosphere. Some of these boundaries are difficult to observe, and progress will require development of new techniques to image, for example, the base of the lithosphere in considerable detail.

Study of the seafloor provides one of the primary windows into a multitude of Earth processes, and the linkages between the various parts of the whole Earth system make the new observational data and quantitative models pertinent to a broad spectrum of Earth problems. Therefore, many of these developments will be multi-disciplinary, involving scientists from outside the oceanographic and solid Earth communities.

GROUP #2: PALEOCEANOGRAPHY

Significant advances in the field of paleoceanography have both sharpened the focus of paleoclimate and ocean history research on classic problems and initiated new research directions. The classic problems driving our long-range research program include: (1) relationships between sea level, ice volume, and climate change; (2) interactions between atmospheric carbon dioxide, climate, and its biospheric and geospheric regulation; (3) long-term changes in ocean chemical composition and geochemical fluxes as related to geological and biological evolution; (4) solar and magnetic field variability and their role in climate change and affect on cosmogenic nuclides; and (5) changes in modes of ocean circulation in relation to climate change and the evolution of oceanic basins.

Work proceeding along these paths has led to two important shifts in how we view climate that seem to cut across all time scales: (1) the growing realization that substantial changes in atmospheric CO_2 are likely to have played a large role in both long-term and short-term climate change, and (2) the discovery that nonlinear interactions in the ocean-climate system may have played a key role in determining the sensitivity of climate to both internal and external forcing. Furthermore, these nonlinear interactions can shift climatic variance to both higher and lower frequency oscillations.

In recent years, work on these classical themes has uncovered "bombshells" that have rattled prevailing views: (a) tropical sea-surface temperatures may have been 5°C cooler during glacial maxima, in contrast to CLIMAP reconstructions with stable tropical SST; (b) evidence for cool tropics and low equator-pole thermal gradients also is found for the late Cretaceous and Eocene, again countering prevailing beliefs; (c) ice core isotope paleothermometers appear to have understated glacial cooling at high latitudes by a factor of two; (d) transitions between glacial and interglacial states can occur in only a few decades, and (e) in the late Paleocene (~56 million years ago), there was a sudden input of isotopically light carbon into the ocean-atmosphere system accompanied by global warming lasting for no more than a few thousand years. These findings have invigorated ocean-climate investigations and forced us to reexamine many traditional assumptions.

In the future, we must emphasize the search for a better understanding of processes that affect climate change and cause variability in the climate-ocean system. Through this effort we hope to gain a better understanding of the coupling of the ocean-climate system through the entire range of the Earth's climatic spectrum (Figure 3).

Examples of specific science questions that will drive research in paleoclimates include:

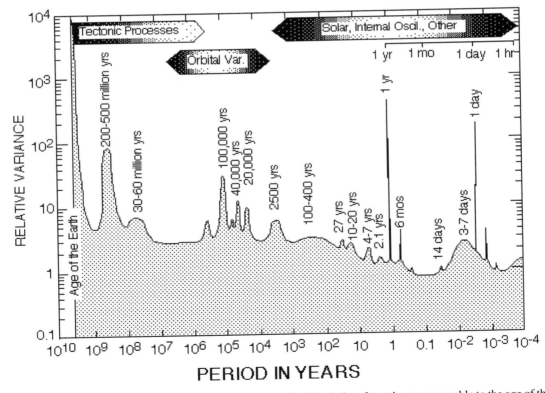

FIGURE 3   Estimate of relative variance of climate over all wavelengths of variation, from those comparable to the age of the Earth to about one hour. Shaded area represents total variance on all spatial scales of variation. Strictly periodic components of variation are represented by spikes of arbitrary width. Modified from Mitchell, Jr., J.M. 1976. An overview of climate variability and its causal mechanisms. *Quaternary Research* 6:481-493, with permission from Academic Press, Inc.

1. What is the cause, nature, and range of climate and ocean variability at the inter-annual to millennial time scale given that there is no obvious external forcing?

2. What processes set the sensitivity of the climate system to external (orbital) forcing, and what processes are responsible for the long-term evolution of this sensitivity?

3. Why are there ice ages in Earth history?

In order to address these questions, integrated model, laboratory, and data studies are essential. The modeling of processes at all levels of complexity—from simple box models to more complex models of mass balance exchange in the ocean-climate system—can provide useful insights into the nature of these processes. Opportunities for model development, including climate, ocean, and biogeochemical models, and accessibility of various types of models for research must be maintained.

Collaboration between researchers across all subdisciplines that study Earth system history must be encouraged and developed. Advances in the field via data-model integration may arise from collaboration of subdisciplines that are not traditionally combined. A critical element of such collaboration will be to have an interdisciplinary peer-review process for interdisciplinary proposals.

GROUP #3: SHELF AND SHOREFACE SEDIMENTS

The coastal-shelf system of the oceans is a critical environmental interface—a fundamental Earth discontinuity—where terrestrial, marine, and atmospheric processes converge and mutually influence one another across a spectrum of spatial and temporal scales. Society relies on the coastal system for its rich biological diversity, extensive mineral resources, and its fulfilling scenic and recreational opportunities. This system satisfies needs for waste disposal, transportation, and a climate moderated by the heat engine of the oceans. It is these attributes that have led to a massive increase in population along the world's shoreline, a pattern that has stressed available resources and exposed development to marine hazards.

Media reports of storm damage, sea-level rise, coastal erosion, and declining nearshore water quality sound a clarion call from the American constituency for the development of a scientific focus on the nation's shelf and shoreface system. As we dam rivers, armor coastlines, disperse pollutants, and mine the shoreface we are forever altering the flux and partitioning of sediments through a sensitively linked series of littoral and marine ecosystems. Human alteration of the coastal system, in fact, constitutes a series of large-scale experiments that are disturbing the natural variability of the environment. Unfortunately, we take these actions without a full understanding of the fundamental processes that provide for the natural health and viability of the afflicted system.

- How do human actions impact natural variability?
- What are the fundamental processes that unify the multiple temporal and spatial scales constituting the dynamic behavior of the shelf and shoreface?
- Are there overarching physical/biochemical processes governing natural variability in the spectrum from microseconds to millennia?

These fundamental and fascinating questions can only be answered with multi-disciplinary and multi-scale investigations of sedimentary dynamics, and resulting environmental and stratigraphic imprints, across the land-sea interface of the continental and insular margin.

Investigations of sediments at the ocean margin range widely in both the time scales of the processes considered and in the spatial scales of the resulting morphologies or stratigraphic record. One investigator might obtain measurements of orbital velocities under waves in the nearshore, and relate those to the resulting transport rates of sediments or to the dimensions of ripple marks formed on the bed. Another investigator could be considering the processes of tides on the mid-shelf and the formation of huge sand waves. Longer time scales and larger spatial considerations apply to the investigator who relates the cycles of sea-level change to the resulting stratigraphy or architecture of deposits that span the entire ocean margin, crossing the shoreline and extending onto the coastal plain.

This breadth of consideration is illustrated by the accompanying diagram (Figure 4) that graphs the time scales of processes (dynamics) versus the scales of the sedimentary features (morphology). In the dynamics domain, the shortest time scale is represented by the rapidly-fluctuating turbulent eddies within currents that are important to the entrainment and transport of sediments. Beyond that are the variations due to wind-generated waves that generally range between 5 and 20 seconds, and the hourly changes in water levels due to tides and the associated currents they generate. Also important are the occurrences of storms, where "normal" storms generally occur a few times each year at a specific coastal site, while a "major" storm such as a hurricane may occur only once in a decade or longer. Such storms have profound effects on the sediments of the nearshore, and even on the seabed sediments across the entire shelf.

Even longer time-scale processes shown in the accompanying diagram are represented by sea-level variations. Tide gauges along our coasts provide a record of relative sea-level change during roughly the past 100 years, the change in global sea level "relative" to the land. Sea-level change can also include punctuated, millennial-scale sea-level events due to shifts in global ice volume (these may influence shelf sediment exchange and seafloor morphology during periods of rapid global change), and transgression/regression cycles that have occurred with glacial-interglacial changes in Earth's climate (Milankovitch cycles). Of critical interest is the knowledge gained from investigations

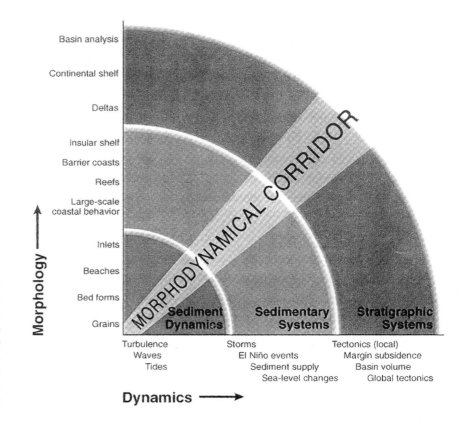

FIGURE 4   The dynamics and morphology of sedimentary environments prescribe specific regions of the *morphodynamical corridor* that tend to define the extent of our research efforts. Workshop members believe that better communication among and between regions along the corridor will enhance the progress of marine sedimentary research. SOURCE: FUMAGES (1998), p. 52.

of the Holocene and last interglacial episodes of transgression. These are specifically important since they have been profoundly influential in governing the present configuration of our coastal plains, coasts, and shelves. The longer-term processes included in the graph are the tectonics of crustal movement at the coastal interface, processes such as continental margin subsidence, changes in basin configuration, and global tectonics that govern the degree of continental freeboard and the fundamental timing of shelf evolution.

The second axis of Figure 4 shows the spatial scales of sedimentary features involved in research investigations. This list is only suggestive of the range of scales and is not an exhaustive account of the many sedimentary bodies found at the ocean margin. At the smallest scale are the sediment particles, obviously important in studies of sediment transport, but also important in the record of grain-size distributions of particles within the resulting deposits that reflect the transport processes. Accumulations of sediment grains form sand ripples or the bars that are an important part of the overall beach morphology, or the large-scale sand waves found in some shelf environments. These morphological features combine to form the entirety of deltas, estuaries, barrier islands, and the present-day shelf. Recorded within the sediments of the margins are ancient shelves, stranded sand bod-

ies, fossil reef tracts, and a stratigraphic record of former changes in sea level.

To a degree, individual research efforts can be placed within the temporal-spatial scale of the accompanying graph. These tend to congregate along the 45° zone shown. For instance, investigations of sediment-transport processes focus on the time scales of waves and currents and down to the scale of turbulent fluctuations, while considering the movement of sediment grains and the effects of sand ripples on that transport. Other investigators document the response of beach morphology to the occurrence of storms, or the formation and migration of sand waves on the continental shelf where sediment transport is due to tidal currents. Yet another group of investigators is focusing on the effects of sea-level change, with the impacts ranging from the present-day changes in coastlines and estuaries to the long-term record within the stratigraphy of the continental margin. The unique ratio of morphologic and dynamic integration that constitutes research on the shelf and shoreface falls within this "morphodynamical corridor."

There is a notable lack of continuity and overlap among separate groups that study discrete morphodynamical ranges. A challenge to our science is to improve the linkage among these research subdisciplines. We must learn to talk to one another more often, and more effectively. Any one investigator tends to be limited to a small range of time scales of processes and resulting spatial scales of sedimentary features. The investigator is familiar with the next lower time-space scale of research, since his/her investigation likely uses the tools of that research (i.e., sediment-transport equations) or relies on its conclusions; the investigator is likely also aware of the implications of his or her research to the next higher time-space scales. While it is seldom that an individual can meaningfully cover an appreciable area of the time-space graph, it is through collaborative research efforts and modeling that connections can be made that lead to a more comprehensive understanding of the processes that are presently important, and were important in the past, to the sediments and sedimentary record of the margin system. A deeper understanding in the future will therefore depend on increased support for such collaborative research efforts.

As we enter the twenty-first century, seventy percent of the world's population will live and work alongside the sedimentary province of the ocean margins. The deposits of the coastal plain, shoreface and shelf, slope, and rise consist of particles and pore fluids arranged in a complex architecture of bedforms, layers, wedges, aprons, and lenses that define the anatomy of the province.

For over a century, human activity has perturbed the coastal sedimentary province by the filling of wetlands and the reclamation of estuaries, through the construction of sea-walls and breakwaters, and by starvation of the sediment supply as a consequence of damming rivers for hydroelectricity, irrigation, and flood control. Many of the environments of this province are "diseased" from thoughtless use, over-exploitation, and even well-intentioned but ignorant attempts at mitigating the problems that come with human influences. Millions of dollars are spent annually to pump sand back to the shore to replenish beaches, only to have these grains disappear into the sea again following northeasters, hurricanes, or typhoons. Much of this province has been alternately submerged and exposed in the past. Some of it will experience renewed flooding if global warming predictions are correct. Some sectors, such as the extremely populated Nile Delta and portions of the Mississippi Delta, are currently in a state of crisis in the wake of severe coastal erosion. The megalopolis of Bangkok is sinking at an alarming rate of one meter in a human lifetime due to a combination of diminished sediment supply and aquifer depletion.

True stewardship of the sedimentary resource will require an improved understanding of its anatomy. The piecemeal, ad hoc examination typical of past research will not provide a sufficiently integrated understanding of the character of the system. A new approach must incorporate the next generation of Earth and environmental scientists trained with greater engineering skills and an integrated knowledge of the physical/chemical/biological metabolism of the sedimentary environment. This is necessary preparation for the challenge of quantitative modeling and measurement that lies ahead. A new, *integrative* science must be the hallmark of

future sedimentological research. In order to achieve new goals of understanding the ocean margin, the science of sedimentology must evolve into a systems approach that integrates theories and concepts of biochemistry, geophysics, meteorology, climatology, population statistics, ecology, and hydrodynamics.

GROUP #4: FLUIDS IN THE OCEANIC LITHOSPHERE AND MARGINS

The importance of water to essentially all aspects of Earth science probably cannot be overstated. Water is a ubiquitous agent of geological creation and transformation and of life. The charge of Working Group 4 was to construct a vision for the next 10+ years of marine geological and geophysical research involving fluids, fluid processes, and fluid products. This vision is to include an assessment of the most important unresolved questions, the means by which these questions should addressed, and the infrastructure and facilities that will be required to do so. The completion of our assignment was complicated by the nature of fluids themselves: they exist at an astonishing array of temperatures, pressures, chemistries, and physical properties, within many different geological systems.

While many questions associated with lithospheric fluids may be addressed successfully through the use of steady-state assumptions, it is becoming increasingly clear that fluid flow and associated processes and properties are inherently transient and interdependent. We suggest that issues associated with fluid flow and resulting reactions, chemistry, biology, and physical properties can be considered in the context of conservation of mass and conservation of energy. The simplest example of this concept is illustrated through examination of the one-dimensional, steady-state diffusion equation (used for chemical, thermal, electrical, and fluid transport): $q = -D\, dP/dl$. This equation states that the flux (q, mass or energy) is a function of the driving force (dP/dl, a gradient in potential) and the properties of the system that govern transmission (D). If one knows any two of the above terms, the third can be calculated. Ideally, all three would be determined independently so that internal consistency can be established.

Other modes of transport can be described in such an equation through inclusion of additional terms (advective, dispersive, reactive, decay, etc.). When such a construction is applied to a volume of the oceanic lithosphere or margin, multidimensional and transient processes can be considered, including the importance of storage terms (for both energy and mass) and tensor properties. With this framework in mind, key questions can be considered:

1. What are the mechanisms influencing hydrothermal fluxes associated with dike injection, transient magma chamber output, and penetration of a cracking front (extent of water-rock reactions, creation and modification of fluid pathways).

2.   What is the flux-frequency distribution for ridge-crest hydrothermal activity (heat, fluid, chemistry)?

3.   To what depth and to what crustal age does significant ridge-flank hydrothermal circulation extend, and what is the influence of this flow on crustal evolution and ocean chemistry?

4.   What are the roles of fluids in the earthquake cycle?

5.   What role does fluid flow play in gas hydrate accumulation and how important are hydrates to climate change, slope stability, and energy resources?

6.   What are the extent and consequences of interactions between terrestrial ground water and marine systems (fluxes, diagenesis, biology, slope stability, canyon formation)?

7.   How much of a role does the microbial community play in subsurface chemical and physical transformations?

These, and a host of other exciting questions, remain mostly conjectural as there have been few quantitative microbiological studies specifically addressing these geological and geochemical problems. We therefore recommend that a small series of highly-focused studies be carefully prepared and executed, within several distinct seafloor environments, as soon as possible. The initial experiments should provide first-order information that will allow an assessment as to whether a significant initiative, and concentration of resources, is justified for more complete, long-term exploration of subseafloor microbial communities.

COMMON THEMES IN INFRASTRUCTURE

The main goal of the workshop meeting was to identify the important scientific questions that would be driving research in marine geology and geophysics in the next 10 to 20 years. However, it was not possible to discuss the science that we wish to accomplish without mentioning the new equipment or changes in the funding infrastructure that would either facilitate or enable researchers to address these questions. Some of these needs are pointed out directly in the thematic group reports, but others are so overarching that it made more sense to call them out in this separate section.

Common Use Equipment

Marine geology and geophysics is an observationally based science and will continue to rely on ocean-going observational capabilities. A significant amount of the technology used in MG&G studies is needed by a wide spectrum of the community (e.g., high resolution seismics for paleoceanographic, geohazard, and sediment processes studies; equipment to sample fluids and sediments; moored arrays for long-term observations for fluid and ridge crest processes; geophysical imaging equipment for studies of lithospheric and mantle dynamics, etc.).

The present funding model for much of the "common use equipment" has resulted in a gradual degradation of many MG&G capabilities. It is not unusual for equipment systems to be maintained as part of a specific ship operation. In some cases, systems are supported as independent cost centers. In either case, a use hiatus results in system degradation, loss of technician expertise, and ultimately a complete loss of the system capability. This problem affects "standard" shipboard equipment as well as portable equipment.

We recommend a community-wide effort to come to a consensus about (1) what instrumentation is broadly needed; (2) how it should be maintained and managed in a way that is appropriate for each facility; (3) how to build in appropriate funding feedbacks so that outdated or poorly managed equipment pools are discontinued and new instrumentation can be added; and (4) what funding structure best supports this equipment. While there will be some short-term costs associated with development of a reliable instrument pool that will be available to the community, in the long run, a cost-effective solution to the problem of deteriorating and unreliable facilities is essential to the health of marine geosciences.

The list of equipment that might be candidates for a shared pool is large: MCS equipment, ocean bottom seismometers, new-generation magnetometers and gravimeters, coring and sampling devices, autonomous underwater vehicles, tethered vehicles, submersibles, etc. We discuss below just a few of these capabilities that we believe are good candidates for placing in a common-use facilities pool.

Navigation

There is a general need for a community-supported facility to (1) position ships, water column instruments, and seafloor instruments in a relative reference frame with a precision of several meters; (2) position ships and instruments in an absolute reference frame with a similar accuracy; and (3) have this information available on the bridge and within scientific facilities in real time. This capability has significant implications for work across the full range of geological environments and scientific problems. Traditionally, precise navigation in a relative frame has been accomplished with a combination of seafloor transponders, shipboard and relay transponders, and shipboard computer hardware and software (long-baseline navigation), all of which must be provided and integrated with the shipboard environment by a project investigator, team member, or consultant. Differential or P-code GPS similarly has not been routinely available, although this situation is changing aboard the larger UNOLS ships. Experience has proven that many scientific programs are difficult or impossible to complete without precise navigational capabilities. In addition, the captains and mates of ships of all sizes are much better able to locate and hold target positions (for both ships and instruments) when they get direct graphical feedback of relative and absolute locations. This capability should be made broadly availability within the MG&G community, and support capabilities (including personnel) should be established for the installation and use of this equipment.

Geophysical Arrays

To address a wide variety of problems in marine geosciences large, portable geophysical arrays of ocean bottom seismometers, magnetometers, and electrometers will be required. Short period geophone and hydrophone instruments will be required in large numbers (e.g., 500) for both active and passive tomography experiments, and for monitoring of microseismicity in tectonic settings ranging from mid-ocean ridges to mid-plate hotspots to the accretionary wedge above a subducting slab. These instruments need to be relatively inexpensive to build and operate, and should be capable of deployment times of up to a year. A smaller number (ca. 50) of portable broadband seismometers will be required for long-term teleseismic tomography studies. All of these instruments must be openly available to investigators throughout the geosciences community. Related to this is a need for shallow-water acoustic mapping instruments of high precision in order to conduct change detection experiments for capturing and quantifying dynamic seabed processes that imprint the sedimentary record. The shallow marine community could also benefit from shared pools of pressure sensors, current meters of various types, optical back-scattering devices, high resolution down-looking *in situ* seabed mapping tools, and the data logging and power units necessary to support such arrays.

Unmanned Vehicles

Tethered and untethered unmanned underwater vehicles have already demonstrated their value in geophysical surveys. They extend the capabilities of conventional surface ships by expanding the area that can be monitored in both time and space and by providing close-up views of events and structures on the seafloor. Advances in technology and design now promise vehicles that are lighter, cheaper, and consume less power. In the future, cost-effective and realistic strategies for underwater event detection and temporal monitoring of systems will likely take advantage of autonomous underwater vehicles (AUVs) and remotely-operated vehicles (ROVs). It is not too soon for the scientific community to begin thinking about how to make these capabilities broadly available and how to manage such a facility. Emphasis should be placed on building some standard "bus" design that can be equipped with mission-specific sensors, without discouraging design improvements in this rapidly evolving field of ocean engineering.

Ocean Drilling Facilities

All thematic groups identified some form of ocean drilling (ODP-like) capability as a long-term requirement of their

sampling and sometimes their monitoring strategies. Sampling and monitoring down-hole conditions in 100-1,000 m sections of zero-age basalts is a priority for both the solid Earth and fluids groups. The solid Earth group also needs sampling capability into older oceanic crust that may eventually require riser capability for deeper sampling. Subduction zone problems require standard ODP capability for flux balance experiments, and riser capability for investigations of the seismogenic aspects of subduction zone systems. The sediments group requires standard ODP capability in a wide variety of sedimentary environments, and in addition, a shallow-water jack-up rig capability. Deep sampling of the thick sedimentary and volcanic sequences of passive margins probably will require riser drilling capability. Paleoceanographers require hydraulic piston coring and good recovery capabilities in a wide variety of lithologies. In particular, they require improved recovery capabilities in difficult sequences such as cherts/chalks and coral.

Archives

The productivity of the entire MG&G community has been greatly enhanced since the introduction of archiving facilities for underway geophysical data and ODP cores. In contrast, there is no uniform archiving procedure for rock samples. These samples retrieved from dredging and submersible operations are a critical long-term resource with which to explore new ideas using ever more sophisticated analytical techniques. Although NSF requires samples to be made available by principal investigators after two years, there is no formal mechanism to implement this requirement, nor any clearly defined long-term repository available and accessible to the entire MG&G community. Repositories at several institutions are beginning to serve this need, but a long-term financial commitment to a sample archive would be beneficial. In addition, we need to encourage investigators to place carefully documented and packed samples into this archive.

Education and Public Relations

Public awareness and support for science, which has always been highly desirable, has become essential in the current national fiscal climate. Furthermore, the science education of the American population is an important part of NSF's federal mandate. We recommend that the Division of Ocean Sciences take a more active role in communicating the excitement of cutting edge scientific discovery to the public. It should be recognized that to be effective a sustained, focussed, long-term effort to develop the needed expertise and experience within NSF and within the science community will be required. Various models for this education and outreach activity should be examined, but one possibility is to work through a publicity office at JOI or CORE. We believe that the costs of such an effort, if it were effective, would be highly worthwhile for the long-term public support of our science.

Some specific educational outreach activities that could be highly effective include:

1. The creation and maintenance of Web-site-based learning sites shared among institutions or agencies. NSF could support the development of teaching modules (electronic workbook) of current oceanographic concepts (e.g., El Niño) for primary and secondary school teachers that would include concise description of the scientific concepts, "classic" data that can be downloaded and manipulated by students, and a list of related projects and reading.

2. A program to support lecture to students in primary and secondary schools by graduate students.

3. Participation of high school teachers in seagoing cruises. Science teachers represent a large reservoir of well-educated, scientifically aware communicators who uniformly welcome the exposure to scientific experiences. They enthusiastically transmit their experiences to their classes. They also interact with the parents of their students, their schools, their districts, and the general public. As a result, including teachers in field studies provides a major multiplier of the research experience. CORE could serve as a center for publicizing this program, and as clearing house for teachers wishing to join a cruise.

WORKSHOP PARTICIPANT LIST

I. Solid Earth

Continental Margins

Dixon, Jacqueline (Univ. of Miami)
Goldstein, Steven (Lamont-Doherty Earth Observatory [LDEO])
Plank, Terry (Univ. of Kansas)
Taylor, Brian (Univ. of Hawaii [UH])

Mid-Plate

Farley, Ken (California Institute of Technology)
Hauri, Erik (Carnegie Institution of Washington)
Larson, Roger (Univ. of Rhode Island)
McNutt, Marcia (Massachusetts Institute of Technology [MIT])

Rifting

Detrick, Bob (Woods Hole Oceanographic Institution [WHOI])
Kelemen, Peter (WHOI)
Langmuir, Charlie (LDEO)

Lonsdale, Peter (Scripps Institution of Oceanography [SIO])
Morgan, JP (SIO)
Orcutt, John (SIO)
Parmentier, Marc (Brown Univ.)
Sawyer, Dale (Rice Univ.)
Sinton, John (UH)
Solomon, Sean (Carnegie Institution of Washington)
Tréhu, Anne (Oregon State University [OSU])

II. Climate

Arthur, Michael (Pennsylvania State Univ.)
Boyle, Edward (MIT)
Curry, William (WHOI)
Delaney, Peggy (University of California, Santa Cruz [UCSC])
Moore, Ted (Univ. of Michigan)
Raymo, Maureen (MIT)
Pisias, Nick (OSU)

III. Sedimentary Processes

Beach, Reg (OSU)
Cacchione, David (U.S. Geological Survey)
Fletcher, Chip (UH)
Ginsburg, Robert (Univ. of Miami)
Hine, Albert (Univ. of San Francisco)
Holman, Robert (OSU)
Kineke, Gail (Univ. of South Carolina)
Komar, Paul (OSU)
Mayer, Larry (Univ. of New Brunswick, Canada)
McCave, Ian (Cambridge Univ.)
Milliman, John (Virginia Institute of Marine Sciences)
Pilkey, Orrin (Duke Univ.)
Ryan, Bill (LDEO)

IV. Fluids/Deformation

Baker, Paul (Duke Univ.)
Davis, Earl (Pacific Geosciences Center, Canada)
Delaney, John (Univ. of Washington [UW])
Deming, Jody (UW)
Fisher, Andy (UCSC)
Fisk, Martin (OSU)
Kastner, Miriam (SIO)
Lilley, Marvin (UW)
Moore, Casey (UCSC)
Shipley, Tom (Univ. of Texas)
Urabe, Tetsuro (Geological Survey of Japan)
Von Damm, Karen (Univ. of New Hampshire)

Additional Participants

Claflin, Lynne (Consortium for Oceanographic Research and Education)
Dauphin, J. Paul (National Science Foundation [NSF])
Elthon, Donald (NSF)
Epp, David (NSF)
Fox, Paul J. (Texas A&M Univ.)

Haq, Bilal (NSF)
Kinder, Tom (Office of Naval Research)
Malfait, Bruce (NSF)
Morell, Virginia (*Science* Magazine)
Mottl, Mike (UH)
Purdy, Michael (NSF)
Sancetta, Connie (NSF)
Shor, Alexander (NSF)

Out Far and in Deep:
Shifting Perspectives in Ocean Ecology

Peter A. Jumars

Darling Marine Center, University of Maine

ABSTRACT

The pace of scientific advance in ocean ecology since the Ocean Ecology: Understanding and Vision for Research (OEUVRE) workshop is impressive. New food-web models reveal the stabilizing influence of weakly interacting species. Hence a reason for population instability becomes the absence or disappearance of these stabilizers. Time-series analysis of the planktonic community and nutrients in the Central North Pacific gyre similarly have led to a much clearer focus on controls of patterns and rates of change in ecosystem structure. This focus on change lends itself also to the extraction of anthropogenic effects through potentially powerful statistical methods such as intervention analysis—if timing of onset of an anthropogenic perturbation is known. Realizing this power for past events, however, requires extending time series backward through paleontological evidence. Advances in understanding of form and function in the organisms that produce microfossils would greatly accelerate such progress and can be expected to follow from impressive recent gains in understanding of how body sizes and shapes interact with bacterial chemotactic capabilities and how copepods distinguish prey from predators hydromechanically. Ocean ecology approaches are accelerating toward Gordon Riley's goal for biological oceanography of having parsimonious equations effectively describing in predictive fashion the interactions of populations of organisms with their abiotic environments as well as with each other.

> The people along the sand
> All turn and look one way . . .
> They cannot look out far.
> They cannot look in deep . . .
> —Robert Frost
> "Neither out Far nor in Deep"

INTRODUCTION

Oceanographers pride themselves in taking a perspective different from the one embodied in Robert Frost's classic poem about human proclivity to stare out to sea but not to see. Scientists of all types pride themselves in being able to adopt multiple, tentative perspectives simultaneously in order to design measurements and experiments that can help to determine which perspectives reflect more accurately the rules that nature follows. It is difficult therefore to summarize perspectives, even within a subdiscipline of oceanography. Approximately 40 scientists met in February 1998, however, to make just such a summary in the form of past successes and future directions of biological oceanography, as supported by the National Science Foundation. The group chose the acronym "OEUVRE" for "Ocean Ecology: Understanding and Vision for Research" to describe its effort. Its report is available at http://www.joss.ucar.edu/joss_psg/project/oce_workshop/, and its separate publication is planned, so I do not repeat it here.

Several months later, I continue to be surprised by the outcome of the meeting. The first surprise was the pleasant one of seeing how much progress has been made in 30 years. Barber and Hilting in a companion paper in this volume have covered some of the major "Achievements in Biological Oceanography," so I will deal primarily with "the vision thing." I cannot claim to have originated the ideas because I was present both for discussions of future research by all 40 of the participants and for the task of writing them up. The vision necessarily is clouded, however, by the second and far less pleasant surprise of the meeting, that anthropogenic effects on marine ecosystems are ubiquitous and probably have been since the advent of commercial whaling. The third surprise on which I focus here is how quickly perspectives in all areas of ocean ecology have been changing since the meeting—and in the direction of making long-difficult problems suddenly more tractable.

For OEUVRE we chose to summarize research opportunities in subject matter categories that are cross-cutting and untraditional. I use the same headings here to facilitate the reader's testing of my assertion of rapid progress against the OEUVRE report. Immediately after the headings, I reproduce the questions that OEUVRE identified under these headings as being both pressing and well poised for progress in the next two decades, and I spend the first few paragraphs in each section explaining the topic heading. I devote most space, however, to developing one or two examples of striking progress since the February 1998 meeting. I make no pretense of balanced coverage of the topic area or questions; scientific progress rarely is even across all fronts. Further, each participant or other ocean ecologist would be likely to choose somewhat different examples. My examples take highly variable space and referencing to develop, depending on the background provided in the OEUVRE report.

The exercise reveals several symptoms of substantial shift in perspectives. Unsuccessful search for a superstable marine ecosystem has ended. Along with this failure comes a new focus on how and why marine ecosystems change over time—and on which changes may contain an anthropogenic component. Coherent, succinct models are emerging of sensory systems and behaviors at spatial scales and Reynolds numbers for which humans have no native intuition at the same time that high-technology sensor systems are being deployed that allow unprecedented spatial and temporal resolution in human exploration of the sea. For many reasons, some of the most revealing exploration now is in time rather than space. Indeed this essay focuses more on how perspectives are changing than on summarizing old or new perspectives and is clearly derivative in that sense as well.

FUNCTIONAL SIGNIFICANCE OF BIOLOGICAL DIVERSITY IN BIOGEOCHEMICAL DYNAMICS

- How do environmental and biotic factors determine the distributions and activities of key species or functional groups important to biogeochemical cycles in space and time?
- What are the important interactions among marine biota, global climate, and biogeochemistry?

Species diversity has received broad attention for many good reasons. The intent of the topic heading was to focus on function and the mapping of biological diversity onto functional diversity in biogeochemical transformations. Analytic and predictive models of ecosystem function for the foreseeable future require some aggregation of organisms into functional groups (e.g., bacterivores or sulfate reducers). The level of aggregation that is useful depends on the question and discriminatory ability at hand, but it is clear that much effort remains to be spent on assigning organisms to biogeochemically functional groups, with due attention to taxonomy and physiology. Excitement is palpable about the maturation of DNA methodologies for both species identification and identification of potential to catalyze specific reactions (e.g., presence of genes that code for nitrogenase) and the maturation of RNA technologies that can assess whether that potential is being realized.

Anticlimax

The central gyres of the ocean present many interesting questions. Collectively they constitute the largest habitat type on Earth and one of the oldest. Among them, the Central North Pacific Gyre (CNP) individually is the largest ecosystem on Earth, and my focus is primarily on this example. A small mistake in understanding of geochemical processes in such habitats can integrate into a large problem with global budgets. One geochemical problem of long standing is an inability to balance the fixed nitrogen budget for the global ocean. In most summaries, the loss terms exceed production substantially. Central gyres present a problem both individually and collectively, in that they seem to use more nitrogen in new production than can be accounted for (e.g., McGillicuddy and Robinson, 1997).

A parallel, long-running argument between geochemists and biologists has been about whether the oceans ultimately are limited in primary production by nitrogen or by phosphorus. The argument is partly a semantic one about what is meant by "ultimately" (i.e., time scale) and by "limited" (i.e., abundance or production of any or most phytoplankton), but it revolves around the issue that phosphorus has no substantial atmospheric source, whereas nitrogen is the largest component of the atmosphere. At its most simplistic, the assertion sometimes is made that because nitrogen-fixing organisms exist, the oceans cannot ultimately be nitrogen limited. Nitrogen fixation is notoriously expensive in energy and phosphorus, however (partly because it must be done anaerobically), and more sophisticated arguments revolve around whether rates of nitrogen fixation are slow enough to make phosphorus the effectively limiting nutrient because its availability restricts nitrogen fixation. There has long been evidence of phosphate as well as nitrate limitation in gyres, even of less demanding taxa than nitrogen fixers (e.g., Perry, 1972, 1976). Moreover, nitrogen fixers like *Trichodesmium* may well be limited or co-limited by iron

(Falkowski, 1997) and their vertical migration may be critical in obtaining nutrients.

The classic view of the gyres is as superstable, very species-rich but biomass-poor ecosystems, both in the pelagic realm and on the seabed (Hessler and Jumars, 1974). Early in the exploration of the CNP, climax communities were a popular ecological concept, and the gyres looked like end-member examples. For this reason, early exploratory cruises and station locations were named "Climax." Species diversity is high, and zooplankton samples taken years apart were as similar in species composition as samples from the same cruise (McGowan and Walker, 1985).

Flaws in the idea of oligotrophic gyres as nutrient impoverished appeared in the form of evidence that some phytoplankton were growing at high rates (e.g., Laws et al., 1984). Flaws appeared in the idea of constancy or static stability after the number of visits grew (Venrick et al., 1987), but disintegration of the idea came from time series funded under JGOFS (Joint Global Ocean Flux Study). The CNP showed two seasons of enhanced new production, one in winter based on enhanced physical mixing and one in summer based on enhanced nitrogen fixation, and these pulses showed substantial interannual variability (Karl et al., 1996). Further analysis of the time series and integration with all prior data suggest a doubling of primary production and a shift from dominance by eukaryotes to dominance by prokaryotes in the mid-1970s (Karl, in review). Some evidence links decadal-scale change in the Central North Pacific to the same large-scale ocean-atmosphere interactions that drive El Niño-Southern Oscillation (Karl et al., 1995). Not even this diverse community can resist basin-scale changes imposed by physics and chemistry.

Wind and eddy activity that can be important in bringing nitrate closer to the surface intuitively is unsteady and difficult to integrate over scales in time and space appropriate to the balance of nutrient budgets, and it is not hard to imagine that this component has interannual variability. The perspective shift underway, however, is that unusual lack of physical mixing also leads to enhanced new production, which is based instead on nitrogen fixation (Karl, in review) and on nitrate transport through vertical migration by mats of the diatom *Rhizosolenia* (Villareal et al., 1999). That is, the CNP's new production is minimized at some intermediate and probably "typical" input of physical energy, and lack of energy input leads to important biological "events." The theme of physical control of functional groups that effect drawdown of nutrients (including CO_2) extends to the Southern Ocean (Arrigo et al., 1999). Margalef (1978) must be pleased.

Time series clearly have power in exploring patterns of temporal variation and cross-correlation and have been key in shifting perspective away from stable, steady climax. They have made central gyres obvious places to improve the global ocean nitrogen budget, making the notion of oligotrophic seas as deserts even less tenable. To what extent nitrogen fixation is limited by phosphorus and trace metals (Falkowski, 1997), and to what extent it occurs in heterotrophic bacteria as well as cyanobacteria, is unclear (Karl, in press). Grazer influences on rates of nitrogen fixation and on the food-web fates of microbially fixed nitrogen beg for exploration. The extent to which and reasons why higher trophic levels are more (or less) stable in composition than primary producers are unquantified. The success of Ironex II and newspaper reports of parallel successes in the Southern Ocean make large-scale manipulation of phosphorus and trace-metal concentrations a tantalizing prospect for oligotrophic gyres, and the existing time series can suggest the season and duration that would be effective. Making explicit, mechanistic, a priori preditions of consequences and their time scales is certain to enhance greatly the knowledge gained from any discrepancies observed.

FUNCTIONAL ECOLOGY (OF INDIVIDUALS, WITH RAMIFICATIONS FOR POPULATIONS, COMMUNITIES, AND ECOSYSTEMS)

- How are mass and momentum transfer and other environmental forces integrated with information to influence behavior?
- How does performance change with size and form?

Functional ecology is not a well-established term in the popular ecological lexicon, despite the fact that a journal of the British Ecological Society bears this name. Loosely, by contrast to numerical ecology, it refers to the performance of individuals in the context of environmental features, including other organisms. Perhaps the best-known functional responses of individuals are "filtering" and ingestion rates as functions of food concentration: rectilinear, hyperbolic, and sigmoidal responses have been described. As a consequence of rapid advances in understanding of mechanics of food encounter and handling, it can now be argued that it is better to use arrival rate of food items in the sensory field of the forager as the independent variable (instead of food concentration) in predicting ingestion rates. Ambient fluid motion, for example, can alter rate of encounter without any change in food concentration.

Tactile Senses of Copepods: Discriminating Food from Foe

Among the most difficult phenomena about which to gain intuition are ones outside human sensory experience. Mechano- and chemosensing at low to intermediate Reynolds numbers may be among the most alien; hence they require accurate description before they can be appreciated and succinct, logical description before they can be intuited. Advances on the front of understanding mechanosensory ecology recently have been stunning; data with broad scatter suddenly have collapsed onto simple curves defined by sys-

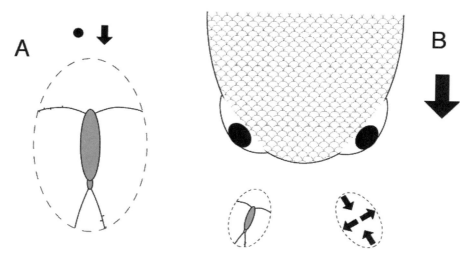

FIGURE 1. The two perspectives of copepod mechanosensing as predator and prey, respectively. A few mechanosensory hairs are sketched in on one antenna and the opposite caudal furca. The parcel of water within which the copepod is embedded is indicated by the dashed ellipse. A. Detection of small prey is via local perturbation of the velocity field. B. Detection of a predator by deformation of the water parcel in which the copepod is embedded. Qualitative features of the deformation are shown in the ellipse containing arrows. Based on analyses in Kiørboe and Visser (1999).

tematic decomposition of fluid dynamic phenomena into their constituent motions.

Kiørboe and Visser (1999) have idealized the motion of both prey and predator as flow around a sphere and have provided clear intuition for the mechanical perspective of a copepod in detecting and distinguishing moving prey and predators (Figure 1). This intuition comes not only from the calculated decomposition, but also from clever experiments that expose copepods to simplified and easily quantified fluid dynamic components of this decomposition. Arrays of mechanoreceptors on the antennae detect prey as local velocity variations. Independently performed experiments and numerical simulations (Bundy et al., 1998) show that the flow field generated by a swimming copepod also can contribute to detection of nonmoving, particulate prey.

Predators of copepods, on the other hand, are large in comparison with the copepod and are detected as larger-scale flow-field deformations that influence the whole space monitored by the copepod; the copepod knows that a predator is near when the stimulus affects the full array of sensors but deflects them with spatially varying velocities. Successful fish predators detect the copepod at a distance visually and decelerate as they approach, dropping the deformation produced by their bow wave below the threshold intensity that elicits the copepod's "jump" escape behavior. This threshold sits orders of magnitude above the neurophysiological detection limits of the mechanoreceptors and just above the level of deformation caused routinely by ambient turbulence. Kiørboe and Visser's (1999) analysis brings intuitive understanding of the process.

The qualitative gain in intuition for mechanoreception by copepods itself is compelling, but the gain is even more impressive when stimuli are quantified. Clearance rates by the ubiquitous omnivore *Oithona similis* are predictable over three orders of magnitude and on radically differing food particles (swimming protists and settling fecal pellets). Moreover, detection of settling particles by coprophagous copepods is found in calculations and experiments to be highly nonlinearly but predictably dependent upon particle size and settling velocity, to the point where detection of the most rapidly settling pellets at natural copepod abundances is virtually certain. Since McCave's (1975) seminal analysis, large particles or aggregates have been thought to account disproportionately for the flux of material to the seafloor. Kiørboe and Visser's (1999) analysis suggests instead that particles of intermediate settling velocity that provide less mechanical stimulus for detection may be more successful in running the suspension-feeder gantlet. Particle-type- and settling-velocity-dependent degradation rates are among the most poorly constrained parameters in global carbon budgets, and this new analysis of mechanosensory abilities provides substantial help in the form of new perspectives and predictions from an unexpected direction. Suddenly complicating this range of issues in vertical transport of carbon still further is the documentation of spontaneous assembly of gels (Chin et al., 1998).

Diversity in Bacterial Tactics

Another important part of the carbon cycle is uptake of dissolved organic carbon by heterotrophic bacteria. *Escherichia coli*, a resident of the human large intestine and colon, has provided the universally used model of chemotaxis. Digestion in humans can be expected to yield large-scale gradients relative to the body size of a bacterium (Dade et al., 1990), and the tumble-and-run approach makes good sense in the episodically stirred gut environment. *E. coli* runs in a straight line, then tumbles and goes off at a random heading from the original. Run duration increases with nutrient concentration, biasing the random walk and moving the bacterium up-gradient (Berg, 1993).

Most suggested sources of dissolved organic matter for bacteria are small in length and hence subject to rapid diffusion. In the case of leaking or exuding phytoplankton, the source may be quasi-steady, but in the cases of sloppy feeding, fecal pellet ejection, or autolysis, the source will be not only small but also short lived. An important question is whether marine bacteria show the same chemotactic behaviors as *E. coli*. Small bacteria apparently cannot find their way up gradient, and taxis is not practiced by bacteria smaller than about 0.6 μm in diameter (Dusenbery, 1998a). Below this size, the course of the bacterium would be changed too rapidly by Brownian rotation to allow directed movement. Conversely, the elongate shapes of chemotactic bacteria generally observed are particularly resistant to Brownian rotation and thereby increase the time over which the organism practically can integrate stimulus strength, and hence increase greatly its sensitivity to chemical gradients (Dusenbery, 1998b).

Recently, swimming paths of large marine bacteria have been recorded optically by taking advantage of light scatter under dark-field illumination (Blackburn et al., 1998). Unlike *E. coli*, they do not turn in a random direction at the end of a run but instead double back at close to 180° from the original heading, with some deviation due to Brownian rotation. This visualization was done in still water, but simulations in shear fields expected from decaying turbulence suggest that doubling back is far more effective than the tumble-and-run behavior in staying near a small, spherical source (Luchsinger et al., in review). There clearly is diversity in chemotactic strategies and patterns among marine bacteria (e.g., Barbara and Mitchell, 1996); *E. coli* is no longer an appropriate universal model. Models also suggest yet undocumented tactics. Quite contrary to intuition, bacteria can in principle use spatial sensing (difference in concentration on two parts of the cell) rather than temporal sensing (difference in concentration over time) to detect stimulus gradients (Dusenbery, 1998c).

Letting released enzymes do the searching for particulate material appears to be another useful bacterial strategy in either large aggregates of particles or in sediments (Vetter et al., 1998; Vetter and Deming, 1999) and explains the paradox of oversolubilization of aggregates by bacteria (more soluble products made than used; Smith et al., 1992). Further challenges to biological-physical modeling, to measurement of both bacterial tactics and carbon dynamics and even to the discrimination of dissolved from particulate carbon are the rapid self assembly and state changes of biogenic polymer gels (Chin et al., 1998).

The conceptual simplifications provided by Kiørboe and Visser (1999), steadily increasing abilities to visualize flows and organisms both optically and acoustically, and increasing computational capabilities poise the study of fluid dynamic and chemical interactions with and among organisms for rapid advance. These advances promise in particular to help understand the vast and beautiful morphological diversity of protists and phytoplankton (though some modern classifications include the phytoplankton and even macroalgae with protists) living at low Reynolds numbers. Easier numerical modeling than at high Reynolds numbers is partial compensation in these regimes for lack of intuition about life in a fluid environment filled with dynamical chemical and physical signals.

STRUCTURING DYNAMICS OF BIOLOGICAL ASSEMBLAGES

- Over what spatial scales are marine populations connected via dispersal of early life stages?
- What are the dynamics of marine food webs, and how will they respond to environmental perturbations?
- As models and synoptic data now are used to forecast the weather, can one forecast changes in physical-chemical-biological interactions in the sea that affect fisheries yields, food-web dynamics, and ecosystem services that the sea provides?

This section covers population and community ecology but in the specific context of the ocean. Important facets of this context are the unquantified connectedness of subpopulations and the pervasiveness of fluid transport of propagules. This unquantified connectedness remains the greatest obstacle to rational establishment of marine preserves and management policy in general.

The OEUVRE workshop, to no one's surprise, cited identification of strong indirect effects through experimental manipulation as one of the great successes of marine ecology over the last 30 years and prediction of which interactions would be strong ones among the greatest challenges for the future. For few communities are the majority of interaction strengths known, but it is clear that in communities of even modest diversity the majority are weak (Paine, 1992), and identifying the important ones by manipulating all the species individually is a daunting empirical task. I left the workshop seeing no clear route to progress through predictive theory, either.

Theory suddenly has jumped to the rescue and shifted perspective by 180° by putting focus not on the strong interactors but on the weak ones. McCann et al. (1998) departed from classic food-web models in two ways. They modeled functional responses as saturating rather than linear in prey concentration, and they allowed population abundances to be away from equilibrium values. With these additions to realism, food-web models better reflect the added stability observed as food-web complexity increases. A deep insight from these models is that weakly interacting species in general damp oscillations. Although the paper was not specifically about marine communities, it is worth noting that communities with food webs rather than simple food chains, with omnivores rather than food specialists, with intraguild predation (predation on a competitor, particularly its young), and with allochthonous food supply are particularly stabilized. The majority of these characteristics apply to the majority of marine food webs. Berlow (1999) amplified this advance by demonstrating the efficacy of weak interactions in generating spatial and interannual variation in community structure with a simple mussel-barnacle-whelk system.

This shift in perspective is remarkable. When the goal was identification of strong interactors, they were hard to guess. When the goal instead is to find weak interactors with potentially large damping or amplifying effects, suspects jump to mind. It is easy to predict a cottage industry among ocean ecologists in the manipulation of omnivores, for example. Observations of strong interactions, particularly when the same species interact strongly in one place and not in another (e.g., the starfish-barnacle-mussel triad in Washington State and southeast Alaska; cf. Paine, 1980), translate into questions about what stabilizing species were missing in the former. Anchovies and sardines in this view exhibit such dramatic oscillations (Bakun, 1997) because they live in simplified food chains.

I cannot help but comment on how pleased I think that Gordon Riley would be to see ocean ecologists employ these new equations and a deeper understanding of the physical and chemical processes of the sea to write quantitative descriptions and predictions of marine ecosystem processes. He sought to combine understanding of the chemistry and physics of the oceans with the Lotka-Volterra equations systematically to dissect the workings of marine ecosystems. He supported use of the term "biological oceanography" to get away from the "grab-bag of semi-defined concepts [that he perceived to dominate the ecology of his day] to clear, step-wise analytical approaches to variation in nature" (Mills, 1995, p. 39). OEUVRE coined the term "ocean ecology" in this spirit but also to encompass the pressing need to obtain at least recent paleo-information about the workings of marine ecosystems.

HUMAN IMPACTS AND HABITAT LINKAGES

• How then can one understand the multiple-scale and pervasive human impacts on the sea in the face of the confounding effects of weather and climate change? *Resolving and understanding anthropogenic and natural sources of variability and change on coastal to basin scales is arguably the greatest challenge to oceanographic science for the foreseeable future.*

Anthropogenic effects from injection of nutrients and pollutants and from removal of predators certainly are pervasive. The challenge, given that a fully "natural" community free of anthropogenic effects appears to be a purely theoretical construct in the pejorative sense, is to avoid the trap of scurrying to understand the magnitudes of anthropogenic impacts without making the effort to understand their mechanisms. In the period since the OEUVRE workshop, I have not discovered any comparably perspective-shifting contributions to the ones mentioned under the other headings. This situation makes me continue to support the position taken by the OEUVRE group, that mankind is doing many manipulations without understanding their consequences and that a greater effort needs to be devoted to taking advantage of these manipulations to uncover the consequences and their mechanisms as the consequences arise rather than afterward. The surest method is to predict these consequences and then learn from the errors in the predictions as the perturbation proceeds.

Time-Series and Intervention Analysis

It is possible to look back at the other sections of this paper, however, and to note that time-series analysis has played a central role in gaining understanding that the CNP ecosystem is dynamically stable, and not statically stable, and that time series similarly have played a large role in dissecting food-web interactions. It is not too soon to think about the consequences of increasing atmospheric inputs to the CNP from the industrialization of Asia and to ask how to resolve them from more natural variation. Time series again seem to be prominent in the answers.

One kind of time-series analysis, called "intervention analysis" (Box et al., 1994), appears to hold particular promise for characterizing some anthropogenic effects because it was developed to do so in the context of atmospheric pollutants. The procedure is to collect a long time series that includes the period before a change in policy or other anthropogenic perturbation whose timing is known. For example, one can look for changes in the record of lead deposition after the switch to lead-free fuels. The procedure is to fit an explicit time-series model to the pre-change data set. The model is then used to forecast the post-change data set, and the residuals from this forecast contain the treatment effect. Statistical power of this method depends on the length and simplicity of the pre-change time series (i.e., the ability to fit an explicit time-series model before the perturbation). Many ocean ecologists will be reluctant to trade traditional replication in space for replication in time, but if the whole system has been altered, then replication in space is elusive.

Ocean ecologists also must become as creative in extending time series backward as they are in extending them forward. Geochemists have made great contributions in this regard with chemical proxies for temperature and nutrients and with biomarkers for some taxa. Archaeology through analysis of Indian middens has contributed to dissection of the sea otter-urchin-kelp interaction in the Aleutian Island chain (Estes et al., 1998). Where could additional effort by ocean ecologists produce the greatest extension and resolution back in time? The successes noted under functional ecology give reason to expect dramatic progress soon in understanding form and function in marine microfossils through understanding of form and function in today's fossilizable organisms. As the costs and benefits of simple shapes yield to analysis (e.g., Dusenbery 1998b), costs and benefits of more complex morphologies seem less daunting to study. Body form, spination, and mechanical properties of phytoplankton and protist individuals and chains certainly contain environmental information to be read. Continued development of "biomarker" compounds also certainly will be repaid. More conjecturally, establishing the extent to which buried bacterial communities reflect the conditions above and on the seafloor at some previous time (i.e., while the surface mixed layer of sediments was in contact with the overlying water) versus their environmental conditions at present may allow extraction of other paleoenvironmental information. At issue is the length of time that bacteria can survive in inactive state and be interrogated by molecular means in this biochemically messy medium.

ACKNOWLEDGEMENTS

I thank John Cullen and Dick Barber for constructive criticism of earlier drafts of this paper and Thomas Kiørboe and David Karl for sharing their unpublished manuscripts.

REFERENCES

Arrigo, K.A., D.H. Robinson, D.L Worthen, R.B. Dunbar, G.R. DiTullio, M. VanWoert, and M.P. Lizotte. 1999. Phytoplankton community structure and drawdown of nutrients and CO_2 in the Southern Ocean. *Science* 283:365-367.

Bakun, A. 1997. Radical interdecadal stock variability and the triad concept: A window of opportunity for fishery management science. Pp. 1-18 in T.J. Pitcher, P.J.B. Hart, and D. Pauly (eds.), *Reinventing Fisheries Management*. Chapman and Hall, London.

Berg, H.C. 1993. *Random Walks in Biology*. Princeton University Press, Princeton, New Jersey.

Berlow, E.L. 1999. Strong effects of weak interactions in ecological communities. *Nature* 398:330-334.

Blackburn, N., T. Fenchel, and J. Mitchell. 1998. Microscale nutrient patches in planktonic habitats shown by chemotactic bacteria. *Science* 282:2254-2256.

Box, G.E.P., G.M. Jenkins, and G.C. Reinsel. 1994. *Time Series Analysis: Forecasting and Control*. Prentice-Hall, Englewood Cliffs, New Jersey.

Bundy, M.H., T.F. Gross, H.A. Vanderploeg, and J.R. Strickler. 1998. Perception of inert particles by calanoid copepods: Behavioral observations and a numerical model. *J. Plankton Res.* 20:2129-2152.

Chin, W.C., M.V. Orellana, and P. Verdugo. 1998. Spontaneous assembly of marine dissolved organic matter into polymer gels. *Nature* 391:568-572.

Dade, W.B., P.A. Jumars, and D.L. Penry. 1990. Supply-side optimization: Maximizing absorptive rates. Pp. 531-556 in R.N. Hughes (ed.), *Behavioural Mechanisms of Food Selection*. Springer-Verlag, Berlin.

Dusenbery, D.B. 1998a. Minimum size limit for useful locomotion by free-swimming microbes. *Proc. Natl. Acad. Sci. USA* 94:10949-10954.

Dusenbery, D.B. 1998b. Fitness landscapes for effects of shape on chemotaxis and other behaviors of bacteria. *J. Bacteriol.* 180:5978-5983.

Dusenbery, D.B. 1998c. Spatial sensing of stimulus gradients can be superior to temporal sensing for free-swimming bacteria. *Biophysical J.* 74:2272-2277.

Estes, J.A., M.T. Tinker, T.M. Williams, and D.F. Doak. 1998. Killer whale predation on sea otters linking oceanic and nearshore ecosystems. *Science* 282:473-476.

Falkowski, P.G. 1997. Evolution of the nitrogen cycle and its influence on the biological sequestration of CO_2 in the ocean. *Nature* 387:272-275.

Hessler, R.R., and P.A. Jumars. 1974. Abyssal community analysis from replicate box cores in the central North Pacific. *Deep-Sea Res.* 21:185-209.

Karl, D.M. In press. A sea of change: Biogeochemical variability in the North Pacific Subtropical Gyre. *Ecosystems*.

Karl, D.M., J.R. Christian, J.E. Dore, D.V. Hebel, R.M. Letelier, L.M. Tupas, and C.D. Winn. 1996. Seasonal and interannual variability in primary production and particle flux at Station ALOHA. *Deep-Sea Res.* 43:539-568.

Karl, D.M., R.M. Letelier, D.V. Hebel, L. Tupas, J.E. Dore, J.R. Christian, and C.D. Winn. 1995. Ecosystem changes in the North Pacific subtropical gyre attributed to the 1991-92 El Niño. *Nature* 373:230-234.

Kiørboe, T., and A.W. Visser. 1999. Predator and prey perception in copepods due to hydromechanical signals. *Mar. Ecol. Prog. Ser.* 179:81-95.

Laws, E.A., L.W. Haas, P.K. Bienfang, R.W. Eppley, W.G. Harrison, D.M. Karl, and J. Marra. 1984. High phytoplankton growth and production rates in oligotrophic Hawaiian coastal waters. *Limnol. Oceanogr.* 29:1161-1169.

Luchsinger, R.H., B. Bergersen, and J.G. Mitchell. In review. Bacterial swimming strategies and turbulence. *Biophysical Journal*.

Margalef, R. 1978. Life-forms of phytoplankton as survival alternatives in an unstable environment. *Oceanologica Acta* 1:493-509.

McCann, K, A. Hastings, and G.R. Huxel. 1998. Weak trophic interactions and the balance of nature. *Nature* 395:794-798.

McCave, I.N. 1975. Vertical flux of particles in the ocean. *Deep-Sea Res.* 22:491–502.

McGillicuddy, D.J., and A.R. Robinson. 1997. Eddy-induced nutrient supply and new production in the Sargasso Sea. *Deep-Sea Res.* 44:1427-1450.

McGowan, J.A., and P.W. Walker. 1985. Dominance and diversity maintenance in an oceanic ecosystem. *Ecol. Monogr.* 55:103-118.

Mills, E.L. 1995. From marine ecology to biological oceanography. *Helgoländer Meeresunters.* 49:29-44.

Paine, R.T. 1980. Food webs, linkage, interaction strength and community infrastructure. *J. Anim. Ecol.* 49:667-685.

Paine, R.T. 1992. Food-web analysis through field measurement of per capita interaction strength. *Nature* 355:73-75.

Perry, M.J. 1972. Alkaline phosphatase activity in subtropical central North Pacific waters using a sensitive fluorometric method. *Mar. Biol.* 15:113-119.

Perry, M.J. 1976. Phosphate utilization by an oceanic diatom in phosphorus-limited chemostat culture and in the oligotrophic waters of the central North Pacific. *Limnol. Oceanogr.* 21:88-107.

Smith, D.C., M. Simon, A.L. Alldredge, and F. Azam. 1992. Intense hydrolytic activity on marine aggregates and implications for rapid particle dissolution. *Nature* 359:139-142.

Venrick, E.L., J.A. McGowan, D.R. Cayan, and T.L. Hayward. 1987. Climate and chlorophyll a: Long-term trends in the central North Pacific Ocean. *Science* 238:70-72.

Vetter, Y.A., and J.W. Deming. 1999. Growth rates of marine bacterial isolates on particulate organic substrates solubilized by freely released extracellular enzymes. *Microbial Ecol.* 37:86-94.

Vetter, Y.A., J.W. Deming, P.A. Jumars, and B.B. Krieger-Brockett. 1998. A predictive model of bacterial foraging by means of freely released extracellular enzymes. *Microbial Ecol.* 36:75-76.

Villareal, T.A., C. Pilskaln, M. Brzezinski, F. Lipshultz, M. Dennett, and G.B. Gardner. 1999. Upward transport of oceanic nitrate by migrating diatom mats. *Nature* 397:423-425.

Global Ocean Science:
Toward an Integrated Approach[1]

COMMITTEE ON MAJOR U.S. OCEANOGRAPHIC RESEARCH PROGRAMS

NATIONAL RESEARCH COUNCIL

The rigorous nature of conducting research at sea has been a major challenge facing oceanographers since the earliest research cruises. As interest in understanding large-scale phenomena with global implications began to shape ocean research, the need for greater spatial coverage and near synoptic observations required a change in the way oceanographic research was done. A significant innovation to emerge early this century was the organization of large expeditions that attempted to systematically collect ocean observations across great expanses of the oceans by extended cruises of one or more research vessels. As a result of the need to coordinate these activities, what became known as major oceanographic programs came into being.

In many ways, these major programs are inexorably linked to this nation's ability to understand and protect our environment and the tremendous resources it contains. As will be demonstrated in this report, the health of the ocean science community and the research community it includes is strongly influenced by these large collaborative efforts. With several of the present group of major oceanographic programs now nearing their conclusion, the Ocean Sciences Division of the National Science Foundation (NSF/OCE) has undertaken a number of steps to evaluate the present vitality of oceanographic research in this country, with the hope of developing a comprehensive research strategy to take ocean science forward into the next century. As part of that effort, NSF/OCE asked the Ocean Studies Board of the National Research Council to conduct a study of the role of major programs in ocean research. This request resulted in the formation of the Committee on Major U.S. Oceanographic Research Programs, whose purpose was to evaluate the impact of the past and present programs and provide advice on how these programs should be developed and managed in the future.

IMPACT OF MAJOR OCEANOGRAPHIC PROGRAMS ON OCEAN SCIENCE

The major oceanographic programs have had an important impact on ocean science. Many breakthroughs and discoveries regarding ocean processes that operate on large spatial scales and over a range of time frames have been achieved by major oceanographic programs that could not have been expected without the concentrated effort of a variety of specialists directed toward these large and often high profile scientific challenges. In addition to these contributions, each program has left (or can be expected to leave) behind a legacy of high-quality, high-resolution, multiparameter data sets; new and improved facilities and techniques; and a large number of trained technicians and young scientists. The discoveries, data, and facilities will continue to be used to increase the understanding of fundamental earth system processes well after the current generation of programs have ended.

Scientific Understanding and Education

Scientific advances in several high-profile areas have been brought about by research conducted through the major

---

[1]Executive summary from: National Research Council (NRC). 1999. *Global Ocean Science: Toward an Integrated Approach.* National Academy Press, Washington, D.C. For information about obtaining this report, contact the Ocean Studies Board, National Research Council, 2101 Constitution Ave., NW, Washington, DC 20418. The Ocean Studies Board's Committee on Major Oceanographic Programs was chaired by Rana Fine (University of Miami-RSMAS). Other committee members included Charles Cox (Scripps Institution of Oceanography), William Curry (Woods Hole Oceanographic Institution), Ellen Druffel (University of California, Irvine), Jeffrey Fox (Texas A&M University), Roger Lukas (University of Hawaii, Manoa), James Murray (University of Washington), Neil Opdyke (University of Florida), Thomas Powell (University of California, Berkeley), Michael Roman (University of Maryland), Thomas Royer (Old Dominion University), Lynda Shapiro (University of Oregon), Anne Thompson (National Aeronautics and Space Administration), and Andrew Weaver (University of Victoria, British Columbia). Dan Walker was the study director. Rana Fine presented a summary of this report at the symposium.

oceanographic programs. Examples include increased understanding of the causes of mass extinction, the role of ocean circulation in climate (e.g., El Niño) and in the decline in fisheries, and the ability of the ocean and marine organisms to buffer changes in the concentrations of the so-called greenhouse gases (e.g., carbon dioxide). Also of importance is the wide use of program discoveries and data in the classroom, the availability of program facilities for general community and educational purposes, and the training of graduate students. As discoveries and advances attributable to these programs continue to influence research conducted throughout the ocean science community, the significance of these programs will become even more apparent.

The usually high-quality, global, multiparameter data sets and time series developed by major oceanographic programs will be some of their most important and enduring legacies. **It is essential to preserve and ensure timely access to these data sets. Every effort must be made to facilitate data exchange and prepare for an ever-increasing demand for access to these large data sets.**

Technology and Facilities

Major programs have affected the size and composition of the research fleet, and provided impetus for the development of technology and facilities used by the wider oceanographic community. The programs have contributed to a range of technological developments, facilities, and standardization of sampling techniques. **Similar to what is done periodically for the research fleet, a thorough review of the other facilities, including procedures for establishing and maintaining them, is necessary to set priorities for support of the facilities used by the wider oceanographic committee.** The very long lead times needed for fleet and facilities development require that the oceanographic community be developing plans for facilities requirements for 2008 and beyond. **Strategic planning for facilities (ship and non-ship) should be coordinated across agencies with long-range science plans and should include input from the ocean sciences community.**

Collegiality

Major oceanographic programs account for a significant proportion of the funding resources available to ocean science. As a consequence of these programs, more money has been made available for ocean science research in general. However, the proportion of funds consumed by these programs has tended to heighten concerns about the effect these programs have had on collegiality within the research community. Nevertheless, many scientists recognize positive impacts of major programs on the way ocean scientists work together toward an objective, including greater willingness to share data.

In the future, allocation decisions should be based on wide input from the research community and the basis for decisions should be set forth clearly to the scientific community. By providing the research community with timely access to information regarding these decisions, misperceptions can be avoided and the impact of funding pressures minimized.

SCIENTIFIC AND GENERIC GAPS

Given the extensive involvement of the academic community in recent activities undertaken by NSF/OCE to develop a research strategy for ocean science, the committee determined that attempting to specifically identify scientific gaps would be redundant and unnecessary. Yet, a number of mechanisms can help the ocean science community's planning process by identifying scientific and generic gaps in and among existing programs. Some scientific gaps can be addressed by enhancing communication and coordination. The sponsoring agencies, especially NSF/OCE, should continue to develop and expand the use of various mechanisms for inter-program strategic planning, including workshops and plenary sessions at national and international meetings and ever greater use of World Wide Web sites and newsletters.

Generic gaps that were identified in and among programs are as follows:

- the need for funding agencies and the major oceanographic programs to develop mechanisms to deal with contingencies;
- the need to establish (with broad input from the ocean science community) priorities for moving long time-series and other observations initiated by various programs into an operational mode, in consideration of their quality, length, number of variables, space and time resolution, accessibility for the wider community, and relevance toward meeting established goals;
- the need for modelers and observationalists to work together during all stages of program design and implementation;
- the need to enhance modeling, data assimilation, data synthesis capabilities, and funding of dedicated computers for ocean modeling and data assimilation with facilities distributed as appropriate; and
- the need for federal agencies in partnership with the National Oceanographic Data Center (NODC) to take steps to prepare for a supporting role in data synthesis activities (including, but not limited to, data assimilation).

STRUCTURING PROGRAMS TO MAXIMIZE SCIENCE ORGANIZATION AND MANAGEMENT

The present NSF/OCE structure has made it difficult to get intermediate-size projects funded (as distinguished from major programs), particularly ones that are interdisciplinary. These intermediate-size projects could be solicited, funded, and executed in a way that would ensure a regular turnover of new ideas and opportunities for different investigators. **Federal agencies sponsoring oceanographic research programs, especially NSF/OCE, should make every effort to encourage and support a broad spectrum of interdisciplinary research activities, varying in size from the collaboration of a few scientists to programs perhaps even larger in scope than the present major oceanographic programs.**

There is no one procedure by which principal investigators with good ideas can start new programs. **The sponsoring agencies, especially NSF/OCE, should develop well-defined procedures for initiating and selecting future major ocean programs.** Successful ideas should be brought to planning workshops that are administered by an independent group to ensure that the process is inclusive.

In the past, major oceanographic programs have been administered by a Scientific Steering Committee (SSC) with a chair and sometimes an Executive Committee. However, **there is no one ideal structure that should be used for all programs, and it is important for NSF/OCE and other agencies to maintain flexibility to consider a number of options regarding the design and execution of future programs. Some factors to be considered include the following:**

- The structure of the program should be dictated by the complexity and nature of the scientific challenge it addresses.
- The nature and support of program administration should reflect the size, complexity, and duration of the program.
- The structure should encourage continuous refinement of the program.
- All programs should have well-defined milestones, including a clearly defined end.

IMPROVING SCIENCE BY ENHANCING COMMUNICATION AND COORDINATION

Better communication, planning, and coordination among major oceanographic programs would serve to maximize the efficient use of resources; facilitate interdisciplinary synthesis; and enhance the understanding of ocean systems, their interaction with each other, and with those of the atmosphere and solid earth. In the past, communication among major ocean programs has been ad hoc, and coordination of field programs has been hampered by funding. Beyond field programs, synthesis activities will benefit from coordination. When appropriate, joint announcements of opportunity for inter-program synthesis should be issued. **Communication and coordination can be facilitated**

among the ongoing major ocean programs by considering joint appointments to SSCs, and annual meetings of the SSC chairs. Greater involvement and appreciation for the accomplishments and challenges facing these programs by scientists not funded through the programs can occur if non-program scientists are recruited to participate as members of the SSCs and in other activities when appropriate.

LESSONS FOR THE FUTURE

The large-scale global scientific challenges of the future will continue to require major oceanographic programs. At the same time, the scientific research conducted by individual investigators in the core disciplines must be healthy. The pursuit of these two goals should include complementary activities that strengthen the overall national and international program of ocean science. The strength of many of the major programs and individual initiatives can be directly attributed to the NSF peer-review system and the flexibility of the agency and program managers. Some tools for federal agencies and the scientific community to use to balance these two often competing needs, based on scientific requirements, are presented in this report. In addition, there are opportunities for some course corrections that will enable the federal agencies, including NSF, to better respond to the growing need of the ocean sciences community to conduct multi-investigator and interdisciplinary research. The need to carry out interdisciplinary research through multi-investigator projects will continue to increase in recognition of the emphasis placed on global environmental and climate issues, issues that have largely displaced national security as an underlying motive for funding research in the geosciences.

The committee's recommended approach for achieving the goals described above would be to create a new interdisciplinary unit within the Research Section of NSF/OCE, charged with managing a broad spectrum of interdisciplinary projects. The large-scale global and integrative nature of some of the present scientific challenges, such as environmental and climate issues, will require greater coordination, as will the need for shared use of expensive platforms and facilities. The creation of such a unit could alleviate many of the real and perceived problems identified throughout this report related to coordination, collegiality, and planning, and thus help maximize the scientific return on the considerable investment this nation makes in ocean-related research.

Ocean sciences must reach a new level in order to successfully meet the emerging needs for environmental science. Doing so will require more integration and greater emphasis on consensus building. If the challenges can be met, a new interdisciplinary unit would be well positioned to aid in building partnerships among the agencies, and play a leading role in helping to create focused national efforts in future global geosciences initiatives.

Education in Oceanography: History, Purpose, and Prognosis

Arthur R. M. Nowell

College of Ocean and Fishery Sciences,
University of Washington

INTRODUCTION

My purpose in writing this review and prognosis on education in ocean sciences is to catalyze discussion and hopefully action in areas of education where often neither faculty nor graduate students in oceanography venture. I will first review some history to understand how we come to our present situation. I will then review the functional areas of education from graduate to informal education and finally raise some questions about our future and our values.

SOME HISTORY OF UNIVERSITIES AND OCEANOGRAPHY RELATED TO EDUCATION

The history of the modern research university as it exists in the late 20th century was envisioned first in 1809. In that year, the year Abraham Lincoln was born, Wilhelm von Humbolt published a report on university reform in Germany. Higher education until then had been more a matter of rote learning than of creative scholarship. Humbolt proposed that:

> The idea of disciplined intellectual activity, embodied in institutions, is the most valuable element of the moral culture of the nation. One unique feature of higher intellectual institutions is that they conceive of science and scholarship as dealing with ultimately inexhaustible tasks: this means they are engaged in an unceasing process of inquiry. At the higher level the teacher does not exist for the sake of the student; both teacher and student have their justification in the common pursuit of knowledge. The teacher's performance depends on the students' presence and interest—without this science and scholarship could not grow. (Humbolt, 1809)

But for many in the United States, the university was a foreign concept, and universities existed only in the old eastern seaboard cities. That changed in 1860 with the passage of the Morrill Land Grant Act that permitted states, territories, and local groups to apply for federal land grants, the proceeds of which could be used to support education. Even

among politicians in Washington, D.C., education and learning were valued. In his autobiography, Lincoln wrote:

> He studied and nearly mastered the six books of Euclid since he became a member of Congress. *He regrets his want of education and does what he can to supply the want.* (my italics)

The Land Grant Act encouraged the establishment of state universities and colleges throughout the United States. My own institution in Seattle was founded in 1861, when Seattle barely had a population of 500. The first class had 5 students!

Education links to the federal government were much stronger in the nineteenth century than today. There was a freer exchange of people. Such leaders of science as John Wesley Powell and G.K. Gilbert both held university positions, both became directors of the U.S. Geological Survey, both published some enduring scientific papers while serving in Washington, D.C. and going into the field in Utah! Today, even as the federal government supports immense research infrastructure in research universities, the exchange between the two sides has dwindled. We place people into separate compartments: once you leave the classroom and research lab to administer or support science in Washington, D.C., you are not much welcomed back. As academics at institutions of higher learning, we have much to learn to change this attitude—to change it back to the mid-nineteenth century ideal.

But the history of education in oceanography in the United States was written broadly with the National Academy of Sciences (NAS) report of 1929, often-called F.R. Lillie Report. Lillie's committee made only four comments on education:

> The general paucity of opportunities for instruction in this general field is so obvious that it needs no detailed survey for corroboration.
>
> The graduate student, sufficiently devoted to the subject and

195

fitted for advanced instruction or research, finds far fewer avenues than the importance of this field of science demands.

The advance of Oceanography in America now suffers from one of its greatest handicaps, for progress in this science is a matter not only of ships, laboratories and money, but far more of men, which implies opportunities for education. And it is of men that there is now the most serious shortage.

It is in fact, one of the most serious obstacles to advances in this field that it is not now possible for a student to obtain a course of instruction, properly graded upward from the elementary introduction to advanced research, in any one American University. In America the oceanographer must today be largely self-taught in the basic aspects of his subject.

Regrettably, F.R. Lillie then proceeded to ignore entirely what was obviously the key obstacle to oceanography. He focussed on the creation of a facility for research. The report recommended establishing Woods Hole Oceanographic Institution. The Lillie Report did enhance research facilities that were used by both a small group of resident scientists and a group of visiting scientists and their students, working as was the style of the time in an apprenticeship mode. The report, however, became a model for how education was to be dealt with in future NAS and overview reports in 1960 and 1969! This apprenticeship mode continued and until the 1980s the majority of scientists in the field received their training outside the discipline, most often in chemistry, geology, applied physics, and zoology for example, and after earning a doctorate degree changing fields to enter oceanography.

At the broader federal level, three developments between 1945 and 1950 changed the entire complexion of higher education in the United States and especially in oceanography. In 1945, the GI Bill was passed, which opened higher education to a much wider segment of the population. Prior to 1945, the university was mainly the provenance of the moneyed families: the GI Bill is often referred to as the largest piece of affirmative action legislation ever passed. In 1946, the Office of Naval Research was founded. Its impact on the field of oceanography as a science and on the institutional characteristics of our science has been immense. Finally, in 1950 the National Science Foundation Act was passed.

The establishment of NSF followed the publication of Vannevar Bush's important study, *Science: The Endless Frontier* (Bush, 1945). In that report, Bush noted:

> Before the War in all but a few of the prosperous universities, teaching loads were excessive from the standpoint of optimal research output. During the war, the university scientist had for the first time the facilities and assistance to carry on research. It is of the utmost importance to maintain a favorable competitive position for universities.

When the Bush Report was written only 35 percent of the U.S. population advanced further than Grade Eight at school and less than 10 percent of the U.S. population went to college. Teaching loads were approximately 18 contact hours per week in engineering and 12 hours per week in the sciences. Today, when faculty talk about teaching load, it is instructive to recall the differences! Today, over 95 percent of the population graduates from high school and over 80 percent of high school graduates attend colleges and universities. The role and function of higher education must surely reflect these differences, as must the role and responsibilities of NSF.

The Symposium celebrated the creation of NSF. The Bush Report called for the creation of the National Science Foundation and in 1945 the Kilgore-Magnuson Bill was introduced to create this independent agency. The process of creating and funding NSF now looks like a template of how much of science is funded by NSF today, a process of submit thrice and fund once at less than requested levels, for the 1945 Bill was submitted and rejected. So like you or I with a proposal, it was reworded and re-submitted by Magnuson in 1947. It was again rejected, resubmitted, and yet again rejected. But Magnuson was a tenacious politician, as was Vannevar Bush, so when the bill was submitted after three rejections in 1949 it was passed: NSF was created and funded. On May 10, 1950, President Harry Truman signed the NSF Act on a whistle-stop train pausing at Pocatello, Idaho. Truman referred to the endless frontier of science and the mystique of the western frontier. He said NSF would provide "new frontiers for the mind and a fuller and more fruitful life for all citizens." Regrettably, NSF only funded research. Education in the form of student support and fellowships was not included. Until the late 1950s, ONR was the major supporter of education and human resources in ocean sciences. From 1958 onward, NSF's rapid growth started to have a major impact, and other agencies such as the Atomic Energy Commission, provided research support and assistantships for targeted research related to nuclear waste or the unintended consequences of nuclear power generation.

By the late 1950s, ONR had assisted in the establishment of many of today's oceanographic institutions. The field of ocean sciences was now established at over a dozen universities, whereas in 1946, doctoral education in oceanography existed mainly through other science departments such as chemistry and zoology. Another NAS report was published on oceanography in 1959 and purported to look to the next decade (NAS, 1959). However, its comments on education and the role of universities was much like the Lillie Report, focusing on producing clones for the research enterprise. Again, there were only three major suggestions:

> Universities now providing graduate education for oceanographers should be encouraged to increase numbers and quality of output.

> It is desirable to develop oceanographic education at new centers that should be at universities with strong faculties in the sciences.

Efforts should be made in the research and survey programs to use larger numbers of assistants at the Bachelor's and Master's level to utilize more efficiently the limited number of persons available at the doctoral level.

The third, and arguably most influential report on oceanography, was published in 1969. It is often referred to as the Stratton Report but its full title is *Our Nation and the Sea: A Plan for National Action* (CMSER, 1969). However, this report is as weak as the Lillie Report in its recognition of the role and responsibilities of education on oceanography. It makes just two recommendations in relation to education:

NOAA [should] be assigned responsibility to help assure that the Nation's marine manpower needs are satisfied and to help devise uniform standards for nomenclature of marine occupations. (Note NOAA wisely, or by default, did not achieve this!)

NSF should expand its support for undergraduate and graduate education in the basic marine-related disciplines and plan post-doctoral programs in consultation with academic and industrial marine communities.

Hence, forty years after the Lillie Report, education in ocean sciences was largely ignored at the federal level and the individual institutions devised their own responses to local educational needs as well as the perceived national research agenda. The growth of graduate programs in the 1950s and 1960s was in large measure a response to the lack of trained staff, so that both master's and doctoral programs were created. NAS (1959) made many recommendations about increasing support for students and suggesting stronger ties between oceanographic labs and academic institutions. These recommendations focused almost exclusively on the graduate level.

EDUCATION IN THE OCEAN SCIENCES: WHAT IS THE POINT?

As the NAS reports did not provide any insight into the role and goals of education in oceanography in the first years of the development of our field, it is now appropriate to ask a series of questions. These questions are not new. Since 1979, a group of deans and directors of academic programs in the United States have met biennially to share information and discuss shared issues related to graduate education. Called the "Deans' Retreat," these meetings were catalyzed by Charley Hollister and Jake Peirson of Woods Hole, and I worked with the group for 17 years ensuring the development of a database for our discussions (Nowell and Hollister, 1988, 1990). The data are now available through the Consortium for Oceanographic Research and Education (http://core.cast.msstate.edu/oserintro.html) and the biennial meetings are called the "Ocean Sciences Educators Retreat."

Is Our Role Simply to Produce Professionals? If you attend a meeting of oceanographers you will almost always

TABLE 1 Employment of Ph.D. Degree Recipients

	Ocean Sciences (%)	Physics (%)
Industry	10	35
Government	14	10
University	60	36
Other (FFRDC)	16	19
TOTAL	100	100

NOTE: FFRDC = Federally Funded Research and Development Center.

hear the clarion call that we are producing too many Ph.D.s—new competition who are sometimes out-competing the established scientists. And, if you ever ask in a group why we have doctoral programs, you will hear a uniform response. We want excellent students who are creative and imaginative to assist in transforming the field, and we want the field to expand to absorb these new scientists. But is that the role of academic departments at universities, or is that just one of many responsibilities? In some fields, such as dentistry, the objective of achieving the professional qualification is to practice the art and skills learned. I can think of no other reason for obtaining a DDS degree than to practice dentistry. In the case of a medical degree the objective is overwhelmingly to produce practicing physicians. Does obtaining a law degree mean you will practice law? I would say overwhelmingly yes, unless you choose to enter politics! (Maybe that explains the difference in regard for learning between 1860 and 1999.) But, does an advanced degree in physics mean you are going to become an academic physicist, or does an advanced degree in oceanography mean you can only become a faculty member undertaking research? Table 1 indicates that the answers to these last two questions show a surprising variance. More physicists than oceanographers enter industry than become academics by a factor of three and almost 50 percent more oceanographers enter research in oceanography than engage in research in physics. Is this a consequence of the structure of our field, in which the overwhelming majority of students are supported on research assistantships versus being supported on teaching assistantships or fellowships? Today, a more eclectic vision is emerging among faculty that recognizes and even encourages students to consider careers besides becoming a federally supported researcher. Faculty are recognizing that the relationship of educator to the student is more than the relationship of craftsperson to apprentice.

Is Our Role to Produce Scientifically Literate and Numerically Adept Graduates Who Enter a Wide Range of Professions? One way to answer this question is to look at the employment patterns from our field and compare them with another field such as physics. Tables 2 and 3 compare master's and bachelor's degree employment, respectively, for oceanography and physics.

TABLE 2   Employment of Master's Degree Recipients

	Ocean Sciences (%)	Physics (%)
Industry	25	40
University	18	16
Government	29	18
Self Employed	>1	22
Other (unknown)	28	4
TOTAL	100	100

TABLE 3   Employment of Batchelor's Degree Recipients

	Ocean Sciences (%)	Physics (%)
Industry	50	55
College/University	15	3
Government	25	11
Military	5	19
High School	>1	10
Other	5	2
TOTAL	100	100

What is evident is the strong role that industry plays in employment in physics all the way from bachelor's (55 percent) to master's (40 percent) to doctoral degree (35 percent) holders. Another surprising difference is the very small numbers of bachelor's degree holders in oceanography that enter high school teaching. Is it possible that the paucity of students interested in the Earth sciences is due to the low number of high school teachers who have a formal education in the Earth sciences?

I conclude from these data that oceanography has yet to capitalize on the value of a master's degree. Too often, a master's student is regarded as a failed doctoral student: but economic data show that a master's degree is the most economically advantageous degree. Bachelor's degrees in oceanography are exceptionally rare: very few academic institutions offer such degrees because the usual argument is made that you can't be an oceanographer with just a bachelor's degree and that a grounding in basic science is crucial for entering graduate school. Such arguments ignore the fact that fewer than 20 percent of undergraduates proceed to graduate school and those faculties at universities have a responsibility to educate undergraduates. Oceanography as a field has missed out on the chance to lead the burgeoning interest in interdisciplinary education even though oceanography is inherently interdisciplinary. It is a problem that could be addressed readily at the local level.

Is Our Role to Provide an Intelligent Basis for Public Decisionmaking About Marine and Coastal Issues as Well as the Larger Context of Global Environmental Issues and

Issues of Science in Public Policy?   A survey of public understanding of science by NSF three years ago (NSB, 1996) found that only 2 percent of respondents understand science as the development and testing of theory. About 13 percent understand that science involves careful measurement and comparison of data and 21 percent understand the concept that an experiment may involve the use of a control group. Approximately 64 percent do not understand science at any of these levels.

A survey of public understanding of environmental concepts is slightly more encouraging. While only 7 percent can list the cause of acid rain, over 17 percent can identify the location of the ozone hole and 32 percent can list harms that result from the ozone hole. I would suggest that oceanographers, whether they be faculty or students, whether they be aquaria employees or research institution staff, have a shared responsibility to ensure the public can make informed choices when environmental issues reach the ballot. Apportionment of water in the U.S. West and the threat of salmon species extinction are part of this suite of issues for which we, as scientists, have a public responsibility.

WHAT ARE OUR VALUES? THE PURPOSES OF EDUCATION IN OCEAN SCIENCES

The responsibility of scientists in the arena of public education cannot be underestimated but a question closer to the hearts of academics is what are the purposes of education in oceanography? I would list specific purposes. They apply equally to undergraduate as well as post-graduate education.

- We strive to teach students the language of ocean sciences and some things of the disciplines that are its underpinnings.
- We introduce students to the ways of science that imply familiarity with the tools and methodologies of inquiry and with the conceptual as well as practical problems of ocean sciences.
- We help students learn critical thinking skills including the methods of reasoning logically, deductively, inductively, of accuracy and precision and the limitations of data and of models.
- We help students become effective communicators and strive to persuade students to teach others.
- We inculcate a personal love of learning that will last a lifetime so that internal scholarly standards and a continuing curiosity become the basis for living.

I would challenge each academic department in ocean sciences to evaluate its curricula and its educational programs and ask how many of these goals are achieved. Although this challenge is partly being addressed as more and more university departments evaluate their responsibility to undergraduate teaching, it is still far from universal that faculty in the ocean sciences perceive their role as educators, and not just master craftspersons.

WHAT CAN WE DO AS A SCIENCE?

There are three arenas in which we can make a difference. The one most central to the hearts of most faculty is graduate education. Although we may do a superb job in training students for research careers in oceanography, the variability in funding, the decline in national interest in science, and the decreasing numbers of students interested in becoming teaching faculty all suggest we have some work to do to better prepare our students for a life other than that of a research scientist. We should begin by asking what roles the differing degrees play and what are the expectations and rewards for different levels of academic effort. We must ensure that graduate education is more than a research apprenticeship. One question we might consider is what role a thesis should play in the degree and what role it should play in relation to student learning and subsequent employment. One question that can be asked is: How do we envision doctoral thesis research? In the past, the thesis was seen as piece of lone scholarship developed by the individual student working as much as possible independently of everyone else. But today many of the problems that are being investigated require multi-disciplinary teams and teams that have programs that last longer than the duration of a student's thesis years. How do we develop team-based collaborative research and teach students how to make significant creative contributions to shared societally relevant problems?

A survey carried out at Stanford University (Massey and Goldman, 1995; see also Golde and Fiske, 1997) revealed recently that 60 percent of doctoral students are looking for careers outside academia or not involving research. In other words, a minority of students are considering traditional careers in academia. In addition, 70 percent of the students claimed they had changed their career goals while at graduate school. But I surmise that an overwhelming majority of the present faculty believe that the only good students are those who are planning to become faculty! That is certainly the perception of the students in the survey who assert that faculty are considerably more supportive if they perceive the student to be pursuing a research career. The Stanford survey showed that students overwhelmingly (73 percent) felt that the doctoral degree takes too long to obtain and that 80 percent claimed advising was the most important aspect of doctoral studies. It would seem then that we must, as faculty address our responsibilities to explain to prospective and in-residence students the differing career paths and do so in a supportive manner. We must also develop better methods of providing students information on differing career tracks (e.g., through professional societies). The American Physical Society already does this. Their Web page is a good example of how to be supportive of beginning scholars in a field.

As a field, oceanography has been significantly absent from undergraduate education. In part, this is because many universities have their oceanography programs located at a distance from the center of mass of their undergraduate programs, but in part, it is a self-sustaining result. We didn't have undergraduates, so we don't have undergraduates, so we don't want undergraduates. But the entire field of Earth sciences has changed. Global environmental science has become of more immediacy to local and national politics and as Earth system science has recently become possible through structured and linked models and global observing networks. The future for oceanography may lie in much stronger linkages to other geosciences including atmospheric sciences, geo-hydrology, environmental chemistry, and sustainable biospheres. The isolation from undergraduate education may then become a major handicap to future university programs. The emerging integration of the global geosciences offers a stellar opportunity for oceanographers to become more actively involved in undergraduate education. It offers the chance to encourage smart students to enter graduate school, learn geosciences and then teach, and learn about the integration of the sciences and the role of collaborative studies in important societal problems. If we do not avail ourselves of this opportunity, oceanography could become marginal to many universities and thus, become even more dependent on federal research funds.

The last area in which we must examine our values and our responsibilities is in the area of societal education. In the past year the American public has been inundated by stories that involve the ocean. The movie *Titanic*, the novel *The Perfect Storm*, the widespread coverage of the impacts of warm Pacific waters through El Niño are perfect examples of a strong base on which we could build public interest and support for our science. Faculty or research scientists alone cannot undertake this responsibility. It is the shared responsibility of public and private universities and research institutions, of scientists at federal and state agencies and aquaria. While individuals look at what contributions they are making, higher educational institutions are increasingly re-evaluating their role in undergraduate education and K-12 education and outreach. Now would be an excellent time for NSF to re-evaluate its organization that separates education from research. Increasingly, we tie these together: leadership by the agency to integrate them at the funding level would be a strong signal of change.

To return to the beginning again to the words of Abraham Lincoln, "Public opinion is everything. With public opinion nothing can fail; without it, nothing can succeed."

In oceanography, in science, and in education, we must recognize, and we must realize, and we must respond to this concept.

Oceanography, and all of us who are committed to the field, will succeed when the public shares and supports our goals. We can succeed best when our goals and public goals are one and the same. This requires listening to an audience that we in academia rarely consider. While we educate the public about oceanography, we should also listen to the challenges that the public believes are important.

REFERENCES

Bush, V. 1945. *Science-The Endless Frontier. A Report to the President on a Program for Postwar Scientific Research.* U.S. Office of Scientific Research and Development, Government Printing Office, Washington, D.C.

Commission on Marine Science, Engineering and Resources (CMSER). 1969. *Our Nation and the Sea: A Plan for National Action.* U.S. Government Printing Office, Washington, D.C.

Farrington, J.W., and A.L. Peirson, III. 1996. Undergraduate college faculty workshop. *Oceanography* 9:135-139.

Golde, C.M., and P. Fiske. 1997. Graduate school and the job market of the 1990s: A survey of young geoscientists. AGU On-Line Discussion http://earth.agu.org/eos_elec/97117e.html

Massey, W., and C. Goldman. 1995. The Production and Utilization of Science and Engineering Doctorates in the United States. Stanford University, California.

National Academy of Sciences (NAS). 1929. *Oceanography: Its Scope, Problems, and Economic Importance.* Houghton Mifflin Company, Boston.

National Academy of Sciences (NAS). 1959. *Oceanography 1960 to 1970.* National Academy Press, Washington D.C.

National Science Board (NSB). 1996. *Science and Engineering Indicators.* NSB 96-2,Washington D.C.

Nowell, A.R.M., and C.D. Hollister. 1988. Graduate students in oceanography: Recruitment, success and career prospects. *EOS* 69:834-843.

Nowell, A.R.M., and C.D. Hollister. 1990. Undergraduate and graduate education in oceanography. *Oceanus* 33:31-38.

Evolving Institutional Arrangements for U.S. Ocean Sciences

WILLIAM J. MERRELL, MARY HOPE KATSOUROS, AND
GLENN P. BOLEDOVICH

The H. John Heinz III Center for Science, Economics, and Environment

INTRODUCTION

Our understanding of the oceans has changed markedly since the creation of the National Academy of Sciences (NAS) in 1863. In assessing where federally supported ocean research and science are going in the next millennium, it is instructive to understand where we have been. This paper highlights some of the major institutional forces that have influenced national ocean research in the past 135 years. We discuss the evolution of federal agencies, with emphasis on how the NAS, the Office of Naval Research (ONR), and the National Science Foundation (NSF) have influenced the development and direction of ocean science and oceanographic institutions. Based on this historical understanding we next discuss emerging national needs for ocean research in light of the principal factors that drive ocean science.

Factors That Drive Ocean Science

There are five principal factors driving ocean science: basic research, national pride, national defense, economic benefits, and environmental concerns. Changing national needs influence the weight and priority given to these factors and their significance shifts over time.

Basic Research—Curiosity and Understanding

The quest to improve our understanding of the world around us is the driving force for scientific inquiry. Basic research, as it applies to oceanography, has the primary goal of understanding ocean phenomena. The nature of basic research is such that its purposes are broad and the results often are not readily predicted. Although basic research often supports other objectives, there is an inherent value in improved knowledge and understanding of fundamental systems.

National Pride

National pride is a sense of national consciousness exalting one nation's accomplishments above others. A prime example was the race with the former Soviet Union for the exploration of space. Support for research and exploration often increases when the connection between science and national prestige becomes apparent to policy makers and the public.

National Defense

National defense typically means the protective steps taken by a country to guard against attack, espionage, sabotage, or crime. The development of oceanography in the United States grew in large part because of national security interests during World War I, World War II, and the Cold War. In the Department of Defense, the Navy has responsibility for marine research and has supported extensive scientific investigations to provide a more complete understanding of the ocean environment as necessary for our national security.

Economic Benefits

Nations have always viewed the seas as a source of wealth. Scientific knowledge and technical capabilities in the marine environment often have been supported to maintain and expand our national economy. Two overriding concerns are (1) avoiding being confronted with a shortage of raw materials and (2) developing marine resources to advance economic growth. Principal economic uses of the sea are energy, mineral and fishery production, transportation, and recreation. In the United States, these activities are carried out largely by the private sector; however, federal, state, and local authorities often regulate the industries.

Environmental Concerns

In the 1950s, research into the impacts of marine pollutants flourished after the incident of mercury poisoning in Minamata, Japan. In the 1960s, a series of alarming events raised our national environmental consciousness. For example, the discovery that dichlorodiphenyltrichloroethane (DDT) was the agent responsible for the inability of pelican eggs to hatch verified Rachel Carson's (1962) warning in *Silent Spring* of chemical dangers lurking in the environment. At about the same time, oil from an offshore drilling rig blowout coated beaches in Santa Barbara, California.

THE NATIONAL ACADEMY OF SCIENCES AND U.S. OCEANOGRAPHY

The Early Years

On March 3, 1863, as its last act on its last day, the 37th Congress passed legislation establishing "an independent organization to address scientific issues critical to the defense of the country." That evening, Abraham Lincoln signed this bill—creating an organization that would be known as the National Academy of Sciences—into law.

Its charter mandated that "whenever called upon by any department of the government" the NAS was to "investigate, examine... and report upon any subject of science or art." Federal agencies made ten requests to NAS in the first year. Three were ocean and defense related:

1. The Committee on Protecting the Bottom of Iron Clad Ships from Injury by Saltwater: On May 8, 1863, the Navy Department through the chief of its Bureau of Navigation, Admiral Charles H. Davis, asked the Academy to investigate protection for the bottoms of iron ships from injury by salt water. Wolcott Gibb's committee, appointed the next day, reported that a metallic coating or alloy was commonly used to prevent or arrest corrosion of metals and that substances in paints often were used to destroy accumulations of plants or animals on ship bottoms. The committee provided its report in seven months and was discharged early the next year.

2. The Compass Committee: Also on May 8, 1863, the Academy was asked to conduct an investigation of magnetic deviations in iron ships and means for better correction of their compasses. Alexander Bache chaired the committee appointed on May 20 and made his report with seven subreports on January 7, 1864.

3. The Committee to Examine Wind and Current Charts and Sailing Directions: The third request was for recommendations regarding the proposed discontinuation of Matthew Fontaine Maury's *Wind and Current Charts* and *Sailing Directions*. The committee's view was less than favorable, finding the charts to be "a most wanton waste of

TABLE 1 Era of Early Institution Building

Dates of Origin	Institutions
1853	California Academy of Sciences, California
1885	U.S. Fish and Wildlife Biological Lab in Woods Hole, Massachusetts
1888	Marine Biological Laboratory, Woods Hole, Massachusetts
1892	Hopkins Marine Station, California
1903	Scripps Institution of Oceanography, California
1904	University of Washington, Friday Harbor Labs, Washington

valuable paper" that "embrace much, which is unsound in philosophy, and little that is practically useful." It recommended that they be discontinued in their current form. In Maury's defense, his charts did, in fact, reduce sailing times, and a simplified version was republished 20 years later.

These early ocean committees set the tone for the Academy's future role in advancing ocean science in support of national security. But, for the next fifty years, federal agencies made no major marine research requests to NAS. During this time, however, a number of small marine laboratories were established and were used by biologists and their students from nearby universities during the summer months. Some of these were and still are supported by state funds, whereas others received funds from private foundations. As these seaside biological stations grew so did the scope of their investigations and the interests of the scientists using them. Some of them grew to become oceanographic laboratories. Table 1 indicates the dates when some of these early oceanographic institutions began.

In 1916, the National Academy of Sciences formed the National Research Council (NRC) to improve cooperation among government, academic, industrial, and other research organizations. The principal aims in creating the NRC were to encourage investigations of natural phenomena, increase the use of research to develop U.S. industries, strengthen national defense, and promote national security and welfare.

World Wars Spur Investment and Advances in Ocean Science

With the outbreak of World War I, the federal government sought the assistance of the NAS-NRC to support the national defense. From 1916 to 1918, three committees were formed:

1. The Committee on Physics chaired by Robert A. Millikan,

2. The Submarine Investigations Subcommittee chaired by Robert A. Millikan, and

3. The Committee on Navigation Specifications for the Emergency Fleet chaired by Lewis S. Bauer.

Also during this time, NAS proceedings found that basic marine research was "a realm in unsurpassed promise for the fruits of investigation." But these fruits proved hard to pick. In 1919, a Committee on Oceanography was formed and chaired by Henry Bigelow. The purpose of this committee was to survey ocean life, but it disbanded in 1923 with "frustrated members feeling they could serve no useful purpose."

In 1927, the Committee on Oceanography was reformed with Frank L. Lillie as chair and Henry Bigelow as secretary. The committee was charged to consider the U.S. role in a worldwide oceanographic research program. The committee produced *Oceanography: Its Scope, Problems and Economic Importance* (NAS, 1929) and *The International Aspects of Oceanography* (NAS, 1937). The Lillie Committee and its reports highlighted national pride and economic factors as drivers for increased oceanographic research. They pointed out the lack of U.S. research vessels and shore facilities, and concluded that the nation was far behind many European nations in the study of physical oceanography and marine biology.

The reports led to an effort to build up national oceanographic institutions, including enhanced facilities at the Scripps Institution of Oceanography and the University of Washington. The Lillie Committee also recommended the establishment of a central oceanographic research institution on the East Coast to promote research and education and to provide a place to integrate the various research activities that were being pursued by private institutions and federal agencies. This led to the establishment of the Woods Hole Oceanographic Institution (WHOI) in 1930 with a $2.5 million grant from the Rockefeller Foundation. Lillie became the chair of the WHOI Board, and Bigelow became the WHOI director.

Post-World War II: A Golden Age for Oceanography

The rapid development of technologies during World War II resulted in an increased appreciation of science and the importance of ocean research for national defense. On August 1, 1946, President Harry S. Truman signed the law creating the Office of Naval Research (ONR). Its primary mission was to secure the collaboration of top-level civilian scientists in all fields of research having a bearing on national security. The Navy worked out a contract arrangement acceptable to the universities that were to undertake the research. The agreements specifically ensured that the scientists involved would retain a large degree of academic freedom by allowing them to initiate projects "in fundamental research without restrictions" — in nuclear physics, medicine, physics, chemistry, mathematics, electronics, mechanics, and oceanography.

At the urging of the Navy, a second Committee on Oceanography chaired by Detlev Bronk was formed in 1948 to assess the state of oceanographic research as part of a larger ONR effort to prepare a long-range national plan. This committee produced *Oceanography 1951* (NAS, 1952) and again reinforced the issue of national pride by describing the United States as far behind other maritime nations in its support of oceanographic research for national defense, transportation, and the exploitation of natural resources. The report found the number of U.S. oceanographers to be fewer than 100. It recommended that $750,000 be allocated to train oceanographers and that additional support be provided for basic research in biological and chemical oceanography.

Meanwhile in 1950, Congress authorized the creation of the National Science Foundation to promote the progress of science; to advance the national health, prosperity, and welfare; to secure the national defense; and to serve other purposes. The President approved the act on May 10, 1950. NSF's support for oceanographic research would prove to be a valuable asset, making important contributions in the advancement of an improved national marine science infrastructure.

It was ONR, however, that continued to take a leadership role. In the 1950s, ONR supported 80 to 90 percent of the oceanographic research occurring in the academic community. ONR concurred with the findings of *Oceanography 1951* and resolved to build up U.S. capabilities including new facilities, ships, and equipment. In 1956, a third NAS Committee on Oceanography (NASCO) was formed at the request of ONR, this time by Art Maxwell who was impressed with the work produced by the Lillie Committee. NASCO became one of the most important, productive, and influential committees in the history of the Academy. It formed more than 20 panels and task groups to examine specific oceanographic challenges and opportunities. The committee produced *Oceanography, 1960 to 1970* (NAS, 1959), an outline for future oceanographic research, and *Economic Benefits from Oceanographic Research* (NAS, 1964), which proved to have a significant effect on oceanographic research as well as on relations between the government and the NAS.

Maxwell personally attended NASCO meetings for ONR. Together with Gordon Lill and Feenan Jennings, Maxwell produced a complementary internal report *The Next Ten Years of Oceanography* (the "TENOC" report; Lill et al., 1959), which was endorsed by the Chief of Naval Operations in 1959 as a plan to increase research funding and provide additional buildings, ships, and pier construction. The additional support was concentrated at ten institutions: Scripps, Woods Hole, and the following universities: Washington, Columbia, Miami, Rhode Island, Oregon State, Texas A&M, New York, and Johns Hopkins.

Oceanography had achieved an importance that was unforeseen at the close of World War II. Outreach efforts triggered interest from leading policy makers, including President John F. Kennedy who in a letter to congressional leaders stated, "We are just at the threshold of our knowledge of the oceans, . . . [This] knowledge is more than a matter of curiosity. Our very survival may hinge upon it." This status had been achieved largely through Navy interest in the oceans. ONR-sponsored oceanography roared into the 1960s, solidifying a strong infrastructure for blue-water oceanographic institutions in the United States. The percentage of NSF funding also increased, resulting in an academic structure based primarily on supporting federal research with little academic and local funding.

Expanding Requirements and Shifting Priorities

As the influence and support of ONR, NSF, and NAS drove oceanographic research into the 1960s, there arose a renewed interest in the economic and environmental aspects of ocean research. The Mansfield Amendment requiring ONR's ocean research to be defense related caused some uncertainty and ultimately shifted responsibility for some types of basic research to NSF. Legislation in support of the International Decade of Ocean Exploration (IDOE) nearly doubled NSF's funding for ocean science. The National Sea Grant College Act was introduced by Senator Claiborne Pell of Rhode Island, and on its passage in 1966 the program initially was placed under NSF. In the same year, Congress passed the Marine Resources and Engineering Development Act of 1966 authorizing a Commission on Marine Science, Engineering and Resources, more commonly known today as the Stratton Commission.

The primary objectives of the Stratton Commission were to support the expanding economy and develop marine resources. In January 1969, the Stratton Commission released its influential report *Our Nation and the Sea* (CMSER, 1969). The report made 126 recommendations spread over 17 categories. From these recommendations a flurry of legislation was enacted: the Coastal Zone Management Act, the National Marine Sanctuaries Act, the Marine Mammal Protection Act, and the Fishery Conservation and Management Act. Based on the report's recommendations, the National Oceanic and Atmospheric Administration (NOAA) was established in 1970 with Robert White as its first administrator. NOAA was tasked with administering these new laws; conducting integrated ocean and atmospheric research, and Earth data collection; and providing related grants for research, education, and advisory services.

Other important environmental legislation such as the Clean Water Act and the Endangered Species Act, was enacted in 1970. The Environmental Protection Agency (EPA) was formed as an independent agency within the Executive branch. At the international level, negotiations of a new treaty on the Law of the Sea were initiated in the early 1970s. As the environmental movement grew, so did the number of ocean-related environmental non-governmental organizations (NGOs), such as Greenpeace and the Center for Marine Conservation. Many coastal laboratories expanded their research into marine pollution and living marine resources; however, the blue-water institutions with longer traditions

of basic research were not as quick to move into these emerging fields.

The NAS remained active following the release of the Stratton Commission report. In the early 1970s, NAS convened the Ocean Affairs Board chaired by Robert Morse. The board produced several important reports on topics ranging from the Law of the Sea, to climate prediction, numerical modeling, and offshore petroleum resources. Later in the 1970s, the NAS established the Ocean Sciences Board and the Ocean Policy Committee. The Ocean Sciences Board, chaired by John Steele, produced reports on NOAA, the need for increased large-scale marine research on climate, and other issues in the 1980s. The Ocean Policy Committee, chaired by Edward Miles and Paul Fye, produced reports on the Law of the Sea, fisheries, and other international marine policy issues.

The expansion of ocean-related research led the academic community to form its own associations, for example, the University-National Oceanographic Laboratory System (UNOLS) and the National Association of Marine Laboratories (NAML). In 1976, ten leading U.S. blue-water oceanographic institutions formed the Joint Oceanographic Institutions (JOI), Inc. to facilitate and foster the integration of program and facility requirements and to bring to bear the collective capabilities of the individual oceanographic institutions on research planning and management of ocean sciences. JOI continues to manage the Ocean Drilling Program and the U.S. Science Support Program.

In 1983, NAS merged its ocean science and policy boards to form the Board on Ocean Sciences and Policy. John Slaughter chaired the board that produced several reports on oil development, ocean dumping, and climate. In 1985, the board was renamed the Ocean Studies Board (OSB). Walter Munk was the first chair, followed by John Sclater, Carl Wunsch, William Merrell, and Kenneth Brink. The OSB has produced more than 50 reports on a broad array of topics ranging from climate to coastal ecosystems and from fisheries and marine mammals to improved integration of science and policy. The number of committees of the board expanded, including the Committees on Major U.S. Ocean Research Programs, U.S.-Mexico Collaboration for Ocean Science Research, Operational Global Ocean Observing System, Fish Stock Assessment Methods, and Ecosystem Management for Sustainable Marine Fisheries.

As the range of OSB committees indicates, support for oceanographic research in the 1980s began to increasingly reflect the demand and need to address a diverse range of issues and problems. This trend continued into the 1990s. The president of JOI, Dr. D. James Baker, was appointed by newly elected President Clinton to head NOAA. The OSB reviewed NOAA and Navy research programs and convened Committees on Science and Policy for the Coastal Ocean, Identifying High-Priority Science to Meet National Coastal Needs, Biological Diversity in Marine Systems, and Low-Frequency Sound and Marine Mammals. Throughout this period of expansion, other NAS boards, such as the Marine Board and the Polar Research Board, investigated additional marine-related issues. The latter's 1996 report on the Bering Sea ecosystem (NRC, 1996) complemented efforts of the OSB and provided a comprehensive analysis of the challenges besetting this productive arctic ecosystem.

Ocean institutions continued to develop and made efforts to improve their capabilities. The long-standing oceanographic centers that made up JOI were complemented by an increasing number of academic centers focusing on nearshore issues including fisheries, coastal pollution, and marine toxicology. Responding to these trends, the president of JOI, Admiral James T. Watkins created a new organization of more than 50 marine research institutions, the Consortium for Oceanographic Research and Education (CORE), encompassing a broader array of marine science expertise.

In addition to changing research needs, the political landscape continued to evolve, opening up new opportunities for joint research and improved integration. OSB recognized that important changes were altering research needs and opportunities. The board itself took on the task of reviewing trends in ocean science and provided its views in the 1992 report *Oceanography in the Next Decade* (NRC, 1992). Working to implement many of the reports' recommendations, JOI and the CORE president, Admiral Watkins, took the case for ocean research to Congress, which subsequently passed the National Oceanographic Partnership Act supporting partnership-based research among federal agencies, academic institutions, and other interests. Led by ONR and with increasing support from NOAA, the National Oceanographic Partnership Program is forging new relationships and cross-cutting approaches to ocean research. It also is worth noting that the recently appointed director of the NSF, Dr. Rita Colwell, has extensive expertise and interest in marine science. The 1998 International Year of the Ocean provided more opportunities to chronicle the importance of continued and increased support for ocean research, although legislation to create an ocean commission died in the final hours of the 105th Congress.

The Factors Driving Ocean Science in the Future

The five factors driving ocean science—basic research, national pride, national defense, economic benefits, and environmental concerns—continue to influence ocean research today. However, in some cases the scope of the factors themselves has evolved. The changing marine environment and our improved understanding of it also are influencing the focus of marine research. Environmental laws and changing concepts of government administration are creating new opportunities and demands for science and research to support responsible decisionmaking. Overriding concepts of sustainability, biodiversity, biocomplexity, and ecosystem management are moving from theory to implementation. Increased emphasis on partnerships, interdisciplinary research, and cross-cutting projects often combines research goals and brings several factors into play simultaneously.

Basic Research

Historically, the federal government has supported basic oceanographic research. This support fostered the development of our ocean research institutions. Some researchers believe basic research warrants continued government support. In a 1998 report prepared under the guidance of Vice Chairman Vernon Ehlers (R-MI), the House Committee on Science issued a report, *Unlocking Our Future: Toward a New National Science Policy* (House Committee on Science, 1998). The report concluded that there is a continuing need for research driven by a need for basic understanding. Some level of basic oceanographic research will continue to be supported.

However, as the scope and number of critical needs for applied research increases, policy makers are under increasing pressure to support science that directly meets these challenges. Arguably, the research community also has a responsibility to respond to such national priorities. Also, today's ability to analyze, model, and communicate information instantly has altered basic research in that various researchers can immediately begin to apply the work of others to meet specific needs. For example, basic research on ocean dynamics today may have immediate bearing on improving our understanding of the impacts that climate events, such as El Niño or global warming, may have on society.

The demand for blue-water oceanography is today being joined with an increasing demand for coastal oceanography where many environmental challenges exist. Whether this means increased competition among researchers for the same pool of funds is unclear. If the larger goal of increasing overall support for ocean research can be attained, then the needs of different ocean disciplines may be met without compromising a range of research interests.

National Pride

U.S. leadership in oceanography became a matter of national pride as people increasingly realized that ocean science was important to our continued prosperity. But since the fall of the Soviet Union, the United States is no longer locked in the kind of one-on-one competition that for so many years made national pride an important driver. As the primary global superpower in an increasingly interdependent economy, U.S. national pride is to some degree giving way to a sense of global responsibility. Pride in being a world leader is supplanting the former sense of pride based on comparison to a competitor. This is certainly true in the arena of ocean research, where less developed countries will continue to rely on the United States and other developed countries to provide much of the science to support important ocean and marine resource policy decisions.

National Defense

The importance of ONR to the advancement of ocean science cannot be underestimated. In the wake of the Cold War, ONR continues to substantially fund ocean research and is playing a leading role in the successful implementation of the National Oceanographic Partnership Program. But our understanding of national defense is itself evolving. Increasingly, it is referred to as national security and includes notions of economic and environmental security. Undoubtedly, national defense will continue to be a major driver of ocean research. However, budget pressures may reduce support for basic research in favor of applied science linked directly to meeting priority security concerns.

Economic Benefits

Despite concerns for environmental integrity and declines in the populations of many fish and other marine resources, economic opportunities in the ocean continue to drive ocean research. In some cases, such as fisheries, the economic revival of a now-compromised industry is driving increased science. At the same time, new economic opportunities are driving ocean science. For example, the discovery of new life forms around thermal vents in the deep ocean is resulting in expanding research into marine pharmacology. The economic importance of fisheries, marine transportation and trade, coastal tourism, and mineral development will continue to drive the need for science to promote wise decision making, balance conflicts among users, and promote sustainable practices.

Environmental Concerns

Clearly, environmental concerns will continue to be an increasingly important driver of ocean science. Just 30 years ago, economics was the primary driver as evidenced by the work of the Stratton Commission. Today, climate change, seasonal events such as El Niño, depleted fisheries, nutrient enrichment, harmful algal blooms, dying coral reefs, coastal water pollution, and other environmental challenges are driving a larger share of the investment in marine research. Implementing principles of sustainable development, that is, the balancing of economic and environmental objectives, will require increased investment in science to provide the basis for resource use decisions. For example, increased interest in science is a fundamental result of employing precautionary practices because science will provide the basis for improved assessments of impacts and for reducing the uncertainties posed by potentially high-risk activities.

CONCLUSION

The outlook for investment in ocean science is bright in part because there are so many critical and emerging national needs for improved information on oceans, marine ecosystems, and marine resources. However, despite the rapid rate of change and technological advancement during the past 30 years, the nation has not updated its ocean policy since the Stratton Commission. For example, 30 years ago we did not have rights and responsibilities for the exclusive economic zone—a national marine area larger than the land mass of the entire country for which we have no research strategy or integrated management plans.

ONR, NAS, NSF, other agencies, and the oceanographic community at large have done a fairly good job of accommodating and adjusting to changing ocean science and policy needs. However, rapid change and growth have made it difficult to keep an eye on the big picture. If ocean research is to result in improved policy making and best serve the public and future generations, there is a need to undertake a review of where we are going and to set a path to get us there. There is a need for a collaborative effort that includes marine scientists, government policy makers, industry, and environmental interests to forge a national ocean strategy.

REFERENCES

Carson, R. 1962. *Silent Spring*. Houghton Mifflin, Boston, Massachusetts.

Commission on Marine Science, Engineering and Resources (CMSER). 1969. Pp. 21-22 in *Our Nation and the Sea: A Plan for National Action*. U.S. Government Printing Office, Washington, D.C.

House Committee on Science. 1998. *Unlocking Our Future: Toward a New National Science Policy*. House of Representatives, Washington, D.C. http://www.house.gov/science/science_policy_report.htm

Lill, G.G., A.E. Maxwell, and F.D. Jennings. 1959. *The Next Ten Years of Oceanography*. Internal Memo, Office of Naval Research.

National Academy of Sciences (NAS). 1929. *Oceanography: Its Scope, Problems, and Economic Importance*. Houghton Mifflin Company, Boston.

National Academy of Sciences (NAS). 1937. *International Aspects of Oceanography: Oceanographic Data and Provisions for Oceanographic Research*. National Academy of Sciences, Washington D.C.

National Academy of Sciences (NAS). 1952. *Oceanography, 1951: A Report on the Present Status of the Science of the Sea*. National Academy of Sciences/National Research Council, Washington D.C.

National Academy of Sciences (NAS). 1959. *Oceanography 1960 to 1970*. National Academy Press, Washington D.C.

National Academy of Sciences (NAS). 1964. *Economic Benefits from Oceanographic Research, a Special Report*. National Research Council, Washington, D.C.

National Academy of Sciences (NAS). 1969. *Oceanic Quest: The International Decade of Ocean Exploration*. National Academy Press, Washington, D.C.

National Academy of Sciences (NAS). 1979. *The Continuing Quest: Large-Scale Ocean Science for the Future*. National Academy Press, Washington, D.C.

National Research Council (NRC). 1992. *Oceanography in the Next Decade: Building New Partnerships*. National Academy Press, Washington, D.C.

National Research Council (NRC). 1996. *The Bering Sea Ecosystem*. National Academy Press, Washington, D.C.

NSF's Commitment to the Deep

Rita Colwell

Director, National Science Foundation

It is a genuine pleasure for me to be here to celebrate fifty years of ocean discovery and to be able to join you in planning for the next fifty years. The oceans have intrigued and attracted human populations for centuries. There isn't an activity more compelling than the search for knowledge and understanding of the oceans.

As you know, I have spent most of my career on or near the ocean, but I also bring a more personal perspective to ocean discovery. My husband Jack and I are racing sailors, therefore the sea is our recreation but also an arena of challenge and discovery in another dimension. And my own research into cholera is directly related to the oceans and their influence on weather and climate.

The history of cholera reveals a remarkably strong association with the sea. El Niño brings rain, an influx of nutrients from land, and warm sea surface temperatures. These conditions are associated with initiating plankton blooms. A single copepod can carry many *Vibrio cholerae* cells.

I will spare you from further details. Suffice to say that those details have engaged me in decades of research. It is especially fulfilling to be here today, in my still new role as Director of the National Science Foundation, to talk about ocean discovery, past, present, and future.

Many of you have been active participants in NSF's wide-ranging and significant history in ocean research. The Foundation is grateful for your commitment and the excellent work you have done. We are also proud of both our primary role in ocean discovery and our role as a collaborator with other institutions, other agencies, and other nations.

The pervasive, powerful, mysterious, and bountiful forces of the oceans on the planet are evident not only in science but also literature. Seneca, one of Rome's leading intellectuals in the mid-first century (AD) spoke of the ocean in one of his plays. He wrote, "An age will come after many years when the Ocean will loose [loosen] the chains of things, and a huge land lie revealed. . . ." This has been variously interpreted in different historical eras. Christopher

Columbus' son, Ferdinand, thought the comment foretold his father's discovery of the New World in 1492.

Those same words today, in relation to ocean discovery, signify to me that the next fifty years in ocean research will be truly extraordinary. Although Columbus' son thought the reference to a "huge land . . . revealed" was the discovery of the Americas, those now working in ocean science might well think of that "huge land" as the abyssal regions of the ocean that we have only begun to explore. Sophisticated tools and technology are changing every aspect of science—from writing a grant proposal, to the data we gather, and to methods of data collection and analysis.

The impact of the new tools and technology on the way we do oceanography has been revolutionary. NSF supports this revolution and its expanding diversity into robotic vehicles, permanent seafloor observatories, new optical and acoustic imaging methods, long-term moorings, and the vast opportunities opening up in satellite communications. These technological leaps and others to be made will transform the ocean sciences.

The academic research vessel fleet, the *JOIDES Resolution*, and the *Alvin* are the precursors of the robotic eyes and ears that mark a whole new era in oceanography. In the late 1970s, researchers, with the help of the workhorse *Alvin*, were able to explore areas like the volcanic terrain of the East Pacific Rise. Their findings changed our entire understanding of the deep ocean. They found unique ecosystems teeming with life supported by the geothermal energy given off by Earth's inner heat. This discovery was possible with the technological capability of *Alvin*.

Today, oceanography continues to evolve from an exploratory endeavor to efforts that require an ability to make observations of ocean processes over periods of years. Recent advances in technology have enabled us to establish the first permanent U.S. deep seafloor observatory that is connected to shore by a dedicated cable. This observatory, called H2O, or the Hawaii-2 Observatory, sensed its first earthquake this week, a magnitude 5.7 quake near Papua, New Guinea. Needless to say, this has generated tremors of excitement on land.

This new ability to receive and record ocean data continuously and to communicate with scientific instruments on the seafloor will greatly advance our knowledge and predictive capabilities in ocean science. Nonetheless, even with this ever-expanding capability, we all know that scientists have explored relatively little of the deep sea and ocean floor.

Given the fact that our planet's predominant geographical feature is ocean, approximately 70%, some have suggested that the name Earth is a misnomer, that we should have more aptly called ourselves Water. No matter the name, it is our interest and investment in the ocean sciences that counts. These have serious implications for our future prosperity and survival.

Historically, civilizations developed at the water's edge. Today, there are 16 cities in the world with a population of over 10 million. Thirteen of these cities are along coastlines. And the present is not unlike the past, at least in the United States. According to a recent survey from *Economist* magazine, ". . . in America, almost half of all new residential development is near the ocean, and people are moving there at the rate of 3,600 a day." The sustainability of urban and altered ecosystems challenges our best scientific knowledge and opens new directions for research.

The many dimensions of an increasing world population, combined with the power of technology, have changed our global environment. There is both opportunity and responsibility for the science community here. Going back to Seneca's lines, "An age will come after many years when the Ocean will loose [loosen] the chains of things." That suggestion, which dates back some 1,900 years, has a certain currency about it today.

A new, multifaceted idea has begun to take shape as a research direction, as well as a social understanding. I refer to this concept as biocomplexity, a word not yet familiar to most of the scientific community. The oceans play a significant role in the biocomplexity concept. Biocomplexity is not a synonym for biodiversity. It includes and reaches beyond biodiversity. When we speak of sustaining biodiversity, we mean primarily maintaining the plant and animal diversity of the planet—a very important goal. However, biocomplexity speaks of a deeper concept. It is not enough to explore and chronicle the enormous diversity of the world's ecosystems. We must do that . . . but also reach beyond, to discover the complex chemical, biological, and social interactions that comprise our planet's systems. From these very subtle, but very sophisticated interrelationships, we can tease out the fundamental principles of sustainability.

Without a doubt, the oceans form the largest, most formidable, and even perhaps the least explored and understood of those systems. However, our survival as a human species, and the ecological survival of the entire planet, will depend on our ability to achieve what is a truly interdisciplinary task.

Over many centuries, ocean travel has allowed us to discover the shape and size of the planet and to acquaint ourselves with its diverse cultures and commodities. Many hundreds of generations have fed themselves from the enormous variety of fish and seafood. We have taken this for granted and, out of ignorance, often abused instead of used, this bounty.

We have only recently begun to discover the hidden understandings and more subtle complexity of the sea. The economic and biomedical potential of the sea is just beginning to be realized. With the discovery of marine organisms that can thrive in extreme environments of heat or cold, all of the old truths about conditions required for life have been tossed to the winds. Results of on-going studies support the possibility that life originated near hydrothermal vents deep in the ocean. The microorganisms found there today appear to be genetically among the oldest on Earth. Our current knowledge of thermophiles and psychrophiles has broad implications for the future. The enzymes produced by thermophiles currently have widespread application in biotechnology.

Although biotechnology is a young field, it has already burgeoned into a $40-billion-dollar-a-year industry. Marine biotechnology has applications in medicine, agriculture, materials science, natural products, chemistry, and bioremediation. It is estimated that aquaculture, just one branch of marine biotechnology, will be relied upon heavily to help meet world food needs. While world fisheries are over-exploited and/or commercially extinct, world population burgeons and world food needs increase. In addition, aquaculture can produce organisms that can be used as biomedical models in research, reservoirs for bioactive molecule production, and useful in bioremediation.

Clearly, we are just at the beginning of an exciting era in ocean discovery and ocean science. What lies ahead will have increasing impact on our daily lives. Our task will be to educate the public about the economic importance and new environmental understanding that continued work in ocean discovery will bring. Ocean science can no longer be viewed as an esoteric, "off-shore" discipline. It is mainland and mainstream. The health and bounty of our oceans are issues of planetary survival.

Centuries ago, the oceans served as our vehicle for learning about distant peoples and distant lands. Only recently have we been learning about the ocean's cyclical control of the planet's climate. The bounty of the oceans has fed human populations since the beginning of time. We are now acutely aware of how fragile that food supply can be. For centuries, human populations have been vulnerable to the forces of weather and climate; many of them are triggered by the oceans.

Today we have tools and knowledge to predict the onset of severe weather cycles months before their occurrence and prepare for the impact. Nineteen hundred years ago, Seneca predicted that many years in the future "the Ocean will loose [loosen] the chains of things." Today we are unlocking the very essence of those words with tools and equipment of unrivaled sophistication. The biocomplexity of the Earth's major systems are the very "chains of things." And so what does it mean to speak of NSF's commitment to the next fifty years of ocean discovery?

It means investment, imagination, and a focus on interdisciplinary work. It means boldly pushing the frontiers of our knowledge while pursuing their economic, social, environmental, and medical applications. It means that understanding and initiating sustainable strategies for the planet will depend heavily on ocean research. It means bringing the importance of the oceans to the forefront of public understanding. It means that we enter our next fifty years of ocean discovery with grand expectations to fulfill.

Fifty Years of Ocean Discovery

RADM Paul G. Gaffney II

Chief of Naval Research

I am very pleased to be with you for this special occasion, and confess to finding myself feeling a bit humbled as I look around this room and see the depth of oceanographic expertise in the audience and on this panel. The National Science Foundation (NSF) and the Office of Naval Research (ONR) share a long history together, and I am honored to be able to help celebrate and congratulate NSF's contribution to our understanding of the oceans, and to share with you some thoughts about the future of ocean discovery.

Before I get started, however, some good news. We now have a signed appropriations bill, and for the first time in many years, the Office of Naval Research gets an upswing in basic research. Our goal for the future is $400 million, and the recent increases give us some cause for measured optimism.

I want personally to thank the entire science and technology community for your involvement over the last few years to help us make this happen—when you tell the story, it's credible (sometimes even charming). It would not have taken place without your engagement and efforts. What a difference a year can make! Last year we were wondering where the floor might be . . . now we can plan a strategy for continued modest growth to restore hope, take prudent risks, and provide for a future as strong as our past. So it is a pleasure to bring some good news to this gathering, and to thank you for the support that helped produce it.

As I contemplated my remarks to you this afternoon, my thoughts turned to the essence of what we are celebrating. We are essentially celebrating a vision. That vision was born more than 50 years ago at the end of World War II, which showed that our nation could not afford to be without a strong, robust basic research program . . . a program that turns predominantly to universities for the substance of that research. It was a vision that has proven critical not only to our national security, but to the overall economic and social well-being of this country.

Both ONR and NSF were born of that vision—and now we are neighbors. Because we were the two first U.S. government agencies dedicated to supporting basic research, we have each had the pleasure of watching and helping one another mature and make our marks. (Since contributions that go unremarked don't get made into marks, permit me to ask you to mark this—two of this year's Nobel laureates were supported by ONR—professor Walter Kohn, who won a Nobel in Chemistry; and Professor Daniel Tsui, who won a Nobel in Physics. I don't have to tell this audience the role the NSF has also played in supporting Nobel-quality research. I think "Father Vannevar" would be proud of his two offspring.)

Since ONR was founded in 1946, we have shared with NSF a leadership role in ocean science. For us that leadership is critical, because ocean science is a core science, it is a Naval science, and it is one we choose not because we like it, because it is interesting, or because it is our charter. We choose it because we need it—this maritime nation needs it and our Naval Service needs to understand it.

Prior to the mid-1960s, ONR was the only major funder of ocean research. Since then, NSF's ocean research program has made its own mark, as has the National Oceanic and Atmospheric Administration (NOAA) . . . and oceanography has become a substantial part of your portfolio in part because the national priorities for ocean research have grown. And why not? This is the world's greatest nation and it is a nation virtually surrounded by water and dependent on water.

ONR retains the leadership role in some areas—areas such as ocean acoustics that are critical to our national security needs—but NSF has assumed responsibility for other areas of broader national importance such as global climate change research and polar science. This is as it should be for siblings who are responsible for the overall health of the research field.

This division of responsibility has proven to be a successful arrangement for both of us . . . and, I believe, has ensured a strong and healthy community. As a matter of fact, this year, we examined our ocean acoustics investment very carefully and as a result are re-invigorating the program with additional resources that restore funding levels to those of a few years ago. We will also provide a few special programs to ensure that we have a sufficient pipeline of students, post-docs, and young faculty to address future acoustic issues, and we hope to send more acoustical research to sea! (Isn't that a novel idea!) I look forward to collaborating with NSF here as well.

Together, we have made a significant impact on ocean sciences. For example, long-term investments in the studies of ocean thermal, chemical, and acoustical properties, and bottom topography ultimately helped submariners process and assess the many noise sources in the sea, and enabled them to discriminate undersea threats. This capability was one of many that helped end the Cold War.

That research investment also led to a significant improvement in understanding ocean dynamics and the processes that control how the ocean responds to atmospheric forcing and internal changes in ocean structure.

Together, we have shared in the development and improvements in manned vehicles such as *Alvin* and remotely operated vehicles such as *Jason* that have allowed us to "look" at 98 percent of the ocean floor, and to discover (another novel concept . . . is discovery science?) new geological features and life forms.

Together, we have provided the oceanographic community with the tools they need to do their work. ONR has built a first-class armada of ships for global research in both shallow and deep waters . . . while NSF's normal role in the partnership is a major supporter of these national assets to address leading national ocean science challenges. We are working closely here as we take advantage of the good news about our funding to make sure we optimize both science and these assets for the good of the nation over the long-term. Together, we have undertaken significant work with these tools—the Arabian Sea process study, for example—and will continue to advance the state of the ocean sciences.

In the polar areas, NSF is clearly the leader—and we

salute you. I had a chance to personally thank Joe Bordogna last month for NSF's major support of SCICEX (Scientific Ice Expeditions) research, while the Navy provided a unique platform of opportunity.

I think we should be very proud of our accomplishments—they form a strong foundation for the work we will do in the next millennium.

Ladies and gentlemen, it is expensive to do work in the oceans, we must cooperate—if not in shared funding, then at least in planning. No overlap is affordable. And what is really exciting is knowing there is so much we haven't even imagined yet!

This year was the International Year of the Ocean—a celebration that has placed ocean science both on the national and international agendas. Hurricanes, droughts, floods, coastal erosion, El Niños, La Niñas, and national security problems during the last few years have also brought home why it is imperative that we understand the oceans. Secretary of the Navy Dalton has taken the opportunity afforded by the International Year of the Ocean to remind not only the public, but also the Navy itself, of the importance of understanding the oceans, and how they affect our lives.

In closing, I believe that we are in a time of rich opportunity for research in ocean science. (No, "rich" is not right . . . it is imperative that we understand the ocean now, broadly; it is the most American of all sciences as we are the world's greatest maritime nation.)

So permit me to offer a vision for the future of our partnership in ocean observing and exploration that may prepare us to move into areas yet unimagined. The vision is one of a maritime nation whose well-being is seen against the broader background of this planet's waters. Let us regard the future of research with alert and open minds prepared for fresh insights, and take care that we not lose our vision. In the future, there will be a place for ocean science in the interest of national security . . . and there will be a place for ocean science in the interest of national need. Let's go forth together and do great things.

Argo to ARGO

D. James Baker

Under Secretary for Oceans and Atmosphere and Administrator,
National Oceanic and Atmospheric Administration, U. S. Department of Commerce

A while back, Russ Davis came in to give a seminar at the National Oceanic and Atmospheric Administration (NOAA) about ARGO—the Array for Real-Time Geostrophic Oceanography. I'm very excited about this plan for an array of 3,200 profiling drifting floats distributed globally at 1500 meters. It is clearly the next logical step towards a true global ocean observing system.

The name ARGO made me think back to my first long cruises in the 1960s in the Indian Ocean aboard the Scripps R/V *Argo* as part of the International Indian Ocean Expedition. This is where I first learned oceanography, with long talks in the evening under the tropical stars with Henry Stommel, Jule Charney, Allan Robinson, John Knauss, Bruce Taft, and others who became life-long friends. Equipment and navigation then seems crude compared to what we have now, but those were happy days when someone else was responsible for funding my research.

It was then that I began to get interested in global observing systems. Since then it has been great fun to participate in the kaleidoscope of acronyms: IDOE (International Decade of Ocean Exploration), GARP (Global Atmospheric Research Program), MODE (Mid-Ocean Dynamics Experiment), ISOS (International Southern Ocean Studies), JGOFS (Joint Global Ocean Flux Study), a host of equatorial programs leading to TOGA (Tropical Ocean Global Atmosphere) and WOCE (World Ocean Circulation Experiment); and now to see today's CLIVAR (Climate Variability and Prediction Program), BECS (Basin-wide Extended Climate Study), GODAE (Global Ocean Data Assimilation Experiment), and ARGO. In spite of an initial reluctance to come together in cooperative programs, we oceanographers, led by Feenan Jennings and his IDOE band at NSF led by Worth Nowlin and Bill Merrell, have learned to do that while preserving individuality and freedom of research.

This is a fitting time for such a celebration of ocean science—because I believe we are on the verge of getting the necessary pieces to understand the ocean. Why do I think that?

First, the results of NOAA's TOGA TAO (Tropical Atmosphere-Ocean) array in the successful forecast of the 1997-1998 El Niño show the wisdom of the ENSO (El Niño Southern Oscillation) community in putting this together. So from the early ideas of Sir Gilbert Walker, sitting in New Delhi wondering about the cause of the 1899 Monsoon failure and its affect on crops to Jacob Bjerknes seeing the need to involve the ocean, we now have the rudiments of an operational system in place. What's more, armed with the success of forecasts and with the considerable help of the academic community, we at NOAA have been able to achieve operational status for the TOGA array. That means long-term funding as part of our operational budget. This is a major step—the first new operational funding for ocean monitoring in decades.

Second, the success of the altimeter satellite TOPEX (Ocean Topography Experiment)/Poseidon. Once again the community came together—Carl Wunsch and his colleagues in the United States and France working with Stan Wilson at the National Aeronautics and Space Administration (NASA) and its Jet Propulsion Laboratory (JPL), to put together a mission that has proved successful beyond original plans in giving a precision measurement of the shape of the ocean surface for tides, waves, and currents. The data are also being used for heat content and were a key element in helping us understand the 1997-1998 El Niño. TOPEX/Poseidon was my very first lesson in talking to Congress about a project—I knew I was successful when staffers said to me, "Baker, we don't need to talk to you. We've heard enough about TOPEX." The final cooperation was also a good lesson for all of us. Now we are embarked on the follow-on missions, and NOAA will play a role there.

Finally, undergirding all of this is the new scientific understanding of the ocean that has been achieved by scientists supported by NSF, the Navy, NOAA, and NASA. I started my physics career supported by the Air Force, then my ocean career was supported by the Navy, then NSF, then I worked for NOAA in the Pacific Marine Environmental Laboratory. This is where I learned the difference in approaches in funding—NSF being peer review and results-oriented, while the Navy is willing to take longer chances with instrumentation projects that would not have survived peer review. All of this support for the oceans community has led to phenomenal new understanding of physical, chemical, and biological processes in the sea. As Administrator of NOAA, and as the first administrator who worked as a scientist in the agency, I've had a special interest in the practical application of science and technology to better observations, and a unique view of all of these changes.

This past year has been declared the International Year of the Ocean. At the beginning, several of us met and asked how we could get more attention to oceans issues—more than just posters. We ended up doing a lot of things, including the National Ocean Conference where the President announced a special focus on ocean observations. He is the first President to do so. If we think back to President Kennedy and his 1962 announcement about GARP and geostationary satellites, we can see how important this is. Today, the geostationary satellites operated by NOAA are fundamental to our observing system. Hopefully, the initial commitment by President Clinton is a down payment on a fully operational ocean observing system. The first step will be full deployment of ARGO, together with satellite altimeter and scatterometer systems, and data assimilation and modeling. These contributions to GODAE and CLIVAR will provide the information we need to make available real practical applications of ocean understanding. In fact, I can note that the latest development in operational oceanography is wave forecasts for surfers—an application of the ideas of Walter Munk and his colleagues during World War II now being used today in a very different mode.

Physical oceanography is not the only side of this issue. I believe that changing ocean chemistry is as great or greater a human public concern than changing atmospheric chemistry leading to global warming. We're just starting to see this, as we experience the global impacts of non-point source pollution. A coastal global ocean observing system is one way to address this.

So from *Argo* to ARGO, I can see a progression of ocean science to operational oceanography. I wouldn't have missed it for anything, and I'm pleased to have interacted with so many colleagues at sea, in the academic community, in endless but productive meetings and conferences, and in achieving funding and results.

Congratulations to the National Science Foundation as it celebrates Fifty Years of Ocean Discovery.

The Importance of Ocean Sciences to Society

Admiral James D. Watkins, U.S. Navy (Ret.)

President, Consortium for Oceanographic Research and Education

Congratulations to the National Science Foundation for 50 years of research excellence and delivery of valuable products from discovery that enhance human understanding.

To tackle the challenges associated with the topic given to this panel, let me start out with a quote:

> Yet the present achievements, exciting though they are, must be considered only a beginning to what is yet to be achieved by probing the vast depths of water that cover most of the surface of the earth. [Some years ago] a group of distinguished scientists . . . of the National Academy of Sciences declared that "Man's knowledge of the oceans is meager indeed compared with their importance to him."

This statement is as true now as when it was first made by Rachel Carson in the 1960s.

The United States will have focused incredible resources during the last 50 years of this century—the precise period of scientific accomplishments by NSF that we celebrate at this symposium—on exciting space exploration looking outward toward the frontier of the "Big Bang." As we turn the corner into the next millennium, I sincerely hope we can also energize the American public and our national leadership to look, more thoroughly, inward toward Earth's last frontier, the oceans, to help solve the countless growing challenges to humankind, investing the necessary resources to significantly enhance our knowledge of the greatest natural resource on Earth that may house answers to these challenges.

To energize the public and elected officials, we will need new vision, new strategies, much better scientist-to-citizen communication techniques, and much more aggressive follow-through on the part of our scientific community and other stakeholders to carry out the strategy. NSF, as a nearly unique non-mission science support agency of the federal government, has a key part to play in defining this needed new vision for science's role for tomorrow's society.

The exciting thing about the ocean is that its science is virtually *all* relevant to societal needs—quality of life, economic development, national security, education—and hence more potentially salable to our society, one that demands to know what's in it for them. So what must this new vision of ocean sciences include? I'll only focus briefly on four elements, which although not all-inclusive, are seemingly lacking today:

- Integrated Science Education—Marriage between all levels of science education and the science researcher.
- Human Health and the Oceans—Incorporation of the ocean's impact on human health into community thinking when addressing ocean science and technology drivers.
- Product Delivery—Establishment of the paradigm of research and development (R&D) as a business that, in certain cases, can predictably lead from basic research all the way to "products."
- International Coordination.

Let me touch on each one of these individually.

EDUCATION

Arthur Nowell has covered this area well in an earlier paper, but let me just add a few comments.

- Science should proactively involve and integrate education, and education should incorporate the most current science. Education means both formal (classroom) and informal (e.g., science centers and aquariums). It includes everything from kindergarten to doctoral levels.
- Ocean science lends itself uniquely and ideally to the new initiatives for education and the new National Science Education Standards produced by National Research Council (NRC, 1996) in that it is:

1. Interdisciplinary and integrated by nature
2. Inquiry-based
3. Hands-on

And, what better medium can we find than the oceans: (1) for kids to be motivated to learn something about all scientific disciplines, and (2) for ocean researchers to actively help out in the national educational reform effort underway today. This education reform is desperately needed to convert a scientifically illiterate society to one that can better understand the changing world around them. Decisions of the Ocean Research Advisory Panel (ORAP) related to education have been made as a forethought, not an after-thought, and adopted as an integrating concept with all our ocean partnership programs.

HUMAN HEALTH AND THE OCEANS (NRC, 1999)

In 1997, the NRC's Governing Board approved a project proposed by the Ocean Studies Board (OSB), in cooperation with the Institute of Medicine's Division of Health Science Policy entitled "The Ocean's Role in Human Health," which included a workshop in 1998. This workshop was sponsored jointly by the National Oceanic and Atmospheric Administration (NOAA), the National Institute of Environmental Health Services (NIEHS) of National Institutes of Health, and the National Aeronautics and Space Administration, and was tasked to examine a variety of ways in which the oceans play a role in human health. Specifically, the invited speakers addressed the following topics:

1. Marine natural disasters and public health: Can we better model and predict marine natural disasters? Can we better anticipate effects on public health?
2. Climate and the incidence of infectious diseases: Are waterborne diseases detectable and predictable? How do changes in climate—both regional and global—impact disease vectors?
3. Hazards associated with toxic algal blooms: What causes toxic algae to bloom? Can their outbreak be predicted, mitigated, and prevented? Why has the incidence of these blooms been increasing?
4. The therapeutic potential of marine natural products: What are the implications of the discovery of life a thousand meters below the seafloor? Have we adequately examined marine biotechnology for medically important products?
5. Marine organisms as models for biomedical research: Are there marine species that could serve as useful medical models? Can marine species offer new understanding of human development or physiology?

In relation to the report, I can say this:

1. This has been the first time that a panel has been organized to address this subject at such a high level; and the breadth of opportunity was surprising to all participants. Initiation of a dialogue between the medical and ocean science communities has been a most valuable outcome of this exercise.

2. The committee was able to identify specific areas of cooperation—with high potential payoff—that the ocean and medical research communities should pursue jointly. For a range of applications, from mitigating natural disasters to minimizing the outbreak and spread of epidemics, from keeping our recreational beaches and seafood safe to extracting life-saving products from the sea, there is an exciting spectrum of interdisciplinary and doable research that is either unfunded or undersupported.

The NRC has identified these research opportunities. We marine policy-makers must make them happen.

PRODUCT DELIVERY

R&D is a business with a $75 billion bottom line in the federal government alone. But society seems to be demanding more identifiable products for its continued investment. One product can be *understanding*, but we need to sell that product better. Like any business, the science community needs a business plan, a market analysis, and a sales department if it is moving from the product of *understanding* toward potentially valuable and more conventional long-range product objectives, which often demand enhanced resources to move from research into application. The Frank Press Report (NRC, 1995) lays out a proposed approach for allocating federal R&D funds in a manner consistent with this thinking about science, technology, and product delivery.

Fortunately, from a salability standpoint, *ocean* science is one broad area that is inherently product-oriented. We have recently recognized the sales potential in sectors such as coastal zone management, hazard mitigation, agriculture, and public health. Let's design a stronger plan for product delivery of our ocean research and not get too heavily bogged down arguing the merits of basic research versus application. As in so many instances, the mission agencies, for example, already do this. Obviously, when products cannot be foreseen, such as in the case of the Superconducting Super Collider and its search for new discoveries regarding the make-up of matter, it's a much harder "sell" to assert that product objectives are all clear and relevant to societal needs.

INTERNATIONAL COORDINATION

Scientists and those who develop and manage scientific programs must think globally and act globally. We cannot afford, either financially or intellectually, to maintain policies of isolation in the research arena.

Ocean science is inherently international in nature. We have seen the effectiveness of this approach in the Ocean Drilling Program, a model for defining mutually beneficial research objectives among many nations. The current NAS study, chaired by Bob Frosch and requested by the Secretary of State, looks to ocean sciences as one good test-bed for new approaches to using science and technology as tools of diplomacy.

In this connection, I urge you to read the preliminary report of this study signed out to Secretary Albright by the President of the National Academy of Sciences, Bruce Alberts. Let me just quote one vitally important paragraph:

> The opportunities that the areas of science, technology, and health offer in foreign policy are dramatic. . . By forming partnerships with foreign scientists, we enhance their status and support their values, which can do a great deal to promote democracy. In addition, spreading access to new scientific and technical advances is of course essential for providing a decent life and an acceptable environment for the world's expanding population, thereby reducing the potential for destabilizing violent conflicts. (NRC, 1998)

I will close by saying that when the NSF holds its 100th Anniversary Celebration—and I understand that John Knauss has already accepted NSF's invitation to give its keynote address—it is my fond hope that they will look back to the turn of the millennium and say "thanks to those scientific visionaries 50 years ago who set a new and visionary course for ocean science and technology that added such incredible value to the United States and the world."

REFERENCES

National Academy of Sciences (NAS). 1995. *Allocating Federal Funds for Science and Technology*. National Academy Press, Washington, D.C.

National Research Council (NRC). 1996. *National Science Education Standards*. National Academy Press, Washington, D.C.

National Research Council (NRC). 1998. *Improving the Use of Science, Technology, and Health Expertise in U.S. Foreign Policy (A Preliminary Report)*. National Academy Press, Washington, D.C.

National Research Council (NRC). 1999. *From Monsoons to Microbes: Understanding the Ocean's Role in Human Health*. National Academy Press, Washington, D.C.

Appendices

PLATE 7   Between symposium sessions, participants viewed posters in the Academy's Great Hall (top photo; see Appendix C for a list of poster titles).

Another feature of the symposium was an exhibition of photos of marine organisms (center and bottom) in the NAS Auditorium Gallery entitled *Watery Beauties: Discovering Ocean Life*. Contributors to the exhibition were Alice Alldredge, James King, Jan Rines, Larry Madin, and Marsh Youngbluth.

PLATE 8 An important feature of the symposium was the participation of several generations of ocean scientists and both current and past staff of the National Science Foundation (NSF). 32 students from 28 U.S. ocean science institutions participated in the symposium. The top photo shows five of the student participants (from left to right - Cristin Conaway, Dana Lane, Kara Lavender, Juli Dyble, and Janelle Fleming).

Several retired NSF ocean science staff and former rotators participated in the symposium. The bottom photo shows Sandra Toye (left), Jennieve Gillooly (middle), and John Morrison (right).

PLATE 9   John Knauss (left) and Dick Barber (below) gave the
first two presentations of the symposium. Both have been
influential in ocean science and policy over the past decades.

PLATE 10   RADM Paul Gaffney (Chief of Naval Research) and Dr. Rita Colwell (Director of the National Science Foundation) were among the agency heads who closed the symposium with their thoughts about the future of ocean sciences in the United States.

Capt. Diann Lynn (left) speaks with Walter and Judith Munk during one of the breaks.

The audience listens as Sandra Toye introduces the featured speaker of the symposium, Dr. Robert Ballard (President of the Institute for Exploration in Mystic, Connecticut). Dr. Ballard and Dr. John Knauss are seated in the front row.

APPENDIX

A

Symposium Program

Fifty Years of Ocean Discovery
National Academy of Sciences
Washington, D.C.

AGENDA

Wednesday, October 28, 1998

8:30 a.m.
Welcoming Remarks - NAS Auditorium
Dr. William Wulf, President, National Academy of Engineering
Dr. John Steele, Woods Hole Oceanographic Institution, Chair of Organizing Committee

8:45 a.m.
Keynote Lecture: The Emergence of the National Science Foundation as a Supporter of Ocean Sciences in the United States
Dr. John Knauss, University of Rhode Island/Scripps Institution of Oceanography

LANDMARK ACHIEVEMENTS OF OCEAN SCIENCES

Moderator: **Dr. Kenneth Brink**, Woods Hole Oceanographic Institution and Chair, Ocean Studies Board

9:30 a.m.
Achievements in Biological Oceanography
Dr. Richard Barber, Duke University

10:30 a.m.
Break

11:00 a.m.
Achievements in Chemical Oceanography
Dr. John Farrington, Woods Hole Oceanographic Institution

12:00 noon
Lunch

1:00 p.m.
Achievements in Physical Oceanography
Dr. Walter Munk, Scripps Institution of Oceanography

2:00 p.m.
Achievements in Marine Geology and Geophysics
Dr. Marcia K. McNutt, Monterey Bay Aquarium Research Institute

3:00 p.m.
Break

3:30 p.m.
Deep Submergence: The Beginnings of ALVIN
Ms. Sandra Toye, National Science Foundation (ret.)

3:45 p.m.
The History of Woods Hole's Deep Submergence Program
Featured Speaker: **Dr. Robert Ballard**, President, Institute for Exploration, Mystic, Connecticut

4:45 p.m.
Poster Session on the History of Oceanography, Institutions, and Major Programs in the Great Hall
Art Exhibit in the Auditorium Gallery—*Watery Beauties: Discovering Ocean Life*

Thursday, October 29, 1998

CREATING INSTITUTIONS TO MAKE SCIENTIFIC DISCOVERIES POSSIBLE

Moderator: **M. Grant Gross**, Chesapeake Research Consortium

8:30 a.m.
Origins of the Ocean Science Division (panel)
Dr. Michael Reeve, National Science Foundation
Ms. Mary Johrde, National Science Foundation (ret.)
Ms. Sandra Toye, National Science Foundation (ret.)

9:30 a.m.
Development of the Academic Fleet: A Community Perspective
Dr. John Byrne, Oregon State University (ret.)

10:15 a.m.
Break

10:45 a.m.
Drilling Programs from Mohole to the Ocean Drilling Program
Dr. Jerry Winterer, Scripps Institution of Oceanography

11:30 a.m.
Impacts of Technology on Ocean Sciences and NSF's Role in Technology Development
Dr. Larry Clark, National Science Foundation

12:15 p.m.
Lunch

LARGE AND SMALL SCIENCE PROGRAMS: A DELICATE BALANCE

Moderator: **Dr. G. Michael Purdy**, National Science Foundation

1:45 p.m.
Major International Programs in Ocean Science: IGY to CLIVAR (panel)
Mr. Feenan Jennings, Texas A&M University (ret.)
Dr. Richard Lambert, National Science Foundation
Dr. Peter Brewer, Monterey Bay Aquarium Research Institute
Dr. John Delaney, University of Washington

2:45 p.m.
The Importance of Individual Science and Encouragement of Young Scientists (panel)
Dr. Cynthia Jones, Old Dominion University
Dr. Susan Lozier, Duke University

3:15 p.m.
Break

3:45 p.m.
Continuation of Panel
Dr. Maureen Raymo, Massachusetts Institute of Technology
Dr. Miguel Goñi, University of South Carolina

4:15 p.m.
Discussion of the Day's Issues

5:00 p.m.
Symposium Adjourns for the Day

Friday, October 30, 1998

OCEAN SCIENCES TODAY AND TOMORROW

Moderator: **Dr. John Knauss**, University of Rhode Island/Scripps Institution of Oceanography

8:30 a.m.
Opportunities and Challenges in Ocean Science (panel)
Discussion Leader: **Dr. Donald Heinrichs**, National Science Foundation

Physical Oceanography (APROPOS) - **Dr. William Young**, Scripps Institution of Oceanography
Chemical Oceanography (FOCUS) - **Dr. Cindy Lee**, State University of New York, Stony Brook
Marine Geology and Geophysics (FUMAGES) - **Dr. Marcia McNutt,** Monterey Bay Aquarium Research Institute
Biological Oceanography (OEUVRE) - **Dr. Peter Jumars**, University of Washington
The Future of Major Ocean Programs - **Dr. Rana Fine,** University of Miami

10:00 a.m.
Ocean Science Education and Careers
Dr. Arthur Nowell, University of Washington

10:30 a.m.
Break

11:00 a.m.
Evolving Institutional Arrangements
Dr. William Merrell, The H. John Heinz III Center for Science, Economics, and the Environment

11:30 a.m.
The Importance of Ocean Sciences to Society
Discussion Leader: **Dr. Bruce Alberts**. President, National Academy of Sciences

Dr. Rita Colwell, Director, National Science Foundation
RADM Paul Gaffney, Chief of Naval Research
Dr. D. James Baker, Under Secretary for Oceans and Atmosphere and NOAA Administrator, Department of Commerce
Adm. James Watkins, U.S. Navy (ret.), President, Consortium for Oceanographic Research and Education

1:00 p.m.   Symposium Adjourns

APPENDIX

B

Symposium Participants

Carmen Aguilar
Great Lakes Water Institute

Frank Aikman III
National Oceanic and Atmospheric Administration

Arthur Alexiou
Intergovernmental Oceanographic Commission

Helen Almquist-Jacobson
University of Maine

Marc Alperin
University of North Carolina, Chapel Hill

Mary Altalo
Scripps Institution of Oceanography

Evelyn Aquino
Hood College

E. Virginia Armbrust
University of Washington

Thomas Arnold
University of Delaware

Jeff Ashley
University of Maryland

E. Esat Atikkan

Larry P. Atkinson
Old Dominion University

James A. Austin, Jr.
University of Texas

Arthur Baggeroer
Massachusetts Institute of Technology

Rodger Baier
National Science Foundation (retired)

Stephanie Bailenson
Senate Commerce Committee

Megan D. Bailiff
University of Washington

D. James Baker
National Oceanic and Atmospheric Administration

Robert Ballard
Institute for Exploration

Nathan Bangs
University of Texas

Richard Barber
Duke University

Peter Barnes
U.S. Geological Survey

Walter Barnhardt
U.S. Geological Survey

Jack Bash
University-National Oceanographic Laboratory System

Rodey Batiza
University of Hawaii

Sara Bazin
Scripps Institution of Oceanography

Lisa Beatty
Hood College

Peter Betzer
University of South Florida

Brian Blanton
University of North Carolina, Chapel Hill

Jerry Boatman
Naval Meteorology and Oceanography Command

Donald F. Boesch
University of Maryland

Allison Bonner
Center for Marine Conservation

Jack Botzum
Nautilus Press, Inc.

David Bradley
Pennsylvania State University

Jay Brandes
Carnegie Institution of Washington

Stephen Brandt
National Oceanic and Atmospheric Administration

John F. Bratton
Woods Hole Oceanographic Institution

Peter Brewer
Monterey Bay Aquarium Research Institute

Kenneth H. Brink
Woods Hole Oceanographic Institution

David A. Brooks
Texas A&M University

Mark J. Brush
University of Rhode Island

Tammy Bryant
University of Delaware

Ken O. Buesseler
Woods Hole Oceanographic Institution

Karen L. Bushaw-Newton
National Oceanic and Atmospheric Administration

John Byrne
Oregon State University

Louis W. Cabot
Cabot-Wellington

Elizabeth A. Canuel
Virginia Institute of Marine Sciences

Craig Carlson
Bermuda Biological Station

Robert Carney
Louisiana State University

Michael Carr
U.S. Geological Survey

Fei Chai
University of Maine

Gail Christeson
University of Texas

David M. Christie
Oregon State University

James M. Clark
Woods Hole Oceanographic Institution

Larry Clark
National Science Foundation

M. Elizabeth Clarke
National Oceanic and Atmospheric Administration

Mary E. Clutter
National Science Foundation

Kim Cobb
Scripps Institution of Oceanography

Paula Coble
University of South Florida

J. Kirk Cochran
State University of New York, Stony Brook

W. Thomas Cocke
U.S. Department of State

Millard F. Coffin
University of Texas

Muriel Cole
National Oceanic and Atmospheric Administration

Rita Colwell
National Science Foundation

Christin Conaway
University of North Carolina, Chapel Hill

Reide Corbett
Florida State University

Barry A. Costa-Pierce
Mississippi-Alabama Sea Grant Consortium

Nancy Craig
National Oceanic and Atmospheric Administration

Russel Cuhel
UWM Great Lakes Water Institute

Vicky Cullen
Woods Hole Oceanographic Institution

Christopher D'Elia
Maryland Sea Grant College Program

G. Brent Dalrymple
Oregon State University

Margaret A. Davidson
National Oceanic and Atmospheric Administration

Curtiss Davis
Naval Research Laboratory

Ian Davison
University of Maine

Deborah Day
Scripps Institution of Oceanography

Marie DeAngelis
Humboldt State University

Cynthia Jane Decker
Consortium for Oceanographic Research and Education

E. R. "Dolly" Dieter
National Science Foundation

Emanuele Di Lorenzo
Scripps Institution of Oceanography

Jacqueline E. Dixon
University of Miami

Kim Donaldson
University of South Florida

Linda E. Duguay
National Science Foundation

Maria Duva
Hood College

Daniel J. Dwyer
University of Maine

Julianne Dyble
University of North Carolina, Chapel Hill

Sonya Dyhrman
Scripps Institution of Oceanography

James E. Eckman
Office of Naval Research

Margo Edwards
University of Hawaii

Chris Elfring
National Research Council

Olaf Ellers
Bowdoin College

Winford "Jerry" Ellis
Office of the Oceanographer of the Navy

Lisa Etherington
North Carolina State University

Chunlei Fan
University of Maryland

John Farrington
Woods Hole Oceanographic Institution

Drew Ferrier
Hood College

John Field
House Committee on Resources

Rana Fine
University of Miami

Janelle V.R. Fleming
University of North Carolina, Chapel Hill

Maureen Foley
Hood College

Tom Forhan
House Committee on Appropriations

Valerie Franck
University of California

Dirk Frankenberg
University of North Carolina

Carl Friehe
University of California, Irvine

Paul Gaffney, II, USN
Office of Naval Research

Robert B. Gagosian
Woods Hole Oceanographic Institution

Christina Gallup
University of Maryland

Patricia E. Ganey-Curry
University of Texas, Austin

Nikola Marie Garber
Gulf Coast Research Laboratory

Patricia Geetes
Office of Senator Akaka

Jennifer Georgen
Woods Hole Oceanographic Institution

Deirdre Gibson
Skidaway Institute of Oceanography

Jennieve Gillooly
National Science Foundation (retired)

Robert Ginsberg
University of Miami

Pat Glibert
University of Maryland

Linda Glover
Office of the Oceanographer of the Navy

Janice Goldblum
National Research Council

Miguel Goñi
University of South Carolina

Morgan Gopnik
National Research Council

Elizabeth Gordon
University of South Carolina

Jacqueline I. Gordon
Viz.Ability, Inc.

Louis I. Gordon
Oregon State University

David Graham
Sea Technology Magazine

Michael Graham
Scripps Institution of Oceanography

J. Frederick Grassle
Rutgers University

Albert G. Greene, Jr.

Kenneth Grembowicz
University of Southern Mississippi

Gordon Grguric
Richard Stockton College

Jay Grimes
University of Southern Mississippi

Tonya Grochoske
QUEST Explorations

M. Grant Gross
Chesapeake Research Consortium

Peter Hacker
University of Hawaii

Gordon Hamilton
Office of Naval Research

Jennifer Hanafin
Hood College

R. Taber Hand
University of Maryland

Pamela C. Hart
Woods Hole Oceanographic Institution

Paul Hartmann
University of Rhode Island

Susan Haynes
Virginia Institute of Marine Science

Marge Hecht
21st Century Magazine

Carol Hee
University of North Carolina, Chapel Hill

Eli Hesterman
Woods Hole Oceanographic Institution

Dexter Hinckley
Ecological Society of America

W. Hodgkiss

Eileen Hofmann
Old Dominion University

Deane Holt
U.S. Army Corps of Engineers

Raleigh Hood
University of Maryland

Hartley Hoskins
Woods Hole Oceanographic Institution

Edward Houde
University of Maryland

Debra T. Hughes
Office of Congressman Curt Weldon

Anitra Ingalls
State University of New York, Stony Brook

Suzanne V. Jacobson
The H. John Heinz III Center for Science, Economics, and
   Environment

Carol Janzen
University of Delaware

Steve Jayne
Woods Hole Oceanographic Institution

Feenan Jennings
National Science Foundation (retired)

Amy S. Johnson
Bowdoin College

David Johnson
National Oceanic and Atmospheric Administration

Mary Johrde
National Science Foundation (retired)

Cynthia Jones
Old Dominion University

Christy Jordan
University of Maryland

Peter Jumars
University of Washington

John R. Justus
Congressional Research Service

Mary Hope Katsouros
The H. John Heinz III Center for Science, Economics, and
   Environment

Steven Keith
University of North Carolina, Chapel Hill

Charles Kennel
Scripps Institution of Oceanography

William E. Kiene
Smithsonian Institution

Kimani Kimbrough
College of William and Mary

Gail Kineke
Boston College

John Knauss
University of Rhode Island/Scripps Institution of
    Oceanography

Robert Knox
Scripps Institution of Oceanography

Chester Koblinsky
National Aeronautics and Space Administration

Jennie Kopelson
Johns Hopkins University

Fae L. Korsmo
National Science Foundation

Richard Lambert
National Science Foundation

Dana Lane
Scripps Institution of Oceanography

Tom Langman

Shelley Lauzon
Woods Hole Oceanographic Institution

Kara Lavender
Scripps Institution of Oceanography

Cindy Lee
State University of New York, Stony Brook

Margaret Leinen
University of Rhode Island

Don Levitan
Florida State University

Jonathan Lilley
University of Washington

Eric Lindstrom
National Aeronautics and Space Administration

Lawrence Lipsett
Woods Hole Oceanographic Institution

Mike Lomas
University of Maryland

Steven Lonker
U.S. Department of Energy

Susan Lozier
Duke University

Doug Luther
University of Hawaii

Danielle Luttenberg
National Oceanic and Atmospheric Administration

James R. Luyten
Woods Hole Oceanographic Institution

Patricia Sullivan Lynch
The Maritime Aquarium at Norwalk

Diann K. Lynn
U.S. Navy

Diane Lynne
U.S. Environmental Protection Agency

Kimberly Mace
Texas A&M University

Laurence P. Madin
Woods Hole Oceanographic Institution

Christophe Maes
Ocean Forecasting, NCEP

Chris Mann
House Resources Committee

Nancy H. Marcus
Florida State University

Roberta Marinelli
National Science Foundation

John Marr
Mississippi-Alabama Sea Grant Consortium

Ellen Martin
University of Florida

Liz Maruschak
Consortium for Oceanographic Research and Education

Marcia K. McNutt
Monterey Bay Aquarium Research Institute

Jay C. Means
Western Michigan University

David Menzel
National Science Foundation

William Merrell
The H. John Heinz III Center for Science, Economics, and
   Environment

Brian Midson
University of Hawaii, Manoa

Peter Mikhalevsky
Science Applications International Corporation

Stephen Miller
Science Applications International Corporation

Joan Mitchell
National Science Foundation (retired)

M. Moriarty

Sherrie Morris
Hood College

John Morrison
North Carolina State University

John Morse
Texas A&M University

Graham Mortyn
Scripps Institution of Oceanography

Rebecca G. Moser
National Oceanic and Atmospheric Administration

Frank Mueter
University of Alaska

Walter Munk
Scripps Institution of Oceanography

Kurt Mutchler
National Geographic Society

John Mutter
Columbia University

Steven Nadis
Science Writer

Cecily Natunewicz
University of Delaware

Dave Nemazie
University of Maryland

A. Conrad Neumann
University of North Carolina

Scott W. Nixon
University of Rhode Island

Arthur Nowell
University of Washington

Worth D. Nowlin, Jr.
Texas A&M University

Bridgette O'Connor
University of Miami

R. Olivieri

Curtis Olsen
National Oceanic and Atmospheric Administration

Tun Liang Ong
University of Rhode Island

Michael Orbach
Duke University

John A. Orcutt
Scripps Institution of Oceanography

Joseph Ortiz
Lamont-Doherty Earth Observatory

Thomas Osborn
Johns Hopkins University

Gote Ostlund
University of Miami

Janna Owens
Hood College

G.A. Paffenhöfer
Skidaway Institute of Oceanography

Robert Palmer
House Science Committee

Dennis Peacock
National Science Foundation

Ann Pearson
Woods Hole Oceanographic Institution

Mary Jane Perry
University of Washington

Georgia Persinos
Washington Insights

Christine Phillips
University of Maryland

Jonathan Phinney
Center for Marine Conservation

Leonard Pietrafesa
North Carolina State University

Cynthia H. Pilskaln
University of Maine

Richard F. Pittenger
Woods Hole Oceanographic Institution

Elka Porter
University of Maryland

Ellen Prager
Science Writer

John Preston
Pennsylvania State University

Bill Pritchard
Business Publishers

Donald Pryor
Office of Science and Technology Policy

G. Michael Purdy
National Science Foundation

John A. Quinlan
University of North Carolina

Frank R. Rack
Joint Oceanographic Institutions, Inc.

C. Barry Raleigh
University of Hawaii

Steve Ramberg
Office of Naval Research

Mac Rawson
University of Georgia

James Ray
Equilon Enterprises LLC

John Rayfield
House Resources Committee

Maureen Raymo
Massachusetts Institute of Technology

Michael Reeve
National Science Foundation

David Rogers
Scripps Institution of Oceanography

Neil Rolde
University of Maine

Lisa Rom
National Science Foundation

Mike Roman
University of Maryland

Jill Rooth
University of Maryland

Helen Rozwadowski
Georgia Institute of Technology

Stephany Rubin
Lamont-Doherty Earth Observatory

John D. Rummel
National Aeronautics and Space Administration

Lydia R. Runkle
J.A.W.S. Labs

Louis E. Sage
University of Maine

Eric Saltzman
University of Miami

Dana Savidge
University of North Carolina, Chapel Hill

David Scala
Rutgers University

Terry L. Schaefer
National Oceanic and Atmospheric Administration

Terry Schaff
Consortium for Oceanographic Research and Education

Kyra Schlining
Monterey Bay Aquarium Research Institute

Astrid Schnetzer
Bermuda Biological Station

Catherine L. Schuur
University of Texas

Rick Schwabacher
Calypso Log Magazine

Lisa Shaffer
Scripps Institution of Oceanography

Kipp Shearman
Oregon State University

Wilbur G. Sherwood
National Science Foundation (Retired)

Kasey Shewey
American Geological Institute

Hsing-Hua Shih
National Oceanic and Atmospheric Administration

Rebecca Shipe
University of California, Santa Barbara

Randy Showstack
EOS, American Geophysical Union

Bruce D. Sidell
University of Maine

E.A. Silva
Bison Marine and Materials

Kyla Simmons

Maxine F. Singer
Carnegie Institution of Washington

Gabriela Smalley
SERC/UMCES

Brenda Smith
Naval Meteorology and Oceanography Command

Erik Smith
University of Maryland

Sharon Smith
University of Miami

W.D. Smyth
Oregon State University

Evelyn Soucek
Achievement Rewards for College Scientists

A.F. Spilhaus, Jr.
American Geophysical Union

Richard Spinrad
Consortium for Oceanographic Research and Education

Debra Stakes
Monterey Bay Aquarium Research Institute

Edward Stewart
University of Delaware

Karen Stocks
Rutgers University

Paul L. Stoffa
University of Texas, Austin

Glenn Strait
World & I Magazine

Woody Sutherland
Scripps Institution of Oceanography

Tracey Sutton
University of South Florida

Kam Tang
University of Connecticut

Laura Tangley
U.S. News & World Report

Mark Teece
Carnegie Institution of Washington

Anne Tenney
National Science Foundation

David Thistle
Florida State University

Florence I.M. Thomas
Marine Environmental Science Consortium

Ed Thompson
Office of Senator Akaka

Carolyn A. Thoroughgood
University of Delaware

Ronald Tipper
Office of Naval Research

Sandra Toye
National Science Foundation (retired)

Lamarr Trott
National Oceanic and Atmospheric Administration

Elizabeth Turner
National Oceanic and Atmospheric Administration

Ed Urban
National Research Council

David van Keuren
Naval Research Laboratory

S.K. Varma
Congressional Research Service

Kathleen Wage
Massachusetts Institute of Technology

Sharon H. Walker
National Oceanic and Atmospheric Administration

Helen M. Walkinshaw

Michael Wara
University of California, Santa Cruz

Rita Warpeha
National Science Resources Center

James Watkins
Consortium for Oceanographic Research and Education

Lani Watson
National Oceanic and Atmospheric Administration

Gary Weir
U.S. Naval Historical Center

Eli Weissman
National Sea Grant Fellow

Francisco Werner
University of North Carolina

John L. Wickham
National Oceanic and Atmospheric Administration

William Wilcock
University of Texas

Jeff Williams
U.S. Geological Survey

Margaret Williams
University of Miami

Alan Wilson
Sea Technology Magazine

Stan Wilson
National Oceanic and Atmospheric Administration

Herb Windom
Skidaway Institute of Oceanography

Robert Winokur
National Oceanic and Atmospheric Administration

Jerry Winterer
Scripps Institution of Oceanography

Alexandra Witze
Dallas Morning News

Cecily J. Wolfe
Woods Hole Oceanographic Institution

L. Donelson Wright
College of William & Mary

Virginia Wright
Achievement Rewards for College Scientists

Ian W. Young
Lamont-Doherty Earth Observatory

William Young
Scripps Institution of Oceanography

Susan Zeigler
Carnegie Institution of Washington

Yan Zheng
Lamont-Doherty Earth Observatory

Herman Zimmerman
National Science Foundation

Gregory Zwicker
National Oceanic and Atmospheric Administration

APPENDIX

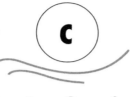

C

Poster Session

INTRODUCTION

To expand community involvement in the symposium, marine science institutions and major ocean programs were given an opportunity to present their roles in the shaping of ocean science in the United States over the past fifty years. Posters (listed below) were displayed in the Great Hall of the National Academy of Sciences Building throughout the symposium. The poster session served as a backdrop and focus for lively discussions and reminiscing (see Plate 7).

BERMUDA BIOLOGICAL STATION
A 50-Year Timeline Guides the Viewer Through BBSR's Scientific Progress

COLLEGE OF WILLIAM AND MARY
Virginia Institute of Marine Science: More Than Fifty Years in Coastal Ocean and Estuarine Science

FLORIDA STATE UNIVERSITY
Oceanography at Florida State University

MONTEREY BAY AQUARIUM RESEARCH INSTITUTE
Underwater Video in Research and Education: A Study of Spot Prawns

NORTH CAROLINA STATE UNIVERSITY
Chaos at Work: A Smooth Noodle Guide to the History of Marine/Ocean Science at North Carolina State University

PENNSYLVANIA STATE UNIVERSITY
Ocean Science Technology at the Applied Research Laboratory of Pennsylvania State University

RIDGE PROGRAM
The RIDGE Program at NSF: Highlights from a Decade of Multi-Disciplinary Science

SCRIPPS INSTITUTION OF OCEANOGRAPHY
Five Decades of Discovery: NSF and the Endless Frontier

SKIDAWAY INSTITUTE OF OCEANOGRAPHY
30 Years of Research: 1968-1998

STATE UNIVERSITY OF NEW YORK AT STONY BROOK
The Marine Sciences Research Center at the State University of New York at Stony Brook: 30 Years of Innovative Coastal Research

TEXAS A&M UNIVERSITY
Fifty Years of Ocean Science Studies at Texas A&M University

UNIVERSITY OF ALASKA, FAIRBANKS
A Brief History of the Institute of Marine Science at the University of Alaska

UNIVERSITY OF CONNECTICUT
The University of Connecticut's Marine Program: Then (1957) and Now (1998)

UNIVERSITY OF DELAWARE
Celebrating a Partnership for Progress: The University of Delaware College of Marine Studies

UNIVERSITY OF MAINE
Seeking Knowledge from the Sea: The History of Marine Research at the University of Maine

UNIVERSITY OF MARYLAND CENTER FOR ENVIRONMENTAL SCIENCE
Leadership at the Ocean Edge

UNIVERSITY OF MIAMI ROSENSTIEL SCHOOL OF MARINE AND ATMOSPHERIC SCIENCE
Salutes NSF's Fifty Years of Ocean Discovery

UNIVERSITY OF RHODE ISLAND
The Graduate School of Oceanography of the University of Rhode Island

UNIVERSITY OF SOUTH CAROLINA
The Belle W. Baruch Institute for Marine Biology & Coastal Research and the Marine Science Program at the University of South Carolina: Partners in Quality Research and Education

UNIVERSITY OF SOUTH FLORIDA
University of South Florida Department of Marine Science: An All Wet World

UNIVERSITY OF SOUTHERN CALIFORNIA
Oceanography Research and Education at the University of Southern California

UNIVERSITY OF SOUTHERN MISSISSIPPI
The University of Southern Mississippi - Institute of Marine Sciences

UNIVERSITY OF TEXAS AT AUSTIN
The History of Geophysical Research at the University of Texas

U.S. GLOBAL OCEAN ECOSYSTEMS DYNAMICS PROGRAM
The U.S. Global Ocean Ecosystems Dynamics Program

U.S. JOINT GLOBAL OCEAN FLUX STUDY
The U.S. Joint Global Ocean Flux Study

WOODS HOLE OCEANOGRAPHIC INSTITUTION
The History of Oceanography at Woods Hole Oceanographic Institution

WORLD OCEAN CIRCULATION EXPERIMENT (WOCE)
Background and Promise of WOCE

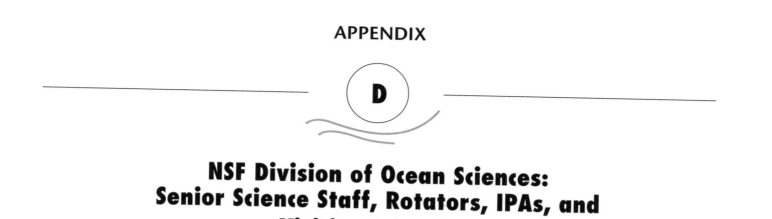

NSF Division of Ocean Sciences:
Senior Science Staff, Rotators, IPAs, and Visiting Scientists

Name	Year(s)	Title	Section/Division
Robert B. Abel	1968-1970	Head	Office of Sea Grant Programs
Sunit Addy	Approx. 1985-1987	Associate Program Director	Marine Geology & Geophysics Program
Frederic A. Agdern	1982	Manager (Rotator)	OSOD Engineering & Operations Section - Design and Construction
Frank R. Alexander	1972-1978	Program Manager for Facilities Acquisitions Program	OOFS
	1979	Assistant Program Manager	OOFS Operations Program
	1980	Associate Program Manager	OOFS Operations Program
	1982-1985	Special Projects Officer	Division of Ocean Sciences
Arthur G. Alexiou	1967-1970	Program Director	Office of Sea Grant Programs
James Allan	1998-present	Associate Program Director (IPA)	Ocean Drilling Program
Mary Altalo	1984-1986, 1989-1990	Associate Program Director	Biological Oceanography Program
Elizabeth Ambos	1992-1993	Associate Program Director (IPA)	Ocean Drilling Program
James Ammerman	1996-1998	Associate Program Director (IPA)	Biological Oceanography Program
Neil R. Andersen	1975 -1981	Program Director	Marine Chemistry Program
	1981-1995	Program Director	Chemical Oceanography Program
Richard G. Bader	1962	Assistant Program Director	Earth Sciences Program
	1963-1965	Program Director	Geochemistry Program - Earth Sciences Section
Rodger W. Baier	1977-1980	Associate Program Director	IDOE Environmental Quality Program
	1981-1998	Associate Program Director	Chemical Oceanography Program
Rodey Batiza	1986	Associate Program Director (IPA)	Marine Geology & Geophysics Program
William R. Benson	1962-1964	Program Director	Earth Sciences Section
	1965-1974	Head	Earth Sciences Section
	1975-1979	Chief Scientist	Division of Earth Sciences
Peter Borella	1982-1983	Associate Program Director (IPA)	Deep Sea Drilling
Garrett W. Brass	1984-1986	Program Director (IPA)	Ocean Drilling Program
Peter G. Brewer	1981-1983	Program Director (IPA)	Marine Chemistry Program

Name	Year(s)	Title	Section/Division
Louis B. Brown	1973-1974	International Affairs Officer	IDOE
	1975-1976	International Liaison Officer	OCE
	1977-1981	International Science Liaison Officer	OCE
	1982-1993	Science Associate	OCE
Ken Buesseler	1996-1998	Associate Program Director (IPA)	Chemical Oceanography Program
Richard Buffler	1986	Associate Director for Science (IPA)	ODP
Glenn A. Cannon	1973	Staff Associate (Rotator)	Physical Oceanography Program
	1973	Acting Program Director	Physical Oceanography Program
	1974	Program Director (Rotator)	Physical Oceanography Program
Robert S. Carney	1978-1980	Assistant Program Director	Biological Oceanography Program
	1981	Program Director	Biological Oceanography Program
James H. Carpenter	1972	Head (Rotator)	Division of Environmental Sciences - Oceanography Section
Daniel Cayan	1975-1977	Associate Program Director (IPA)	Ocean Sciences Research Section
Edward Chin	1968-1969	Acting Program Director & Associate Program Director	Environmental and Systematic Biology Section - Biological Oceanography Program
Thomas M. Church	1975-1976	Program Associate (Rotator)	IDOE Environmental Quality Program
H. Lawrence Clark	1979-1983	Executive Secretary, Advisory Committee	AAEO - OCE
	1984-1987	Program Manager	Oceanographic Centers and Facilities - Oceanographic Technology Program
	1988-present	Program Director	OTIC
Curtis A. Collins	1972-1981	Program Manager	IDOE Environmental Forecasting Program
	1982-1987	Program Director	Physical Oceanography Program
Thomas N. Cooley	1981-1982	Environmental Impact and Communications Officer	Deep Sea Drilling
	1984	Assistant Program Manager	OOFS - Operations Program
A.P. Crary	1970	Acting Head (Rotator)	Research/Oceanography Section
	1971-1974	Division Director	Division of Environmental Sciences
	1975-1976	Division Director	Division of Earth Sciences
David L. Cutchin	1975-1977	Assistant Program Manager	IDOE Environmental Forecasting Program
Kendra Daly	1997-1999	Associate Program Director (IPA)	Biological Oceanography Program
Joseph P. Dauphin	1991-present	Associate Program Director	Ocean Drilling Program

Name	Years	Position	Program
Thomas Davies	1977-1979	Associate Program Director (IPA)	Deep Sea Drilling
Edward M. Davin	1972-1981	Program Manager	IDOE Seabed Assessment Program
Curtiss O. Davis	1982-1984	Associate Program Director (Rotator)	Oceanic Biology Program
Donald B. De Haven	1986	Administrative Specialist (Rotator)	OCE
Jean T. DeBell	1965-1971	Assistant Program Director	Biological Oceanography Program
Christopher D'Elia	1987-1988	Program Director (IPA)	Biological Oceanography Program
Robert F. Devereaux	1971-1975	Special Assistant for Engineering and Technology Program	IDOE
Emma Dieter	1989-1996	Program Director (IPA)	Ship Operations
	1996-present	Program Director	Ship Operations
Clive E. Dorman	1978-1980	Program Associate	IDOE Environmental Forecasting Program
Cathrina Dowd	1986	Assistant Program Director	Biological Oceanography Program
Linda Duguay	1989-1996	Associate Program Director	Biological Oceanography Program
Robert B. Elder	1973	Ocean Projects Manager	Office of Polar Programs - Polar Operations Section
	1974-1978	Program Manager	OOFS - Operations Program
	1979-1980	Program Manager	OOFS Acquisition and Maintenance Program
	1981	Program Manager	OOFS Fleet Maintenance and Upgrading Program
Donald Elthon	1996-1999	Associate Program Director (IPA)	Marine Geology & Geophysics Program
David Epp	1988-present	Program Director	Marine Geology & Geophysics Program
Manuel E. Fiadeiro	1986-1988	Associate Program Director (IPA)	Physical Oceanography Program
Rana A. Fine	1981-1983	Associate Program Director (IPA)	Ocean Dynamics Program
John J. Finnegan, Jr.	1973	Research Program Management Systems Officer (Rotator)	IDOE
Martin Fisk	1994-1996	Associate Program Director (IPA)	Marine Geology & Geophysics Program
Thomas F. Forhan	1978	Staff Assistant	OOFS
	1979	Assistant Program Manager	OOFS Acquisitions and Maintenance Program
	1981	Program Manager	OOFS Acquisitions and Maintenance Program
Dirk Frankenberg	1971	Program Director (Rotator)	Division of Environmental Sciences - Biological Oceanography Program
	1978-1980	Division Director (Rotator)	OCE
John S. Galehouse	1972-1973	Project Officer (Rotator)	Deep Sea Drilling

Name	Year(s)	Title	Section/Division
David Garrison	1993-1994, 1998-present	Associate Program Director	Biological Oceanography Program
Stefan Gartner	1981	Program Associate (IPA)	Deep Sea Drilling
Silvia Garzoli	1993-1994	Program Director (IPA)	Physical Oceanography Program
C.S. Giam	1975	Program Manager (Rotator)	IDOE Environmental Quality Program
Malvern Gilmartin	1970	Program Director	Ecology & Systematic Biology - Biological Oceanography Program
Marcia G. Glendening	1974-1976	Assistant Program Director (Rotator)	Biological Oceanography Program
Harold L. Goodwin	1967-1970	Planning Officer	Office of Sea Grant Programs
Albert G. Greene, Jr.	1972-1973	Program Manager for Operations	OOFS Operations
George Grice	1972	Program Manager (Rotator)	IDOE Environmental Quality Program
M. Grant Gross	1973-1974	Head	Oceanography Section
	1979-1980	Head	IDOE
	1981-1993	Division Director	OCE
Peter W. Hacker	1978-1980, 1982-1985	Program Director	Physical Oceanography Program
William W. Hakala	1981	Systems Integration Engineer (Rotator)	Ocean Drilling Programs - Engineering and Operations Section
Bilal U. Haq	1981	Program Associate (Rotator)	Ocean Drilling Programs - Deep Sea Drilling Science Program
Herman Harvey	1988-present	Program Director	Marine Geology & Geophysics Program
	1981	Staff Associate	Division of Ocean Drilling Programs
	1982	Systems Coordinator	OSOD Planning and Management Section
Michael L. Healy	1973	Staff Assistant	Oceanography Section
	1974-1975	Program Director	Oceanography Section
Donald F. Heinrichs	1975-1985	Program Director	Submarine Geology & Geophysics Program
	1986-1999	Section Head	Oceanographic Centers & Facilities Section
Vernon J. Henry	1967-1969	Program Director (Rotator)	Submarine Geology & Geophysics Program
Deane E. Holt	1972-1981	Program Manager	IDOE Living Resources Program
Edward Houde	1983-1985	Program Director (IPA)	Biological Oceanography Program

Name	Year	Title	Program/Office
Daniel Hunt	1965	Field Operations Chief	Mohole Project Office
David Hurd	1989-1991	Associate Program Director (IPA)	Chemical Oceanography Program
Anton L. Inderbitzen	1981	Program Director	Ocean Margin Drilling Science Program
	1981	Acting Chief Scientist	ODP - Science Section
	1982	Program Director	OSOD - Science Section - Science Program
Eric Itsweire	1991	Associate Program Director (IPA)	Physical Oceanography Program
	1992-1998	Associate Program Director	Physical Oceanography Program
	1998-present	Program Director	Physical Oceanography Program
Feenan D. Jennings	1971-1977	Head	IDOE
David A. Johnson	1979-1980	Program Associate (IPA)	Submarine Geology & Geophysics Program
Ronald E. Johnson	1981	Program Associate (Rotator)	IDOE Environmental Forecasting Program
Mary K. Johrde	1967-1970	Program Director	Oceanographic Facilities Program - Oceanography Section
	1971-1981	Head	OOFS
David Kadko	1999	Associate Program Director (IPA)	Chemical Oceanography Program
Lauriston R. King	1973	Interagency Liaison Officer	IDOE
	1974-1975	Special Assistant for Marine Science Affairs	IDOE
	1976-1977	Program Manager	IDOE Marine Science Affairs Program
Edward J. Kuenzler	1972	Program Director (Rotator)	Biological Oceanography Program
Ronald R. LaCount	1982-1985	Head	OCFS
John Ladd	1988-1990	Associate Program Director (IPA)	Ocean Drilling Program
Richard B. Lambert, Jr.	1975-1977, 1991-1999	Program Director	Physical Oceanography Program
Lawrence H. Larsen	1972	Program Director (Rotator)	Physical Oceanography Program
Michael Ledbetter	1986-1988	Associate Program Director (IPA)	Marine Geology & Geophysics Program
Gordon G. Lill	1964	Head (Rotator)	Mohole Project Office
Carl J. Lorenzen	1971-1972	Staff Associate (Rotator)	Oceanography Section
Robert Lowell	1987-1988	Associate Program Director (IPA)	Marine Geology & Geophysics Program
John Lyman	1960-1962	Associate Program Director	Earth Sciences Section
	1963-1965	Head	Earth Sciences Section - Oceanography Program

Name	Year(s)	Title	Section/Division
Ian D. MacGregor	1982	Chief Scientist (Rotator)	OSOD - Science Section
Bruce Malfait	1975-1977	Assistant Program Manager	IDOE
	1978, 1981	Associate Program Manager	IDOE Seabed Assessment Program
	1978-1980	Acting Section Head	IDOE
	1982-1985	Associate Program Director	Seafloor Processes Program
	1986	Program Director	Marine Geology & Geophysics Program
	1986-present	Program Director	Ocean Drilling Program
W. Bruce McAlister	1971	Program Director (Rotator)	Physical Oceanography Program
Brian K. McKnight	1975-1976	Assistant Project Officer	Ocean Sediment Coring Program
	1977	Program Associate	Ocean Sediment Coring Program
Hugh G. McLellan	1967-1969	Head (Rotator)	Oceanography Section
Archie R. McLerran	1971-1979	Field Project Officer (Scripps)	Ocean Sediment Coring Program
	1981	Field Operations Officer (Scripps)	Division of Ocean Drilling Programs
	1982	Field Operations Manager	OSOD - Field Operations Office
John G. McMillan	1982-1986	Program Manager	OOFS Operations Program
	1987	Ship Operations Manager	Ship Operations
Stephen Meacham	1999-present	Associate Program Director	Physical Oceanography Program
Roberta J. Meares	1972	Assistant Program Director (Rotator)	Biological Oceanography Program
William J. Merrell, Jr.	1972	Systems Officer	IDOE - Research Program Management
	1974-1975	Staff Associate	IDOE
	1976	Executive Officer	IDOE
	1985-1987	Assistant Director	Geosciences Directorate
S. Metz	1999-present	Associate Program Director (IPA)	Chemical Oceanography Program
Joan R. Mitchell	1977-1981	Assistant Program Manager	IDOE - Environmental Quality Program
	1982-1991	Assistant Program Director	Biological Oceanography Program
	1991-1993	Associate Program Director	Biological Oceanography Program
	1993-1998	Associate Program Director	Education
Russell Moll	1994-1996	Associate Program Director (IPA)	Biological Oceanography Program
John M. Morrison	1984	Associate Program Director (Rotator)	Ocean Dynamics Program
John Morse	1979-1984	Associate Program Director (IPA)	Chemical Oceanography Program
Gregory Mountain	1989-1990	Associate Program Director	Marine Geology & Geophysics Program

Name	Year(s)	Title	Program
A. Conrad Neumann	1970	Program Director (Rotator)	Submarine Geology & Geophysics Program
Worth D. Nowlin, Jr	1971	Deputy Head (Rotator)	IDOE
William N. Orr	1978-1979	Program Associate (IPA)	Ocean Sediment Coring Program
Hsien-Wang Ou	1997-1998	Associate Program Director (IPA)	Physical Oceanography Program
Theodore Packard	1987-1989	Associate Program Director (IPA)	Chemical Oceanography Program
Gustav-Adolf Paffenhöfer	1991-1992	Associate Program Director (IPA)	Biological Oceanography Program
P. Kilho Park	1970 1971	Program Director (Rotator) Head (Rotator)	Physical Oceanography Program Oceanography Section
Patrick L. Parker	1973	Program Manager (Rotator)	IDOE Environmental Quality Program
Polly A. Penhale	1982-1985	Associate Program Director (Rotator)	Biological Oceanography Program
Mary Jane Perry	1980-1982	Assistant Program Director (Rotator)	Biological Oceanography Program/Oceanic Biology Program
Morris T. Phillips	1965	Head (Rotator)	Mohole Project
Albert Plueddemann	1995-1996	Associate Program Director (IPA)	Physical Oceanography Program
Bob J. Presley	1973-1974	Program Manager (IPA)	IDOE Environmental Quality Program
G. Michael Purdy	1995-present	Division Director	Division of Ocean Sciences
Richard Ray	1962 1963	Assistant Program Director (Rotator) Head	Earth Sciences Program Earth Sciences Section - Geology Program
Michael R. Reeve	1981-1983 1985-1986 1986-present	Program Director (Rotator) Program Director Section Head	Biological Oceanography Program Biological Oceanography Program Ocean Science Research Section
Donald Rice	1994-1996 1996-present	Associate Program Director (IPA) Program Director	Chemical Oceanography Program Chemical Oceanography Program
Elizabeth Rom	1988-present	Assistant Program Director	Ocean Technology & Interdisciplinary Coordination
William F. Ruddiman	1982-1983	Program Associate (Rotator)	Submarine Geology & Geophysics Program
Raymond Sambrotto	1987-1989	Assistant Program Director	Biological Oceanography Program
Constance Sancetta	1992-1994 1995-1996 1997-1999	Associate Program Director (IPA) Associate Program Director Program Director	Marine Geology & Geophysics Program Marine Geology & Geophysics Program Marine Geology & Geophysics Program
Ronald Schlitz	1989-1991	Associate Program Director (IPA)	Physical Oceanography Program
Detmar Schnitker	1980-1984	Associate Program Director (IPA)	Submarine Geology

Name	Year(s)	Title	Section/Division
Hans Schrader	1985-1986	Associate Program Director (IPA)	Marine Geology & Geophysics Program
Martha Scott	1992-1993	Associate Program Director (IPA)	Chemical Oceanography Program
Gautam Sen	1992-1994	Associate Program Director (IPA)	Marine Geology & Geophysics Program
Wilbur G. Sherwood	1978-1980	Associate Program Manager	Ocean Sediment Coring Program
	1981	Section Manager	Division of Ocean Drilling Programs - Engineering and Operations Section
Allen M. Shinn, Jr.	1982	Director (Rotator)	OSOD
Alexander Shor	1994-1997	Associate Program Director (IPA)	Oceanographic Centers & Facilities Section
	1998-present	Program Director	Oceanographic Instrumentation & Technical Services Program
Miller F. Shurtleff	1964	Executive Assistant (Rotator)	Mohole Project Office
Joe D. Sides	1964	Geophysicist	Mohole Project Office
	1967-1969	Associate Program Director	Submarine Geology & Geophysics Program
	1971	Project Officer	Office of National Centers & Facilities Operations - Ocean Sediment Coring Program
Sergio Signorini	1992-1995	Associate Program Director (IPA)	Division of Ocean Sciences
Sharon Smith	1988-1989	Associate Program Director (IPA)	Biological Oceanography
Thomas Spence	1987-1992	Program Director	Physical Oceanography Program
Harold A. Spuhler	1971-1972	Deputy Head (Rotator)	OOFS
Debra S. Stakes	1982-1984	Program Associate (Rotator)	Submarine Geology & Geophysics Program
Lee Stevens	1986-1989	Ship Operations Specialist	Ship Operations
Alexander L. Sutherland	1982	Operations Manager, Design and Construction	OSOD - Engineering and Operations Section - Design & Construction
	1982-1984	Acting Section Manager	OSOD Engineering & Operations Section
	1984-1988	Associate Program Director	ODP
Philip R. Taylor	1985-1987	Associate Program Director (Rotator)	Biological Oceanography Program
	1987-present	Program Director	Biological Oceanography Program
Fritz Theyer	1980	Program Associate (IPA)	Ocean Sediment Coring Program
Sandra D. Toye	1973-1977	Special Assistant	OOFS
	1982-1983	Executive Officer	OSOD
	1982-1984	Acting Deputy Director	OSOD
	1985-1986	Program Director	ODP

Name	Dates	Position	Program/Section
John R. Twiss, Jr.	1971-1973	Special Assistant	IDOE
Robert E. Wall	1971-1974	Program Director	Submarine Geology & Geophysics Program
	1975-1981	Section Head	Oceanography Section
	1982-1986	Section Head	OSRS
Richard W. West	1982-1998	Program Manager for Oceanographic Facilities Program	OOFS/OCFS
Robert D. Wildman	1967-1970	Program Director (Rotator)	Office of Sea Grant Programs
Peter E. Wilkniss	1976-1980	Project Officer	Ocean Sediment Coring Program
	1981	Acting Division Director	Division of Ocean Drilling Programs
Richard B. Williams	1973-1980	Program Director	Biological Oceanography Program
C. Don Woodward	1964	Head (Rotator)	Mohole Project Office
Marsh Youngbluth	1995-1997	Associate Program Director (IPA)	Ocean Sciences Research Section/Biological Oceanography Program
Herman B. Zimmerman	1979-1980	Associate Program Director (IPA)	Submarine Geology & Geophysics
	1984	Program Associate for Science Coordination (IPA)	ODP

AAEO - Astronomical, Atmospheric, Earth, and Ocean Sciences
IDOE - International Decade of Ocean Exploration
IPA - Intergovernmental Personnel Agreement
OCE - NSF Division of Ocean Sciences
OCFS - Oceanographic Facilities and Centers
ODP - Ocean Drilling Program
OOFS - Office for Oceanographic Facilities and Support
OSOD - Office of Scientific Ocean Drilling
OSRS - Ocean Sciences Research Section
OTIC - Oceanographic Technology and Interdisciplinary Coordination

The names, dates, and position of the individuals who worked in the Division of Ocean Sciences and its precursors in many cases are approximate. Much of the information was compiled from a set of telephone books from 1957-1986 and the memory of those long-time employees of NSF still in residence. Please accept our apologies for any omissions or incorrect information.

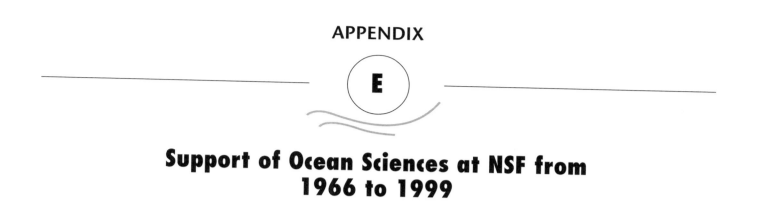

APPENDIX

E

Support of Ocean Sciences at NSF from 1966 to 1999

Expenditures ($ Millions)

	FY 66	FY 67	FY 68	FY 69	FY 70	FY 71	FY 72
Biological	6.6	6.9	6.1	6.3	3.7	3.9	4.4
Chemical						*	*
Geological		1.9	2.9	2.6	3.2	3.4	4.2
Physical	3.7	3.0	1.9	2.2	2.1	2.7	4.0
Ocean Technology							
Education							
OSRS, Subtotals	**10.3**	**11.8**	**10.9**	**11.0**	**9.0**	**10.0**	**12.6**
Living Resources							1.1
Environmental Quality						7.2	5.8
Seabed Assessment						5.3	3.8
Environmental Forcasting						2.4	8.7
General Support						0.3	2.3
IDOE, Subtotals						**15.2**	**21.7**
OFS	**9.3**	**8.4**	**11.6**	**14.0**	**7.4**	**9.2**	**14.5**
Sea Grant			**5.0**	**6.0**	**9.0**	**6.1**	
ODP					**6.6**	**7.1**	**9.3**
OCE, TOTAL	**19.6**	**20.2**	**27.5**	**31.1**	**32.0**	**47.6**	**58.1**

	FY 83	FY 84	FY 85	FY 86	FY 87	FY 88	FY 89
Biological	11.8	12.9	13.9	13.3	14.4	14.8	17.1
Chemical	10.8	12.0	12.4	11.9	13.4	13.7	14.5
Geological	12.6	14.6	15.2	14.6	16.2	16.2	16.0
Physical	14.7	15.6	16.8	17.1	22.5	22.8	23.3
Ocean Technology							
Education							
OSRS, Subtotals	**49.9**	**55.1**	**58.3**	**56.9**	**66.5**	**67.5**	**70.9**
Living Resources							
Environmental Quality							
Seabed Assessment							
Environmental Forcasting							
General Support							
IDOE, Subtotals							
OFS	**31.6**	**32.9**	**35.2**	**33.7**	**37.2**	**37.2**	**43.6**
Sea Grant							
ODP	**21.0**	**26.3**	**27.7**	**28.8**	**30.0**	**30.6**	**31.4**
OCE, TOTAL	**102.5**	**114.3**	**121.2**	**119.4**	**133.7**	**135.3**	**145.9**

This informal 33-year budget history reconstruction was done by Adair Montgomery and Julian Shedlovsky for the period up to the mid-1980s. Please read "A Chronology" by Michael Reeve in this volume for details regarding the sub-components of this table. Program, Section, and Division names are the present-day names. At the end of the International Decade of Ocean Exploration (IDOE), the funds were appropriately merged into the corresponding programs of the Ocean Sciences Research Section (OSRS). The Oceanographic Centers and Facilities Section (OCFS) is tracked in this table by its two sub-

Expenditures ($ Millions)

FY 73	FY 74	FY 75	FY 76	FY 77	FY 78	FY 79	FY 80	FY 81	FY 82
4.3	4.1	4.6	4.5	4.7	5.2	5.3	7.5	10.2	10.7
*	1.6	2.1	2.4	2.6	3.2	3.5	7.7	10.3	10.1
4.3	4.4	5.3	5.7	7.1	6.9	7.2	11.5	12.4	12.0
4.1	3.0	3.3	3.1	3.3	3.7	3.8	14.9	14.7	13.7
12.7	**13.1**	**15.3**	**15.7**	**17.7**	**19.0**	**19.8**	**41.6**	**47.6**	**46.5**
1.8	2.4	2.0	2.1	3.3	2.5	2.6			
4.1	4.5	4.4	3.9	5.1	5.2	5.2			
2.5	3.3	3.0	3.0	3.3	3.8	4.7			
5.3	3.1	5.1	6.0	4.9	6.1	6.8			
3.2	0.5	0.4	0.5	0.6	0.8	0.3			
16.9	**13.8**	**14.9**	**15.5**	**17.2**	**18.4**	**19.6**			
11.0	**18.2**	**20.6**	**16.0**	**18.4**	**20.8**	**23.2**	**24.7**	**27.4**	**28.1**
9.6	**11.1**	**10.5**	**11.8**	**12.8**	**13.4**	**11.6**	**19.5**	**22.0**	**20.5**
50.2	**56.2**	**61.3**	**59.0**	**66.1**	**71.6**	**74.2**	**85.8**	**97.0**	**95.1**

FY 90	FY 91	FY 92	FY 93	FY 94	FY 95	FY 96	FY 97	FY 98	FY 99
17.3	20.3	22.9	21.7	24.3	25.6	26.6	26.7	27.3	30.0
14.9	16.1	17.2	16.2	17.0	17.4	18.3	18.5	18.2	20.1
16.0	17.4	19.0	18.1	20.6	21.4	22.1	22.2	22.5	25.8
24.7	28.3	30.9	29.3	29.9	30.0	30.4	30.4	29.9	32.2
			6.5	8.2	8.2	9.1	9.4	10.3	10.9
							2.1	2.9	2.8
72.9	**82.1**	**90.0**	**91.8**	**100.0**	**102.6**	**106.5**	**109.3**	**111.1**	**121.8**
42.5	**47.7**	**51.2**	**51.7**	**50.2**	**50.4**	**47.5**	**47.7**	**46.7**	**47.3**
32.0	**35.0**	**36.3**	**35.9**	**38.7**	**39.8**	**39.6**	**40.3**	**41.3**	**45.6**
147.4	**164.8**	**177.5**	**179.4**	**188.9**	**192.8**	**193.6**	**197.3**	**199.1**	**214.7**

components Oceanographic Facilities and Ships (OFS) and the Ocean Drilling Program (ODP), and its pre-cursors. The "OCE, total" are the estimated funds of all the program components, which became identified with the present-day Division of Ocean Sciences (OCE). Although these totals represent a rough estimate of NSF expenditures in marine-related fields, they are a minimum estimate because they do not include any expenditures in other branches of NSF, such as Polar Programs or programs within the present-day Directorate for Biosciences in the area of marine biology.

F

Organizational Charts

NATIONAL SCIENCE FOUNDATION
Organization Prescribed in the Enabling Act (1950)

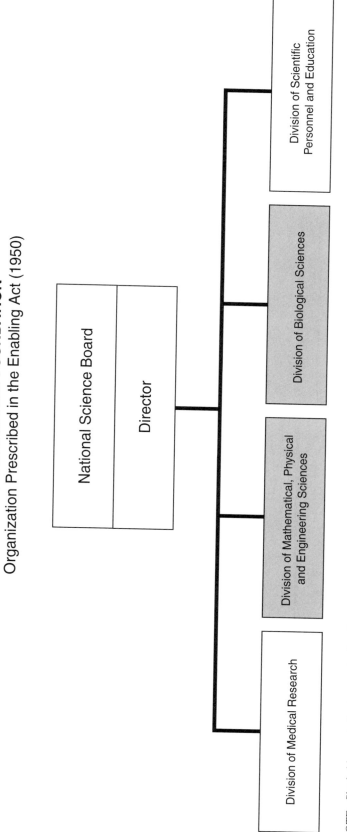

National Science Board

Director

Division of Medical Research

Division of Mathematical, Physical and Engineering Sciences

Division of Biological Sciences

Division of Scientific Personnel and Education

NOTE: Shaded boxes indicate organizational units with significant ocean science responsibilities.

NATIONAL SCIENCE FOUNDATION
Organization as of June 30, 1953

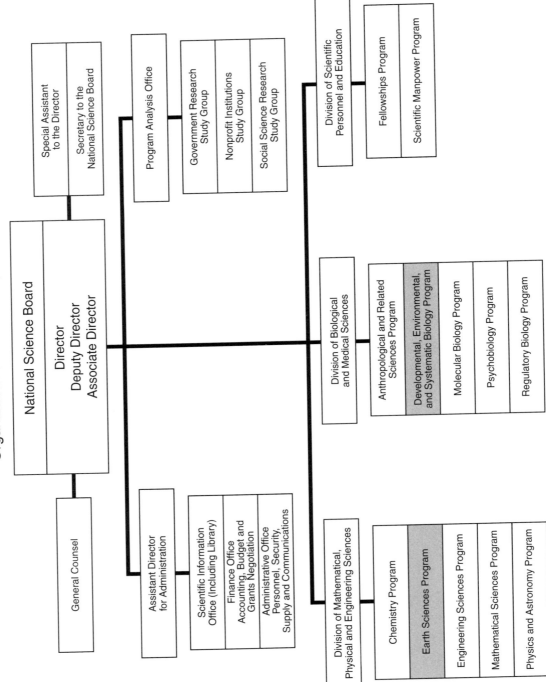

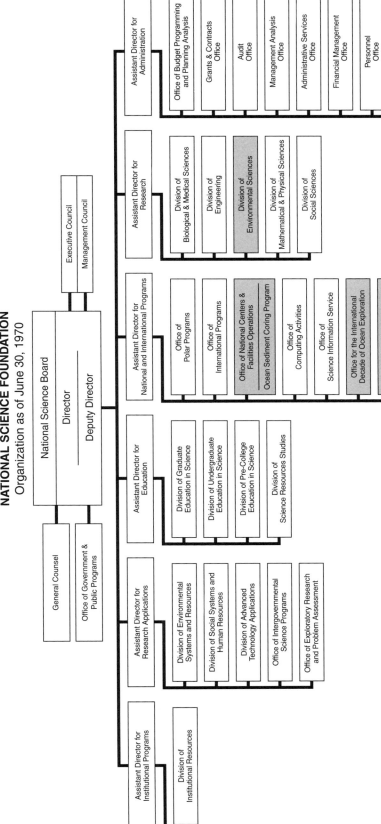

NATIONAL SCIENCE FOUNDATION
Organization as of June 30, 1970

National Science Board

Director

Deputy Director

Executive Council

Management Council

General Counsel

Office of Government &
Public Programs

Assistant Director for Institutional Programs
- Division of Institutional Resources

Assistant Director for Research Applications
- Division of Environmental Systems and Resources
- Division of Social Systems and Human Resources
- Division of Advanced Technology Applications
- Office of Intergovernmental Science Programs
- Office of Exploratory Research and Problem Assessment

Assistant Director for Education
- Division of Graduate Education in Science
- Division of Undergraduate Education in Science
- Division of Pre-College Education in Science
- Division of Science Resources Studies

Assistant Director for National and International Programs
- Office of Polar Programs
- Office of International Programs
- Office of National Centers & Facilities Operations
- Ocean Sediment Coring Program
- Office of Computing Activities
- Office of Science Information Service
- Office for the International Decade of Ocean Exploration
- Office for Oceanographic Facilities and Support

Assistant Director for Research
- Division of Biological & Medical Sciences
- Division of Engineering
- Division of Environmental Sciences
- Division of Mathematical & Physical Sciences
- Division of Social Sciences

Assistant Director for Administration
- Office of Budget Programming and Planning Analysis
- Grants & Contracts Office
- Audit Office
- Management Analysis Office
- Administrative Services Office
- Financial Management Office
- Personnel Office
- Data Management Systems Office
- Program Review Office
- Health Office

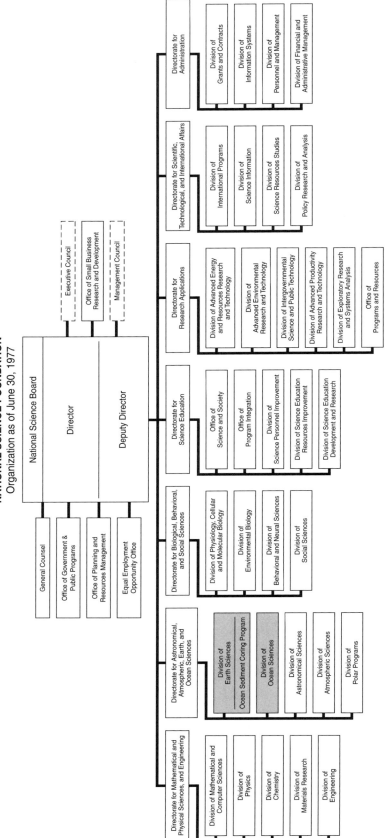

NATIONAL SCIENCE FOUNDATION
Organization as of June 30, 1977

National Science Board

Director

Deputy Director

Executive Council

Office of Small Business Research and Development

Management Council

General Counsel

Office of Government & Public Programs

Office of Planning and Resources Management

Equal Employment Opportunity Office

Directorate for Mathematical and Physical Sciences, and Engineering
- Division of Mathematical and Computer Sciences
- Division of Physics
- Division of Chemistry
- Division of Materials Research
- Division of Engineering

Directorate for Astronomical, Atmospheric, Earth, and Ocean Sciences
- Division of Earth Sciences
  - Ocean Sediment Coring Program
- Division of Ocean Sciences
- Division of Astronomical Sciences
- Division of Atmospheric Sciences
- Division of Polar Programs

Directorate for Biological, Behavioral, and Social Sciences
- Division of Physiology, Cellular and Molecular Biology
- Division of Environmental Biology
- Division of Behavioral and Neural Sciences
- Division of Social Sciences

Directorate for Science Education
- Office of Science and Society
- Office of Program Integration
- Division of Science Personnel Improvement
- Division of Science Education Resources Improvement
- Division of Science Education Development and Research

Directorate for Research Applications
- Division of Advanced Energy and Resources Research and Technology
- Division of Advanced Environmental Research and Technology
- Division of Intergovernmental Science and Public Technology
- Division of Advanced Productivity Research and Technology
- Division of Exploratory Research and Systems Analysis
- Office of Programs and Resources

Directorate for Scientific, Technological, and International Affairs
- Division of International Programs
- Division of Science Information
- Division of Science Resources Studies
- Division of Policy Research and Analysis

Directorate for Administration
- Division of Grants and Contracts
- Division of Information Systems
- Division of Personnel and Management
- Division of Financial and Administrative Management

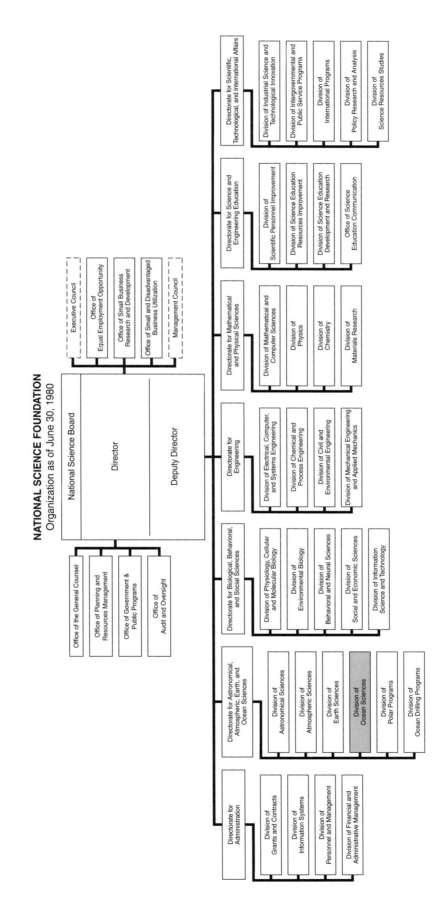

NATIONAL SCIENCE FOUNDATION
Organization as of June 30, 1980

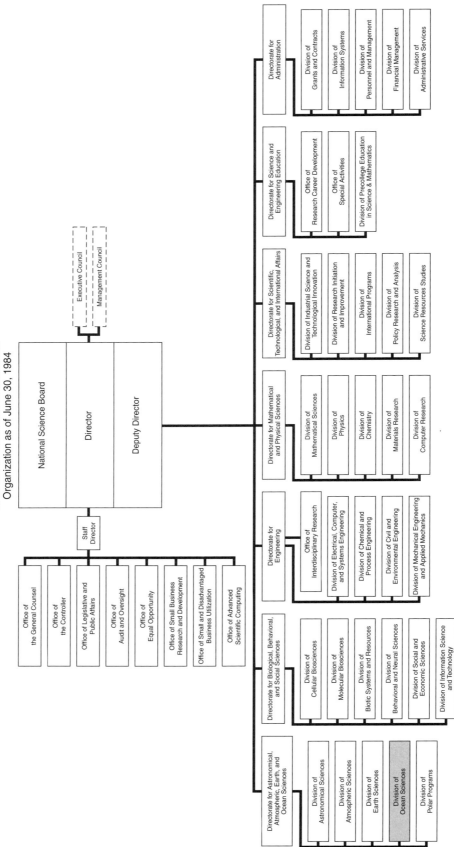

NATIONAL SCIENCE FOUNDATION
Organization as of June 30, 1984

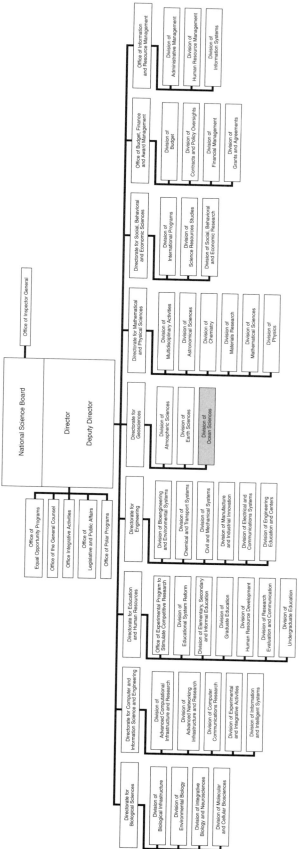

NATIONAL SCIENCE FOUNDATION
Organization as of June 30, 1998

National Science Board

Director

Deputy Director

Office of Inspector General

Office of Equal Opportunity Programs

Office of the General Counsel

Office Integrative Activities

Office of Legislative and Public Affairs

Office of Polar Programs

Directorate for Geosciences
- Division of Atmospheric Sciences
- Division of Earth Sciences
- Division of Ocean Sciences

Directorate for Mathematical and Physical Sciences
- Division of Multidisciplinary Activities
- Division of Astronomical Sciences
- Division of Chemistry
- Division of Materials Research
- Division of Mathematical Sciences
- Division of Physics

Directorate for Social, Behavioral and Economic Sciences
- Division of International Programs
- Division of Science Resources Studies
- Division of Social, Behavioral and Economic Research

Office of Budget, Finance and Award Management
- Division of Budget
- Division of Contracts and Policy Oversights
- Division of Financial Management
- Division of Grants and Agreements

Office of Information and Resource Management
- Division of Administrative Management
- Division of Human Resource Management
- Division of Information Systems

Directorate for Engineering
- Division of Bioengineering and Environmental Systems
- Division of Chemical and Transport Systems
- Division of Civil and Mechanical Systems
- Division of Manufacture and Industrial Innovation
- Division of Electrical and Communications Systems
- Division of Engineering Education and Centers

Directorate for Education and Human Resources
- Office of Experimental Program to Stimulate Competitive Research
- Division of Educational System Reform
- Division of Elementary, Secondary and Informal Education
- Division of Graduate Education
- Division of Human Resource Development
- Division of Research Evaluation and Communication
- Division of Undergraduate Education

Directorate for Computer and Information Science and Engineering
- Division of Advanced Computational Infrastructure and Research
- Division of Advanced Networking Infrastructure and Research
- Division of Computer Communications Research
- Division of Experimental and Integrative Activities
- Division of Information and Intelligent Systems

Directorate for Biological Sciences
- Division of Biological Infrastructure
- Division of Environmental Biology
- Division of Integrative Biology and Neurosciences
- Division of Molecular and Cellular Biosciences

APPENDIX

G

NRC Project Oversight

The organization of the symposium and production of this report were overseen by the Ocean Studies Board of the National Research Council, which is the operating arm of the National Academy of Sciences and National Academy of Engineering.

OCEAN STUDIES BOARD

KENNETH BRINK (Chair), Woods Hole Oceanographic Institution, Massachusetts
ALICE ALLDREDGE, University of California, Santa Barbara
DAVID BRADLEY, Pennsylvania State University, State College
DANIEL BROMLEY, University of Wisconsin, Madison
OTIS BROWN, University of Miami, Florida
CORT COOPER, Chevron Chemical Company LLC
CARL FRIEHE, University of California, Irvine
RAY HILBORN, University of Washington, Seattle
EDWARD HOUDE, University of Maryland, Solomons
JOHN KNAUSS, University of Rhode Island, Narragansett
ROBERT A. KNOX, University of California, San Diego
RAY KRONE, University of California, Davis
CINDY LEE, State University of New York, Stony Brook
ROGER LUKAS, University of Hawaii, Manoa
NANCY MARCUS, Florida State University, Tallahassee
NEIL OPDYKE, University of Florida, Gainesville
MICHAEL K. ORBACH, Duke University, Beaufort, North Carolina
WALTER SCHMIDT, Florida Geological Survey, Tallahassee
GEORGE SOMERO, Stanford University, Pacific Grove, California
KARL K. TUREKIAN, Yale University, New Haven, Connecticut

Staff

MORGAN GOPNIK, Director
EDWARD R. URBAN, JR., Study Director
ANN CARLISLE, Senior Project Assistant

APPENDIX

H

Acronyms

AAEO   Astronomical, Atmospheric, Earth and Ocean Sciences (NSF)
AD   Associate Director
AD/NI   Associate Director for National and International Programs (NSF)
AD/R   Associate Director for Research (NSF)
ADCP   acoustic Doppler current profiler
AEC   Atomic Energy Commission
AEOS   Astronomical, Earth and Ocean Sciences (NSF)
AGU   American Geophysical Union
AMS   Accelerator Mass Spectrometry
AMSOC   American Miscellaneous Society
APROPOS   Advances and Primary Research Opportunities in Physical Oceanography Studies
ARGO   Array for Real-Time Geostrophic Oceanography
ASLO   American Society of Limnology and Oceanography
AUV   autonomous underwater vehicle

BBS   Biological, Behavioral, and Social Sciences (NSF)
BBSR   Bermuda Biological Station for Research
BECS   Basin-wide Extended Climate Study
BMS   Biological and Medical Sciences Division (NSF)
BT   bathythermograph

CCD   calcite compensation depth
CENOP   Cenezoic Paleo-Oceanography Project (IDOE)
CEPEX   Controlled Pollution Experiment
CGO   Coordinating Group on Oceanography (NSF)
CLIMAP   Climate: Long-range Investigation, Mapping and Prediction
CLIVAR   Climate Variability and Prediction program
CMSER   Commission on Marine Sciences, Engineering, and Resources
CNEXO   Centre National pour L'Exploitation de Oceans
CNP   Central North Pacific Gyre
COB   Centre Oceanologique de Bretagne
CODE   Coastal Ocean Dynamics Experiment
CORE   Consortium for Oceanographic Research and Education
CTD   conductivity-temperature-depth profiler
CUEA   Coastal Upwelling Ecosystems Analysis
CZCS   Coastal Zone Color Scanner

DDT   dichlorodiphenyltrichloroethane
DES   Division of Environmental Sciences (NSF)
DISCO   Dissertations in Chemical Oceanography
DOE   Department of Energy
DSDP   Deep Sea Drilling Program

EF   Environmental Forecasting (IDOE)
ENSO   El Niño-Southern Oscillation
EPOCS   Eastern Pacific Ocean Climate Study
EPR   East Pacific Rise
ERDA   Energy Research and Development Administration
EQ   Environmental Quality (IDOE)
ES   Division of Earth Sciences (NSF)

FAMOUS   French-American Mid-Ocean Undersea Study
FOCUS   Future of Ocean Chemistry in the United States
FUMAGES   Future of Marine Geology and Geophysics

GARP   Global Atmospheric Research Program
GCM   general circulation models
GDC   Geologic Data Center
GEO   Directorate for Geosciences (NSF)
GEOSECS   Geochemical Ocean Sections Study
GFD   geophysical fluid dynamics
GLOBEC   Global Ocean Ecosystems Dynamics program
GODAE   Global Ocean Data Assimilation Experiment
GSO   Graduate School of Oceanography (Univ. of Rhode Island)

ICO   Interagency Committee on Oceanography
ICP-MS   inductively coupled plasma mass spectrometer
ICSU   International Council of Scientific Unions (now the International Council for Science)
IDOE   International Decade of Ocean Exploration
IGY   International Geophysical Year
IIOE   International Indian Ocean Exploration
IOC   Intergovernmental Oceanographic Commission
IODP   Integrated Ocean Drilling Program
IPA   Intergovernmental Personnel Agreement
IPOD   International Program for Ocean Drilling
ISOS   International Southern Ocean Studies

JGOFS   Joint Global Ocean Flux Study
JOI   Joint Oceanographic Institutions, Inc.
JOIDES   Joint Oceanographic Institutions for Deep Earth Sampling
JPL   Jet Propulsion Laboratory

LDEO   Lamont-Doherty Earth Observatory
LDGO   Lamont-Doherty Geological Observatory
LOCO   LOng COres drilling program
LR   Living Resources (IDOE)

MANOP   Manganese Nodule Program (Phase 2) (IDOE)
MAR   Mid-Atlantic Ridge
MBL   Marine Biological Laboratory (Woods Hole)

MEDOC	Mediterranean Deep Ocean Convection
MERL	Marine Ecosystems Research Laboratory
MG&G	marine geology and geophysics
MIT	Massachusetts Institute of Technology
MODE	Mid-Ocean Dynamics Experiment
MPE	Division of Mathematics, Physics and Engineering (NSF)
MPES	Mathematical, Physical and Engineering Sciences Division (NSF)
MPS	Mathematical and Physical Sciences (NSF)

NACOA	National Advisory Committee on Oceans and Atmosphere
NAML	National Association of Marine Laboratories
NAS	National Academy of Sciences
NASA	National Aeronautics and Space Administration
NASCO	NAS Committee on Oceanography
NCAR	National Center for Atmospheric Research
NCF	National Centers and Facilities (NSF)
NGO	non-governmental organization
NIH	National Institutes of Health
NOAA	National Oceanic and Atmospheric Administration
NODC	National Oceanographic Data Center (NOAA)
NOLS	National Oceanographic Laboratory System
NORPAX	North Pacific Experiment
NOSAMS	National Ocean Sciences Accelerator Mass Spectrometer
NRC	National Research Council
NSF	National Science Foundation

OAP	Office of Antarctic Programs (NSF)
OCE	Division of Ocean Sciences (NSF)
OCFS	Oceanographic Facilities and Centers (OCE)
ODP	Ocean Drilling Program
OEUVRE	Ocean Ecology: Understanding and Vision for Research
OMD	Ocean Margin Drilling
OMDP	Ocean Margin Drilling Program (NSF)
ONR	Office of Naval Research
OOFS	Office for Oceanographic Facilities and Support (NSF)
OPP	Office of Polar Programs (NSF)
OSOD	Office of Scientific Ocean Drilling (NSF)
OSRS	Ocean Sciences Research Section (OCE)
OT	Oceanographic Technology
OTA	Office of Technology Assessment
OTIC	Ocean Technology and Interdisciplinary Coordination (NSF)

PDR	precision depth recorder
PEQUOD	Pacific Equatorial Ocean Dynamics Experiment
PI	principal investigator
POLYMODE	U.S.-Soviet follow-on to MODE
PRIMA	Pollutant Responses In Marine Animals
PSAC	President's Science Advisory Committee

RIDGE	Ridge Inter-Disciplinary Global Experiments
RISE	RIvera Submersible Experiments
RITA	RIvera and TAmayo (French phase of East Pacific Rise explorations)
ROV	remotely-operated vehicle
RVOC	Research Vessels Operators' Committee

SAR	synthetic aperture radar
SASS	Sound Acoustic Surveillance System
SBA	Sea Bed Assessment (IDOE)
SCICEX	Scientific Ice Expeditions
SEABEAM	shipboard multi-transducer swath echo sounding system
SEAREX	Sea-Air Exchange
SEASAT	Earth Satellite dedicated to Oceanographic Applications
SEATAR	Studies in East Asia Tectonics and Resources (IDOE)
SeaWiFS	Sea-Viewing Wide Field of View Sensor
SES	Seagrass Ecosystem Study (IDOE)
SG&G	submarine geology and geophysics
SIO	Scripps Institution of Oceanography
SSC	Scientific Steering Committee
SSG	Scientific Steering Group
TAC	Technical Assistance Committee
TAO	Tropical Atmosphere Ocean
TH	Tropic Heat (a study of the Eastern Pacific Cold Tongue)
TOGA	Tropical Ocean-Global Atmosphere
TOPEX	Ocean Topography Experiment
TTO	Transient Tracers in the Ocean
TW	Terawatts
UCAR	University Corporation for Atmospheric Research
UNESCO	United Nations Educational, Scientific, and Cultural Organization
UNL	University-National Laboratory
URI	University of Rhode Island
USAP	U.S. Antarctic Program
USGCRP	U.S. Global Change Research Program
USSAC	U.S. Scientific Advisory Committeen (ODP)
VERTEX	VERTical transport and EXchange of materials in the upper waters of the oceans
WEPOCS	Western Equatorial Pacific Ocean Circulation Study
WHOI	Woods Hole Oceanographic Institution
WMO	World Meteorological Organization
WOCE	World Ocean Circulation Experiment

Index

THE NATIONAL ACADEMIES

National Academy of Sciences
National Academy of Engineering
Institute of Medicine
National Research Council

The **National Academy of Sciences** is a private, nonprofit, self-perpetuating society of distinguished scholars engaged in scientific and engineering research, dedicated to the furtherance of science and technology and to their use for the general welfare. Upon the authority of the charter granted to it by the Congress in 1863, the Academy has a mandate that requires it to advise the federal government on scientific and technical matters. Dr. Bruce M. Alberts is president of the National Academy of Sciences.

The **National Academy of Engineering** was established in 1964, under the charter of the National Academy of Sciences, as a parallel organization of outstanding engineers. It is autonomous in its administration and in the selection of its members, sharing with the National Academy of Sciences the responsibility for advising the federal government. The National Academy of Engineering also sponsors engineering programs aimed at meeting national needs, encourages education and research, and recognizes the superior achievements of engineers. Dr. William A. Wulf is president of the National Academy of Engineering.

The **Institute of Medicine** was established in 1970 by the National Academy of Sciences to secure the services of eminent members of appropriate professions in the examination of policy matters pertaining to the health of the public. The Institute acts under the responsibility given to the National Academy of Sciences by its congressional charter to be an adviser to the federal government and, upon its own initiative, to identify issues of medical care, research, and education. Dr. Kenneth I. Shine is president of the Institute of Medicine.

The **National Research Council** was organized by the National Academy of Sciences in 1916 to associate the broad community of science and technology with the Academy's purposes of furthering knowledge and advising the federal government. Functioning in accordance with general policies determined by the Academy, the Council has become the principal operating agency of both the National Academy of Sciences and the National Academy of Engineering in providing services to the government, the public, and the scientific and engineering communities. The Council is administered jointly by both Academies and the Institute of Medicine. Dr. Bruce M. Alberts and Dr. William A. Wulf are chairman and vice chairman, respectively, of the National Research Council.